Course 3

MATH

YOUR COMMON CORE EDITION CCSS

Assessment Masters

Bothell, WA • Chicago, IL • Columbus, OH • New York, NY

connectED.mcgraw-hill.com

STEM McGraw-Hill is committed to providing instructional materials in Science, Technology, Engineering, and Mathematics (STEM) that give all students a solid foundation, one that prepares them for college and careers in the 21st century.

Send all inquiries to:
McGraw-Hill Education
8787 Orion Place
Columbus, OH 43240

ISBN: 978-0-07-662329-7
MHID: 0-07-662329-7

Printed in the United States of America.

7 8 9 QLM 16 15 14

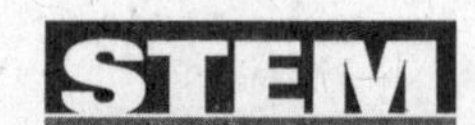

Assessment Masters

Our mission is to provide educational resources that enable students to become the problem solvers of the 21st century and inspire them to explore careers within Science, Technology, Engineering, and Mathematics (STEM) related fields.

Teacher's Guide to Using the *Assessment Masters*

The *Assessment Masters* includes the core assessment materials needed for each chapter. The answers for these pages appear at the back of this book.

Are You Ready? Worksheets

- Use after the Are You Ready? section in the Student Edition.

Chapter Diagnostic Test

- Use to test skills needed for success in the upcoming chapter.
- Retest approaching-level students after the Are You Ready? worksheets.

Chapter Pretest

- Quick check the upcoming chapter's concepts to determine pacing.
- Use before the chapter to gauge students' skill level and to determine class grouping.

Chapter Quiz

- Reassess the concepts tested in the Mid-Chapter Check in the Student Edition

Vocabulary Test

- Includes a list of vocabulary words and questions to assess students' knowledge of those words and can be used in conjunction with one of the Chapter Tests.

Standardized Test Practice

- Assess knowledge as student progresses through the textbook.
- Includes multiple-choice, short-response, gridded-response, and extended-response questions
- Student Recording Sheet corresponds with the Test Practice.

Extended-Response Test

- Contains performance-assessment tasks and includes a scoring rubric

Chapter Tests

- AL 1A-1B Approaching-level students; contains multiple-choice questions
- OL 2A-2B On-level students; contains both multiple-choice and free-response questions
- BL 3A-3B Beyond-level students; contains free-response questions
- Tests A and B are created with parallel format. Use when students are absent or for different rows.

Benchmark Tests

- Contains multiple-choice and short-response questions
- The first three tests provide quarterly evaluations.
- The last test provides a cumulative end-of-year evaluation.

CONTENTS

Chapter 1 Real Numbers

Chapter 2 Equations in One Variable

Chapter 3 Equations in Two Variables

Chapter 4 Functions

Chapter 5 Triangles and the Pythagorean Theorem

Chapter 6 Transformations

Chapter 7 Congruence and Similarity

Chapter 8 Volume and Surface Area

Chapter 9 Scatter Plots and Data Analysis

NAME ______________________ DATE ______________ PERIOD ________

Are You Ready?

Review

Example 1

Find $6 \cdot 3 \cdot 6 \cdot 3 \cdot 6$.

$6 \cdot 3 \cdot 6 \cdot 3 \cdot 6$

$= 3 \cdot 3 \cdot 6 \cdot 6 \cdot 6$ Commutative Property

$= (3 \cdot 3) \cdot (6 \cdot 6 \cdot 6)$ Associative Property

$= 9 \cdot 216$ Multiply.

$= 1{,}944$ Simplify.

Example 2

Find $(-4)(-4)(-4)(-4)$.

$(-4)(-4)(-4)(-4) = 256$ Multiply.

Exercises

Find each product.

1. $9 \cdot 9 \cdot 9$ 1. ________
2. $(-6)(-2)(-2)(-6)$ 2. ________
3. $7 \cdot 4 \cdot 7 \cdot 4 \cdot 4$ 3. ________
4. $8 \cdot 3 \cdot 8 \cdot 8 \cdot 3$ 4. ________
5. $5 \cdot 5 \cdot 5$ 5. ________
6. $(-3)(9)(-3)(9)(-3)(-3)$ 6. ________
7. $2 \cdot 5 \cdot 2 \cdot 5 \cdot 2$ 7. ________
8. $(-4)(6)(-4)(6)$ 8. ________
9. **DOGHOUSE** The doghouse at the local kennel is about $2 \times 4 \times 3 \times 3 \times 4 \times 2$ square feet. About how many square feet is the doghouse? 9. ________
10. **SALES** A furniture store sold $9 \cdot 5 \cdot 9 \cdot 5 \cdot 5$ dollars worth of furniture during a one–day sale. How many dollars was the furniture? 10. ________

NAME ______________________ DATE ____________ PERIOD ________

Are You Ready?

Practice

Find each product.

1. 7 • 7 • 7

2. 8 • 8 • 8 • 8 • 8

3. (−6)(−5)(−5)(−6)(−5)

4. 3 • 3 • 4 • 4 • 4

5. 2 • 9 • 2 • 9 • 9

6. **CALLS** A call center received 9 • 3 • 3 • 9 • 3 calls during a campaign season. How many calls did the call center receive?

7. **RECREATION** The new town recreation center is about 7 • 5 • 7 • 5 • 5 square feet. About how many square feet are in the recreation center?

Find the prime factorization of each number.

8. −96

9. 42

10. 144

11. 17

12. 54

13. −110

14. 16

15. 156

16. **PUMPKINS** The table shows the weight of the winning pumpkins in a contest at the state fair. Find the prime factorization of each weight.

Pumpkin	Weight (lb)
Mr. Smith's	112
Ms. Gonzalez's	98
Mrs. Johnson's	83
Mr. Cyzdin's	72

1. ____________
2. ____________
3. ____________
4. ____________
5. ____________
6. ____________
7. ____________
8. ____________
9. ____________
10. ____________
11. ____________
12. ____________
13. ____________
14. ____________
15. ____________
16. ____________

NAME ______________________ DATE ____________ PERIOD ________

Are You Ready?

Apply

1. DISTANCE The distance from Eli's house to his grandparents' house is $4 \cdot 3 \cdot 4 \cdot 3 \cdot 3$ miles. How many miles away is the grandparents' house?

2. TEMPERATURE The table shows the length and width of Florida at its most distant points. Find the prime factorization of each number.

Measurement	Distance (mi)
Length	500
Width	160

3. DOGS The table shows the weights of the dogs in a local dog show. Find the prime factorization of each number.

Dog	Weights (lb)
Irish Setter	68
Jack Russell Terrier	15
Great Dane	114
Beagle	28

4. HAY Mr. Day feeds the cows on his farm $2 \cdot 7 \cdot 5 \cdot 5 \cdot 5$ pounds of hay per week. How many pounds of hay do they eat per week?

5. NURSERY A nursery owner wants to build a new greenhouse that will have $5 \cdot 2 \cdot 5 \cdot 5 \cdot 5$ square feet. How many square feet is the greenhouse?

6. FUNDRAISER The table shows the amount of money each student raised for the school fundraiser. Find the prime factorization of each number.

Student	Money Raised
Dorsey	$125
Danica	$88
Jazzra	$96
Allen	$150

NAME ______________________ DATE ______________ PERIOD ________

Diagnostic Test

Find each product.

1. $3 \cdot 3 \cdot 3$

2. $6 \cdot 6 \cdot 6 \cdot 6 \cdot 6$

3. $(9)(-5)(-5)(9)(-5)$

4. $7 \cdot 7 \cdot 2 \cdot 2 \cdot 2$

5. $5 \cdot 3 \cdot 5 \cdot 3 \cdot 3$

6. **BIRDS** A bird call manufacturer sold $8 \cdot 4 \cdot 4 \cdot 8 \cdot 4$ calls in two months. How many calls did the manufacturer sell?

7. **RIVER** The St. John's River is $2 \cdot 5 \cdot 31$ miles long. About how many miles long is the river?

Find the prime factorization of each number.

8. −82

9. 38

10. 208

11. 11

12. 60

13. −78

14. 20

15. 124

16. **HEIGHT** The table shows the heights of four sixth grade students. Find the prime factorization of each height.

Student	Height (in.)
Oksana	59
Silvia	62
Joseph	64
Diego	66

1. ____________
2. ____________
3. ____________
4. ____________
5. ____________
6. ____________
7. ____________
8. ____________
9. ____________
10. ____________
11. ____________
12. ____________
13. ____________
14. ____________
15. ____________
16. ____________

Pretest

Write each expression using exponents.

1. $8 \cdot 8 \cdot 8$ **1.** ____________

2. $4 \cdot 4 \cdot 4 \cdot 4$ **2.** ____________

Evaluate each expression.

3. $(-5)^3$ **3.** ____________

4. 3^6 **4.** ____________

Simplify. Express using exponents.

5. $8^2 \cdot 8^4$ **5.** ____________

6. $(7^3)^2$ **6.** ____________

Write each expression using a positive exponent.

7. 9^{-4} **7.** ____________

8. 6^{-3} **8.** ____________

Write each number in standard form.

9. 3.68×10^4 **9.** ____________

10. 7.924×10^{-5} **10.** ____________

Find each square root.

11. $-\sqrt{36}$ **11.** ____________

12. $\sqrt{\frac{16}{25}}$ **12.** ____________

Find each cube root.

13. $\sqrt[3]{27}$ **13.** ____________

14. $\sqrt[3]{512}$ **14.** ____________

NAME ______________________ DATE ____________ PERIOD ________

Chapter Quiz

1. Write $7\frac{1}{5}$ as a decimal. 1. ________

2. Write 4.625 as a mixed number in simplest form. 2. ________

3. Write $-0.\overline{5}$ as a fraction in simplest form. 3. ________

Write each expression using exponents.

4. $4 \cdot p \cdot 4 \cdot p$ 4. ________

5. $x \cdot x \cdot x \cdot x \cdot y \cdot y$ 5. ________

6. Evaluate $k^5 \cdot m$, if $k = 2$ and $m = 5$. 6. ________

Simplify. Express using exponents.

7. $9^3 \cdot 9^5$ 7. ________

8. $\frac{45x^3y}{9x^2y}$ 8. ________

9. $\frac{15m^3n^2}{3mn}$ 9. ________

Simplify.

10. $(6^2)^2$ 10. ________

11. $(5^2)^3$ 11. ________

12. $(-5q^2p)^3$ 12. ________

13. **MONEY** During the school week, Joshua spent $3 each day on lunch. On Tuesday, he bought a $5 ticket to the school play and on Friday he loaned $2 to his friend. When he checked his wallet at the end of the day Friday, he had $3 left. How much money did he start the week with? 13. ________

NAME ______________________ DATE ______________ PERIOD ________

Vocabulary Test

SCORE ________

base	perfect cube	rational number
cube root	perfect square	repeating decimal
exponent	power	scientific notation
irrational number	radical sign	square root
monomial	real number	terminating decimal

Choose the correct term from the two choices given which makes the statement true.

1. In the expression 2^4, 4 is called the (base, exponent). **1.** ____________

2. In the expression a^{-3}, a is called the (base, power). **2.** ____________

3. (Scientific notation, Square root) is a compact way of writing numbers with absolute values that are very large or very small. **3.** ____________

4. A (square root, perfect square) of a number is one of its two equal factors. **4.** ____________

5. A product of repeated factors can be expressed as a (base, power). **5.** ____________

6. A (radical sign, scientific notation), $\sqrt{}$, is used to indicate a positive square root. **6.** ____________

7. Numbers such as 1, 4, 9, 16, and 25 are called (irrational numbers, perfect squares) because they are squares of integers. **7.** ____________

8. Numbers that are not rational are called (base, irrational) numbers. **8.** ____________

Define in your own words.

9. monomial **9.** ____________

10. real number **10.** ____________

NAME ______________________ DATE ______________ PERIOD ________

Standardized Test Practice

Read each question. Then fill in the correct answer on the answer document provided by your teacher or on a sheet of paper.

1. The distance from Earth to the Sun is 92,900,000 miles. What is this number in scientific notation?

   A. 92.9×10^6

   B. 9.29×10^7

   C. 9.29×10^6

   D. 929×10^5

2. **GRIDDED RESPONSE** Ms. Leigh wants to organize the desks in the study hall into a square. If she has 64 desks, how many should be in each row?

3. Between which two whole numbers is $\sqrt{66}$ located on a number line?

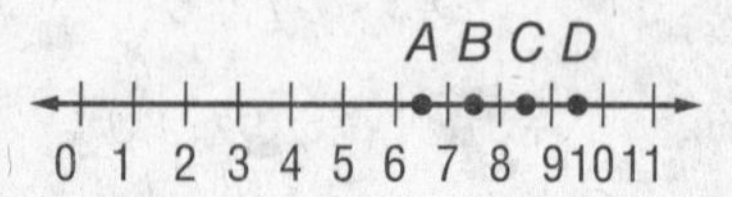

   F. 6 and 7

   G. 7 and 8

   H. 8 and 9

   I. 9 and 10

4. **SHORT RESPONSE** The area of each square in the figure below is 25 square units.

   What is the perimeter of the figure?

5. **GRIDDED RESPONSE** Sarah wants to install new carpet in her living room. The small square is tiled and will not be carpeted. How many square yards of carpet will she need?

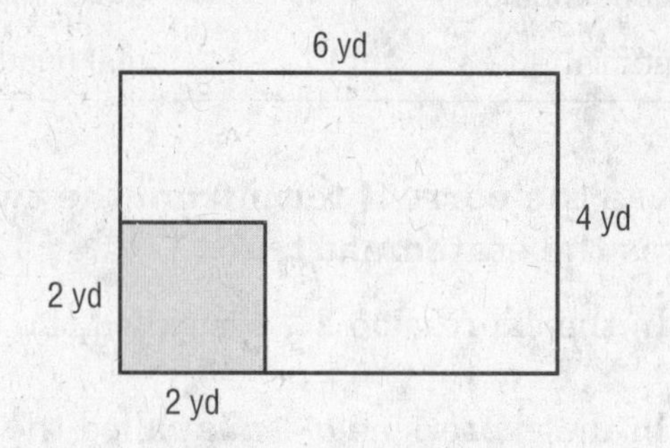

6. **SHORT RESPONSE** The area of a rectangle is $30m^{11}$ square feet. If the length of the rectangle is $6m^4$ feet, what is the width of the rectangle?

7. The table shows the area in square miles of certain states.

State	Area (mi²)
California	1.64×10^5
Ohio	4.48×10^4
Oregon	9.84×10^4
Vermont	9.62×10^3

   Which state has the greatest area?

   A. California

   B. Ohio

   C. Oregon

   D. Vermont

8. The mass of a paper clip is 9.4×10^{-4} kilogram. What is this mass in standard form?

   F. 0.000000094 kg

   G. 0.0000094 kg

   H. 0.000094 kg

   I. 0.00094 kg

9. What is the decimal form of the fraction $-\frac{1}{16}$?

A. 0.0625

B. 0.07

C. −0.07

D. −0.0625

10. Which of the following is equivalent to $(-3)^{-3}$?

F. −9

G. $\frac{1}{-27}$

H. $\frac{1}{27}$

I. 9

11. Which of the following is closest to point *A* on the number line below?

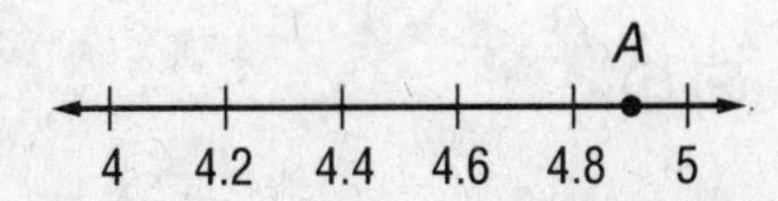

A. $\frac{1}{2}$

B. $\sqrt{20}$

C. $\sqrt{24}$

D. 480%

12. Part of a recipe for one batch of macaroni and cheese is shown below.

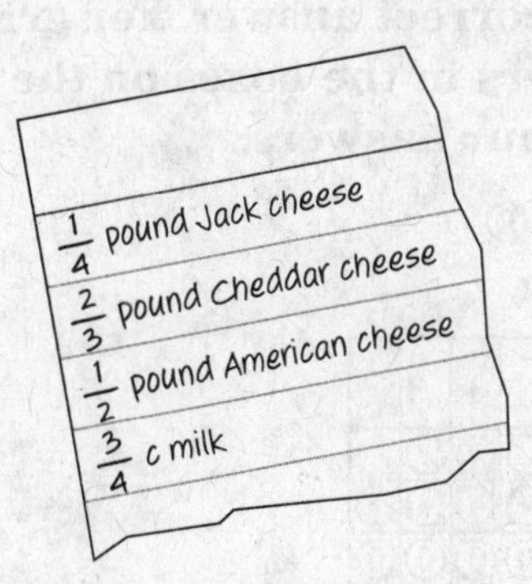

Which of the fractions shown is not a terminating decimal?

F. $\frac{1}{4}$

G. $\frac{1}{2}$

H. $\frac{2}{3}$

I. $\frac{3}{4}$

13. THINK SOLVE EXPLAIN **EXTENDED RESPONSE** The container for a child's set of blocks is 9 inches by 9 inches by 9 inches. The blocks measure 3 inches by 3 inches by 3 inches.

Part A Describe how to determine the number of blocks needed to fill the container.

Part B Write and simplify an expression to solve the problem.

Part C How many blocks will it take?

NAME ________________ DATE ________ PERIOD ________

Student Recording Sheet

SCORE ________

Use this recording sheet with the Standardized Test Practice pages.

Fill in the correct answer. For gridded-response questions, write your answers in the boxes on the answer grid and fill in the bubbles to match your answers.

1. Ⓐ Ⓑ Ⓒ Ⓓ

2.

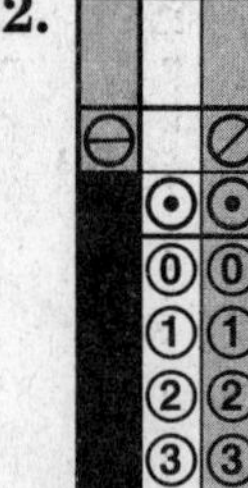

3. Ⓕ Ⓖ Ⓗ Ⓘ

4. ________________

5.

6. ________________

7. Ⓐ Ⓑ Ⓒ Ⓓ

8. Ⓕ Ⓖ Ⓗ Ⓘ

9. Ⓐ Ⓑ Ⓒ Ⓓ

10. Ⓕ Ⓖ Ⓗ Ⓘ

11. Ⓐ Ⓑ Ⓒ Ⓓ

12. Ⓕ Ⓖ Ⓗ Ⓘ

Extended Response

Record your answers for Exercise 13 on the back of this paper.

NAME ______________________ DATE ____________ PERIOD ________

Extended-Response Test

SCORE ________

Demonstrate your knowledge by giving a clear, concise solution to each problem. Be sure to include all relevant drawings and justify your answers. You may show your solutions in more than one way or investigate beyond the requirements of the problem. If necessary, record your answer on another piece of paper.

1. a. Explain what is meant by the *square root* of a number.

b. How many square roots does 36 have?

c. Draw a model to use in estimating $\sqrt{150}$. Explain your reasoning.

d. Explain what is meant by the *cube root* of a number.

e. Explain how you can determine which number is greater, 6 or $\sqrt[3]{220}$.

2. The approximate distance from each planet to the Sun is given below.

Mercury: 35,983,610 miles
Earth: 92,957,100 miles
Neptune: 2,798,842,000 miles

a. For each planet round the distance to the Sun to the nearest million miles. Write each answer in scientific notation.

Mercury:
Earth:
Neptune:

b. About how much farther away from the Sun is Earth than Mercury? Write your answer in standard form and scientific notation.

c. About how much farther away from the Sun is Neptune than Earth? Write your answer in standard form and scientific notation.

3. a. Explain what is meant by the *real number system.*

b. What is the difference between *rational* and *irrational* numbers? Give an example of each.

NAME ______________________ DATE ____________ PERIOD ________

Extended-Response Rubric

SCORE ________

Score	Description
4	A score of four is a response in which the student demonstrates a thorough understanding of the mathematics concepts and/or procedures embodied in the task. The student has responded correctly to the task, used mathematically sound procedures, and provided clear and complete explanations and interpretations. The response may contain minor flaws that do not detract from the demonstration of a thorough understanding.
3	A score of three is a response in which the student demonstrates an understanding of the mathematics concepts and/or procedures embodied in the task. The student's response to the task is essentially correct with the mathematical procedures used and the explanations and interpretations provided demonstrating an essential but less than thorough understanding. The response may contain minor flaws that reflect inattentive execution of mathematical procedures or indications of some misunderstanding of the underlying mathematics concepts and/or procedures.
2	A score of two indicates that the student has demonstrated only a partial understanding of the mathematics concepts and/or procedures embodied in the task. Although the student may have used the correct approach to obtaining a solution or may have provided a correct solution, the student's work lacks an essential understanding of the underlying mathematical concepts. The response contains errors related to misunderstanding important aspects of the task, misuse of mathematical procedures, or faulty interpretations of results.
1	A score of one indicates that the student has demonstrated a very limited understanding of the mathematics concepts and/or procedures embodied in the task. The student's response is incomplete and exhibits many flaws. Although the student's response has addressed some of the conditions of the task, the student reached an inadequate conclusion and/or provided reasoning that was faulty or incomplete. The response exhibits many flaws or may be incomplete.
0	A score of zero indicates that the student has provided no response at all, or a completely incorrect or uninterpretable response, or demonstrated insufficient understanding of the mathematics concepts and/or procedures embodied in the task. For example, a student may provide some work that is mathematically correct, but the work does not demonstrate even a rudimentary understanding of the primary focus of the task.

NAME ____________________ DATE ____________ PERIOD ________

SCORE ________

Test, Form 1A

Write the letter for the correct answer in the blank at the right of each question.

1. What is the fraction $\frac{6}{11}$ written as a decimal?

A. 0.54 **B.** $0.\overline{54}$ **C.** 0.55 **D.** 0.611 **1.** ________

2. What is the value of the expression $(-4)^3$?

F. −64 **G.** −12 **H.** 12 **I.** 64 **2.** ________

3. Which of the following is $0.\overline{7}$ as a fraction in simplest form?

A. $\frac{7}{12}$ **B.** $\frac{7}{11}$ **C.** $\frac{7}{10}$ **D.** $\frac{7}{9}$ **3.** ________

4. Using exponents, what is the simplified form of the expression $\frac{10^{15}}{10^3}$?

F. 10^{18} **G.** 10^{12} **H.** 10^5 **I.** 1^{12} **4.** ________

5. Using exponents, what is the simplified form of the expression $6^5 \cdot 6^2$?

A. 6^7 **B.** 6^{10} **C.** 36^7 **D.** 36^{10} **5.** ________

6. Rory's garden is square in shape. The length of one side of her garden is 5^2 feet. What is the area of her garden in square feet? Express your answer using exponents.

F. 10^4 **G.** 10^2 **H.** 5^4 **I.** 625 **6.** ________

7. What is the simplified form of the expression $(3x^4)^3$?

A. $9x^7$ **B.** $9x^{12}$ **C.** $27x^7$ **D.** $27x^{12}$ **7.** ________

8. What is the next term in the pattern $3^2 = 9$, $9^2 = 81$, $81^2 = 6{,}561, \ldots$?

F. $324 + 2 = 326$ **H.** $6{,}561^2 = 43{,}046{,}721$

G. $324 \times 1 = 324$ **I.** $324 \times 2 = 648$ **8.** ________

9. How is the expression 5^{-3} written using a positive exponent?

A. 3^5 **B.** 5^3 **C.** 15 **D.** $\frac{1}{5^3}$ **9.** ________

10. How is the fraction $\frac{1}{2^3}$ written using a negative exponent?

F. -3^2 **G.** -2^3 **H.** 2^{-3} **I.** 3^{-2} **10.** ________

11. What is 3.471×10^{-5} written in standard form?

A. 3,471,000 **B.** 347,100 **C.** 0.0003471 **D.** 0.00003471 **11.** ________

NAME ______________________ DATE ____________ PERIOD ________

Test, Form 1A *(continued)*

SCORE ________

12. In one 24-hour day there are 86,400 seconds. What is this number written in scientific notation?

F. 8.64×10^4 **G.** 8.64×10^2 **H.** 864×10^{-2} **I.** 864×10^{-4}

12. ________

13. What is the value of the expression below written in scientific notation?

$$(2.5 \times 10^3)(3 \times 10^2)$$

A. 750,000 **B.** 7.5×10^5 **C.** 7,500,000 **D.** 7.5×10^6

13. ________

14. What is the value of the expression below written in scientific notation?

$$(4.7 \times 10^5) - (2.8 \times 10^3)$$

F. 467,200 **H.** 1.9×10^3

G. 4.672×10^5 **I.** 1.9×10^2

14. ________

15. The speed of light is approximately 3×10^8 meters per second, while the speed of sound is approximately 3.4×10^2 meters per second. How many times faster is the speed of light than the speed of sound?

A. 9×10^3 **B.** 9×10^4 **C.** 9×10^5 **D.** 9×10^6

15. ________

16. What is the solution of the equation $y^2 = 64$?

F. 32 **G.** 8 **H.** 8 or −8 **I.** −8

16. ________

17. Which point is closest to $\sqrt{29}$ on the number line?

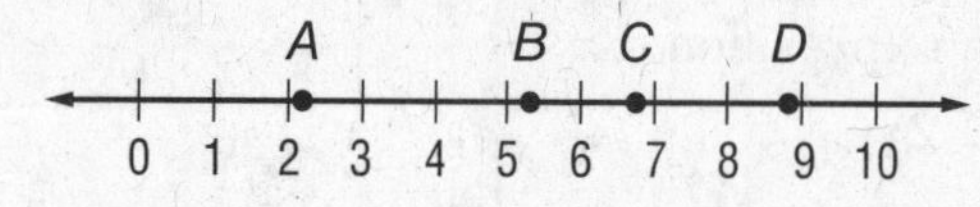

A. *A* **B.** *B* **C.** *C* **D.** *D*

17. ________

18. To which set of numbers $\frac{7}{8}$ belong?

F. rational **G.** integer **H.** irrational **I.** whole

18. ________

19. Which set of numbers is ordered from least to greatest?

A. $\left\{2.82, \sqrt{8}, \sqrt{11}, 3\frac{1}{2}\right\}$ **C.** $\left\{-\sqrt{16}, -\sqrt{17}, -\sqrt{18}, -\sqrt{19}\right\}$

B. $\left\{\sqrt{5}, -\sqrt{6}, 2\frac{1}{2}, -3\right\}$ **D.** $\left\{\sqrt{10}, 4, \sqrt{4}, 1.5\right\}$

19. ________

20. What is the value of $\sqrt[3]{8}$?

F. 2 **G.** 3 **H.** 8 **I.** 24

20. ________

NAME ______________________ DATE ____________ PERIOD ________

Test, Form 1B

SCORE ________

Write the letter for the correct answer in the blank at the right of each question.

1. What is the fraction $\frac{3}{11}$ written as a decimal?

A. 0.27 **B.** $0.\overline{27}$ **C.** 0.28 **D.** 0.311 **1.** ________

2. What is the value of the expression $(-3)^5$?

F. −243 **G.** −15 **H.** 15 **I.** 243 **2.** ________

3. Which of the following is $0.\overline{5}$ as a fraction in simplest form?

A. $\frac{5}{9}$ **B.** $\frac{5}{10}$ **C.** $\frac{5}{11}$ **D.** $\frac{5}{12}$ **3.** ________

4. Using exponents, what is the simplified form of the expression $\frac{5^{10}}{5^5}$?

F. 5^{50} **G.** 5^{15} **H.** 5^5 **I.** 1^5 **4.** ________

5. Using exponents, what is the simplified form of the expression $2^4 \cdot 2^7$?

A. 14^{28} **B.** 2^{28} **C.** 2^{11} **D.** 2^3 **5.** ________

6. The game of checkers is played on a square board. If the length of one side of the board is 4^2 inches, what is the area of the board in square inches? Express your answer using exponents.

F. 8^4 **G.** 4^4 **H.** 8^2 **I.** 32 **6.** ________

7. What is the simplified form of the expression $(4x^3)^3$?

A. $64x^9$ **B.** $64x^6$ **C.** $12x^6$ **D.** $12x^9$ **7.** ________

8. What is the next term in the pattern 100, 88, 76, 64, ...?

F. 12 **G.** 24 **H.** 36 **I.** 52 **8.** ________

9. How is the expression 4^{-2} written using a positive exponent?

A. 2^4 **B.** 4^2 **C.** 8 **D.** $\frac{1}{4^2}$ **9.** ________

10. How is the fraction $\frac{1}{5^2}$ written using a negative exponent?

F. -2^5 **G.** 5^2 **H.** 5^{-2} **I.** 2^{-5} **10.** ________

11. What is 2.1×10^4 written in standard form?

A. 210,000 **B.** 21,000 **C.** 0.0021 **D.** 0.000021 **11.** ________

Test, Form 1B *(continued)*

SCORE ________

12. In one week there are 10,080 minutes. What is this number in scientific notation?

F. 10.08×10^3 **G.** 1.008×10^4 **H.** 10.08×10^{-1} **I.** 1.008×10^{-4}

12. ________

13. What is the value of the expression below written in scientific notation?

$$(4.2 \times 10^2)(2 \times 10^3)$$

A. 840,000 **B.** 8.4×10^5 **C.** 8,400,000 **D.** 8.4×10^6

13. ________

14. What is the value of the expression below in scientific notation?

$$(4.7 \times 10^5) + (2.8 \times 10^3)$$

F. 4.728×10^{-8} **H.** 4.728×10^5

G. 4.7282×10^{-5} **I.** 472,800

14. ________

15. The top speed of a cheetah is approximately 1.2×10^2 kilometers per hour, while the speed of the fastest human is approximately 4×10^1 kilometers per hour. How many times faster is the top speed of a cheetah than the speed of a human?

A. 3×10^0 **B.** 3×10^1 **C.** 3×10^2 **D.** 3×10^3

15. ________

16. What is the solution of the equation $y^2 = 900$?

F. 30 or -30 **G.** -30 **H.** 30 **I.** 450

16. ________

17. Which point is closest to $\sqrt{41}$ on the number line?

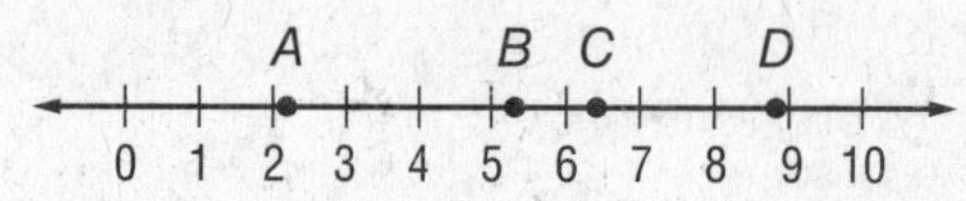

A. *A* **B.** *B* **C.** *C* **D.** *D*

17. ________

18. To which set of numbers does -5.2 belong?

F. rational **G.** integer **H.** irrational **I.** whole

18. ________

19. Which set of numbers is ordered from least to greatest?

A. $\{\sqrt{8}, 3, \sqrt{3}, 1.5\}$ **C.** $\{-\sqrt{21}, -\sqrt{22}, -\sqrt{23}, -\sqrt{24}\}$

B. $\{\sqrt{7}, -\sqrt{8}, 4\frac{1}{7}, -4\}$ **D.** $\{3.31, \sqrt{11}, \sqrt{13}, 3.61\}$

19. ________

20. Which is the value of $\sqrt[3]{27}$?

F. 2.7 **G.** 3 **H.** 9 **I.** 81

20. ________

NAME ______________________ DATE ______________ PERIOD ________

Test, Form 2A

SCORE ________

Write the letter for the correct answer in the blank at the right of each question.

1. What is the value of the expression $(-4)^3$?

A. -64 **B.** -12 **C.** 12 **D.** 64 **1.** ________

2. Using exponents, what is the simplified form of $\frac{12x^5}{6x^2}$?

F. 2^3 **G.** 6^3 **H.** $6x^3$ **I.** $2x^3$ **2.** ________

3. Using exponents, what is the simplified form of $(-3x^4y^2)^2$?

A. $-6x^6y^4$ **B.** $6x^6y^4$ **C.** $-9x^8y^4$ **D.** $9x^8y^4$ **3.** ________

4. How is the expression 10^{-5} written using a positive exponent?

F. -10^5 **G.** $\frac{1}{10^5}$ **H.** 10^{-5} **I.** 0.0001 **4.** ________

5. The Statue of Liberty weighs 450,000 pounds. What is this number written in scientific notation?

A. 4.5×10^{-5} **C.** 4.5×10^4

B. 4.5×10^{-4} **D.** 4.5×10^5 **5.** ________

6. What is the value of the expression $-\sqrt{\frac{144}{100}}$?

F. -120 **G.** $\frac{36}{25}$ **H.** $-\frac{6}{5}$ **I.** $\frac{6}{5}$ **6.** ________

7. To the nearest whole number, what is the best estimate of $\sqrt{214}$?

A. 9 **B.** 15 **C.** 36 **D.** 41.5 **7.** ________

8. Which of the following is equivalent to $-\frac{9}{15}$?

F. -9.15 **G.** -0.6 **H.** 0.6 **I.** 9.15 **8.** ________

9. Which of the following is equivalent to $0.\overline{75}$?

A. $\frac{3}{4}$ **B.** $\frac{75}{100}$ **C.** $\frac{25}{33}$ **D.** $7\frac{1}{2}$ **9.** ________

Test, Form 2A *(continued)*

SCORE ________

10. The area of a square sandbox is 83 square feet. To the nearest foot, what is the perimeter of the sandbox?

F. 9 ft **G.** 9.1 ft **H.** 36 ft **I.** 41.5 ft **10.** ________

11. Which number best represents the point graphed on the number line?

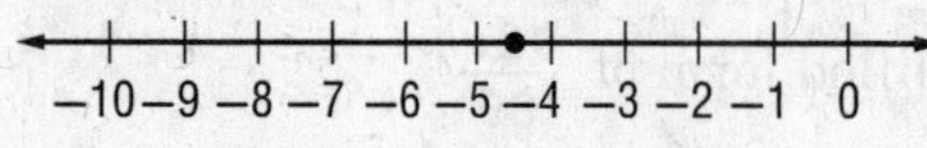

A. $\sqrt{-20}$ **B.** $-\sqrt{20}$ **C.** $-\sqrt{25}$ **D.** $\sqrt{25}$ **11.** ________

12. The band is selling 50 hats for a fundraiser. Each hat is being sold for \$12. The hats cost a total of \$400. If they sell all of the hats, how much money will be raised by the band? Use the *four-step plan*. **12.** ________

13. Recently in the United States, there were about 300,000,000 cell phone users. That same year, there were 5.7×10^9 cell phone users worldwide. About how many times larger was the number of cell phone users worldwide than in the United States? **13.** ________

14. Tito is installing a new kitchen floor. The kitchen is square in shape and has an area of 441 square feet. What is the length of one side of Tito's kitchen? **14.** ________

15. Name one whole number, one integer, one rational number, and one irrational number. Do not use the same number twice.

15. ________
Whole: ________
Integer: ________
Rational: ________
Irrational: ________

16. Find $\sqrt[3]{216}$. **16.** ________

17. Estimate $\sqrt[3]{130}$ to the nearest whole number. **17.** ________

18. Solve the equation $x^2 = 400$. **18.** ________

NAME ______________________ DATE __________ PERIOD ______

Test, Form 2B

SCORE ______

Write the letter for the correct answer in the blank at the right of each question.

1. What is the value of the expression $(-2)^5$?

A. 32 **B.** 10 **C.** −10 **D.** −32 **1.** ______

2. Using exponents, what is the simplified form of $\frac{15x^6}{3x^2}$?

F. $5x^4$ **G.** $5x^3$ **H.** $5x$ **I.** 5 **2.** ______

3. Using exponents, what is the simplified form of $(-2x^2y^3)^3$?

A. $-6x^5y^6$ **B.** $6x^5y^6$ **C.** $-8x^6y^9$ **D.** $8x^6y^9$ **3.** ______

4. How is the expression 10^{-3} written using a positive exponent?

F. 0.001 **G.** 10^{-3} **H.** $\frac{1}{10^3}$ **I.** -10^3 **4.** ______

5. The Washington Monument weighs approximately 90,800 tons. What is this number written in scientific notation?

A. 9.08×10^5 **C.** 9.08×10^{-4}

B. 9.08×10^4 **D.** 9.08×10^{-5} **5.** ______

6. What is the value of the expression $-\sqrt{\frac{196}{81}}$?

F. $\frac{14}{9}$ **G.** $\frac{14}{81}$ **H.** $-\frac{14}{81}$ **I.** $-\frac{14}{9}$ **6.** ______

7. To the nearest whole number, what is the best estimate of $\sqrt{444}$?

A. 21 **B.** 21.1 **C.** 22 **D.** 23 **7.** ______

8. Which of the following is equivalent to $\frac{-13}{40}$?

F. 13.40 **G.** 3.25 **H.** 0.325 **I.** − 0.325 **8.** ______

9. Which of the following is equivalent to $0.\overline{45}$?

A. $\frac{9}{20}$ **B.** $\frac{45}{100}$ **C.** $\frac{5}{11}$ **D.** $2\frac{1}{5}$ **9.** ______

NAME ______________________ DATE ____________ PERIOD ________

Test, Form 2B *(continued)*

SCORE ________

10. The area of a square ice rink is 404 square yards. To the nearest yard, what is the perimeter of the rink?

F. 80 yd **G.** 40 yd **H.** 20.1 yd **I.** 20 yd

10. ________

11. Which number best represents the point graphed on the number line?

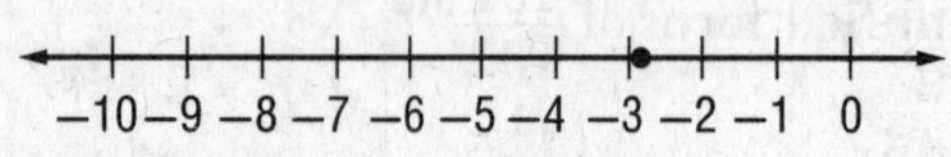

A. $\sqrt{-10}$ **B.** $-\sqrt{10}$ **C.** $-\sqrt{8}$ **D.** $\sqrt{8}$

11. ________

12. For a fundraiser, the basketball team is selling 75 wrist bands for \$3 each. The wrist bands cost a total of \$37.50. If they sell all of the wrist bands, how much money will be raised by the team? Use the *four-step plan*.

12. ________

13. In a recent year, there were about 400,000,000 mobile internet users. That same year, there were about 1.2×10^9 desktop internet users. About how many times larger was the number of desktop internet users than the number of mobile internet users?

13. ________

14. Natasha is seeding her backyard. The backyard is square in shape and has an area of 4,225 square feet. What is the length of one side of Natasha's backyard?

14. ________

15. Name a number that is not a whole number, a number that is not an integer, a number that is not a rational number, and a number that is not an irrational number. Do not use the same number twice.

15.
Not Whole: ________
Not Integer: ________
Not Rational: ________
Not Irrational: ________

16. What is the value of $\sqrt[3]{1{,}000}$?

16. ________

17. Estimate $\sqrt[3]{30}$ to the nearest whole number.

17. ________

18. Solve the equation $x^2 = 900$.

18. ________

NAME ______________________ DATE ____________ PERIOD ________

Test, Form 3A

SCORE ________

1. Evaluate the given expression if $a = 4$ and $b = -3$.

$$a^2 - b^3$$

1. ________

Simplify using the Laws of Exponents. Write each expression using a positive exponent.

2. $\frac{n^7}{n^3}$ — 2. ________

3. $-4x^2y(-3xy^3)$ — 3. ________

4. $[(u^3)^2]^4$ — 4. ________

5. $\frac{42c^4}{-6c^{12}}$ — 5. ________

6. Marta is making a quilt in the shape of a square. The length of one edge of the quilt is $2g^2h^3$. What is the area of the quilt? — 6. ________

7. Write 2.18 as a mixed number in simplest form. — 7. ________

8. Write 7^{-5} using a positive exponent. — 8. ________

9. Find the missing exponent in the equation $3y^5 \cdot y^{\square} = 3y^{10}$ — 9. ________

10. The volume of a drop of water is 0.00005 liter. Write this number in scientific notation. — 10. ________

11. Write 3.07×10^{-4} in standard form. — 11. ________

Test, Form 3A *(continued)*

SCORE ________

12. Evaluate the expression. Express the result in scientific notation. **12.** ________

$$(1.2 \times 10^4)(3.2 \times 10^{-6})$$

13. The closest distance from Venus to Earth is about 40,000,000 kilometers. The closest distance from Saturn to Earth is about 1.2×10^9 kilometers. How many times closer to Earth is Venus than Saturn? Write your answer in standard notation. **13.** ________

14. Evaluate $(2.1 \times 10^4) + (5.68 \times 10^{-2})$. Express the result in standard form. **14.** ________

15. Find $\sqrt[3]{729}$. **15.** ________

16. The area of a square carpet tile is 900 square centimeters. What is the length of one edge of the tile? **16.** ________

17. Without using a calculator, which is greater, 8 or $\sqrt[3]{510}$? Explain your reasoning. **17.** ________

18. Which number(s) in the set listed below are irrational numbers? **18.** ________

$$\left\{-\frac{2}{5}, 0.005, 3.2 \times 10^{-4}, \pi, \sqrt{13}\right\}$$

19. Order the set of numbers from least to greatest. **19.** ________

$$\left\{4.509, \frac{229}{50}, 4.09, \sqrt{21}\right\}$$

20. Graph $\sqrt{32}$ on the number line. **20.** ________

NAME ______________________ DATE ____________ PERIOD ________

Test, Form 3B

SCORE ________

1. Evaluate the given expression if $a = -2$ and $b = 5$.

$$a^3 + b^2$$

1. ________

Simplify using the Laws of Exponents. Write each expression using a positive exponent.

2. $\frac{p^5}{p^3}$

2. ________

3. $5x^4y^2(-2x^2y)$

3. ________

4. $[(p^2)^5]^2$

4. ________

5. $\frac{-28d^3}{7d^{18}}$

5. ________

6. The game of chess is played on a square shaped board. If the length of one edge of the board is $3m^4n$, what is the area of the board?

6. ________

7. Write 5.62 as a mixed number in simplest form.

7. ________

8. Write a^{-6} using a positive exponent.

8. ________

9. Find the missing exponent in the equation $-6x^{10} \cdot x^{\square} = -6x^{14}$

9. ________

10. The volume of a drop of a certain oil is 0.00002 liter. Find and write the volume of 8 drops of the oil in scientific notation.

10. ________

11. Write 2.01×10^5 in standard form.

11. ________

Test, Form 3B *(continued)*

SCORE ________

12. Evaluate the expression. Express the result in scientific notation.

$$(4.3 \times 10^2)(1.1 \times 10^{-7})$$

12. ____________

13. The closest distance from Venus to the Sun is about 46,000,000 kilometers. The closest distance from Neptune to the Sun is about 4.5×10^9 kilometers. About how many times closer to the Sun is Venus than Neptune? Write your answer in standard notation.

13. ____________

14. Evaluate $(3.61 \times 10^{-4}) + (7.8 \times 10^2)$. Express the result in standard form.

14. ____________

15. Find $\sqrt[3]{512}$.

15. ____________

16. The area of a square ceiling tile is 576 square inches. What is the length of one edge of the tile?

16. ____________

17. Without using a calculator, which is greater, 7 or $\sqrt[3]{345}$? Explain your reasoning.

17. ____________

18. Which number(s) in the set listed below are irrational numbers?

$$\left\{-\frac{3}{7}, \pi, 0.03, 2.1 \times 10^8, \sqrt{19}\right\}$$

18. ____________

19. Order the set of numbers from least to greatest.

$$\left\{5.4, \frac{537}{100}, 5.09, \sqrt{29}\right\}$$

19. ____________

20. Graph $\sqrt{56}$ on the number line.

20. ____________

NAME ______________________ DATE ______________ PERIOD ________

Are You Ready?

Review

The integers are the numbers . . . , –3, –2, –1, 0, 1, 2, 3, ... The dots indicate that the numbers go on without end in both directions. On a number line the negative numbers are to the left of zero and the positive numbers are to the right of zero, with the numbers increasing in value from left to right. Zero is neither positive nor negative.

Example 1

Determine whether the statement $-2 > 1$ is *true* or *false*.

The statement is false. On a number line the number –2 is to the left of zero and the number 1 is to the right of zero. Since numbers increase in value from left to right, $-2 < 1$.

Example 2

Determine whether the statement $5 \leq 7$ is *true* or *false*.

The statement is true. Although both numbers are to the right of zero, on a number line the number 5 is to the left of the number 7. Since numbers increase in value from left to right, $5 \leq 7$.

Exercises

Determine whether each statement is *true* or *false*.

1. $-4 \geq 3$ — 1. ______________
2. $7 < 11$ — 2. ______________
3. $-6 > -8$ — 3. ______________
4. $2 \leq 0$ — 4. ______________
5. $-\frac{3}{4} \geq -\frac{1}{2}$ — 5. ______________
6. $\frac{5}{8} < \frac{7}{8}$ — 6. ______________
7. **TEMPERATURE** The extremes in temperature in the continental United States were 134°F recorded in California on July 10, 1913 and –80°F recorded on January 23, 1971. Which of the temperatures represents a smaller value? Explain. — 7. ______________

NAME ______________________ DATE ____________ PERIOD ________

Are You Ready?

Practice

Add, subtract, multiply, or divide.

1. $45 \div (-5)$ **1.** ____________

2. $56 + (-16)$ **2.** ____________

3. $-20(12)$ **3.** ____________

4. $33 - (-35)$ **4.** ____________

Determine whether each statement is *true* or *false*.

5. $\frac{2}{3} < \frac{10}{30}$ **5.** ____________

6. $-44 > 1$ **6.** ____________

7. **GOLF** A golfer's score is measured against what a professional golfer would score for the course, or "par." If a golfer is −1 after the first nine holes, it means he or she is 1 stroke below par. Suppose Stevie scored −2 overall for the first nine holes and −3 overall for the last nine holes. What is his final score after eighteen holes? **7.** ____________

8. **SHOPPING** Corbin is shopping for a pair of jeans and a shirt. What will it cost Corbin to buy one of each?

Item	Unit Cost
Shirt	$27
Belt	$11
Jeans	$38

8. ____________

NAME ______________________ DATE ____________ PERIOD ________

Are You Ready?

Apply

1. **TOURS** Fred and Mary's Ice Cream Company gives tours of their facilities on Mondays and Wednesdays. If 45 people went on the Monday tour and 77 people went on the Wednesday tour, how many people toured the facilities this week?

2. **MOVIES** Mr. Ramiro wants to take his 6 adult children to the movies. If one adult ticket cost $9, how much will it cost Mr. Ramiro to take himself and his children to the movies?

3. **ELEVATION** Death Valley, located in the United States, is 282 feet below sea level. The Sea of Galilee, located in Israel, is 682 feet below sea level. How much further below the lowest point in Death Valley is the lowest point in the Sea of Galilee?

4. **ANGLES** Two angles are said to be supplementary when the sum of their measures is 180°. Find the measure of the supplementary angle to each of the given angle measures.

Angle Measure	Supplementary Measure
35°	
144°	
83°	

5. **POPULATION** In a small town, $\frac{3}{4}$ of the population is older than 55. In a larger town, $\frac{2}{3}$ of the population is older than 55. Which town has the greater fraction of its population over 55, the smaller or the larger town?

6. **AVERAGING** Gabe recorded the low temperature on Saturday to be −13° and on Sunday to be −7°. What was the average low temperature over the weekend?

NAME ______________________ DATE ______________ PERIOD ________

Diagnostic Test

Add, subtract, multiply, or divide.

1. $39 - (-11)$ **1.** ________

2. $7(-12)$ **2.** ________

3. $-44 + 25$ **3.** ________

4. $63 \div 3$ **4.** ________

5. **SAVINGS ACCOUNT** Mr. Swisher made the following deposits to his savings account. How much did he deposit altogether?

Week	Deposit
1	$45
2	$72
3	$36

5. ________

6. **RUNNING** Mia runs 5 miles per day, 6 days per week. How many miles does Mia run each week? **6.** ________

7. **SCUBA DIVING** Reuben and Melina are scuba diving. Reuben is at a depth of −3 meters and Melina is at a depth of −7 meters. Who is farther from the surface of the water? **7.** ________

Evaluate each expression if $x = -5$ and $y = 3$.

8. $x + 6$ **8.** ________

9. $-1 - y$ **9.** ________

10. xy **10.** ________

11. $-8x$ **11.** ________

NAME ______________________ DATE ______________ PERIOD ________

Pretest

Solve each equation. Check your solution.

1. $x - 3 = -17$ 1. ____________

2. $-14 = \frac{q}{3}$ 2. ____________

3. $-4p + 21 = -43$ 3. ____________

4. $\frac{n}{3} - 7 = -17$ 4. ____________

5. **SCHOOL DANCE** Forty-two students arrived at the school dance after it started. In all, 78 students attended. Write and solve an equation to find how many students arrived at the dance before it started. 5. ____________

Name the property shown by each statement.

6. $m + n = n + m$ 6. ____________

7. $(pq)r = p(qr)$ 7. ____________

8. $-8(a - 2) = -8a + 16$ 8. ____________

Solve each equation. Check your solution.

9. $4n - 3 = -2n + 15$ 9. ____________

10. $2(2 - x) = 4(-2 + x)$ 10. ____________

NAME ______________________ DATE ______________ PERIOD ________

Chapter Quiz

Solve each equation.

1. $\frac{5}{6}x = 35$ **1.** ________

2. $\frac{7}{9}x = 42$ **2.** ________

3. $3.5v = 161$ **3.** ________

4. $0.6r = 4.5$ **4.** ________

5. $-\frac{4}{9}x = \frac{2}{3}$ **5.** ________

6. $3g - 5 = 7$ **6.** ________

7. $10 + 6y = 16$ **7.** ________

8. **SPORTS EQUIPMENT** The price of a baseball glove is $8 more than half the price of spikes. The glove costs $54. Solve the equation $\frac{s}{2} + 8 = 54$ to find out how much the spikes cost. **8.** ________

Translate each sentence into an equation.

9. Five more than twice a number is 17. **9.** ________

10. The difference between a number divided by 3 and 4 is 20. **10.** ________

11. **TENNIS** Audrey has $120 to spend on a tennis racket and lessons. The racket costs $45 and the lessons cost $15 per hour. Define a variable. Then write and solve an equation to find how many hours of lessons she can afford. **11.** ________

12. **EXERCISE** A one-year membership to a gym costs $725. The registration fee is $125, and the remaining amount is paid monthly. Define a variable. Then write and solve an equation to find how much new members pay each month. **12.** ________

NAME ______________________ DATE ____________ PERIOD ________

Vocabulary Test

SCORE ________

coefficient	multiplicative inverses	properties
identity	null set	two-step

Choose from the terms above to complete each sentence.

1. ________________ are statements that are true for any number. **1.** ________________

2. Two numbers with a product of 1 are called ________________. **2.** ________________

3. A(n) ________________ contains two operations. **3.** ________________

4. The numerical factor of a term that contains a variable is called the ________________ of the variable. **4.** ________________

Define each term in your own words

5. null set **5.** ________________

6. identity **6.** ________________

NAME ______________________ DATE ______________ PERIOD ________

Standardized Test Practice

Read each question. Then fill in the correct answer on the answer document provided by your teacher or on a sheet of paper.

1. Which of the following equations matches the description below?

Six more than the quotient of a number and three is 14.

A. $14 = \frac{x}{3} + 6$

B. $6 = 14 + \frac{x}{3}$

C. $14 = \frac{x + 6}{3}$

D. $6 = \frac{x + 14}{3}$

2. THINK SOLVE EXPLAIN **SHORT RESPONSE** The sum of a number and 6 is 23. Write an equation to represent this situation.

3. Zach, Luke, and Charlie ordered a large pizza for \$11.99, breadsticks for \$2.99, and chicken wings for \$5.99. If the three friends agree to split the cost of the food evenly, how much will each friend pay?

F. \$20.79

G. \$7.93

H. \$7.32

I. \$6.99

4. Which property is illustrated by the equation below?

$$5(x + 2) = 5x + 10$$

A. Associative Property of Addition

B. Commutative Property of Addition

C. Distributive Property

D. Multiplicative Identity

5. THINK SOLVE EXPLAIN **SHORT RESPONSE** Simplify the expression shown below.

$$(3m^3n^2)(6m^4n)$$

6. **GRIDDED RESPONSE** To approximate the radius r of a circle, you can use the formula $r = \sqrt{\frac{A}{3.14}}$, where A is the area of the circle. Find the radius in feet of the circle below. Round to the nearest tenth.

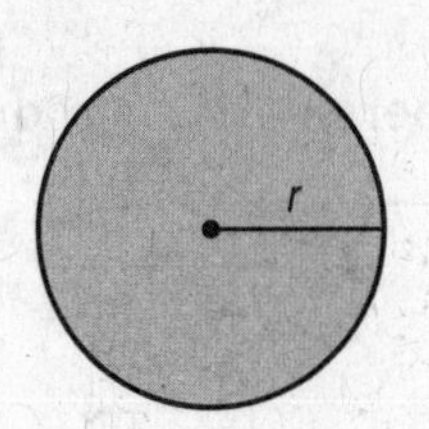

$A = 60$ ft^2

7. What is the value of m in the equation below?

$$4m + 7 = -3m + 49$$

F. −6

G. 6

H. 12

I. 42

8. THINK SOLVE EXPLAIN **SHORT RESPONSE** What value of x makes the polygons below have the same perimeter?

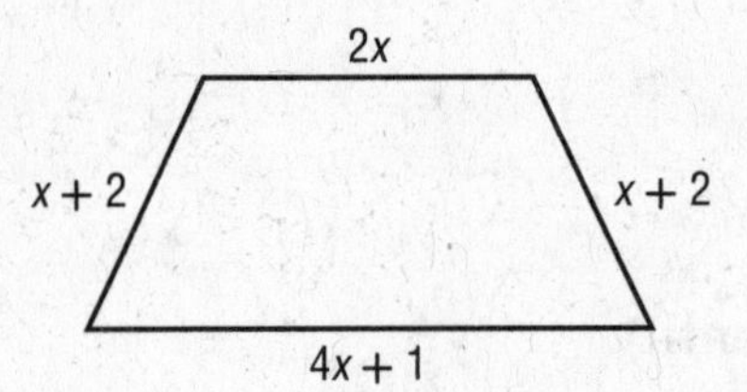

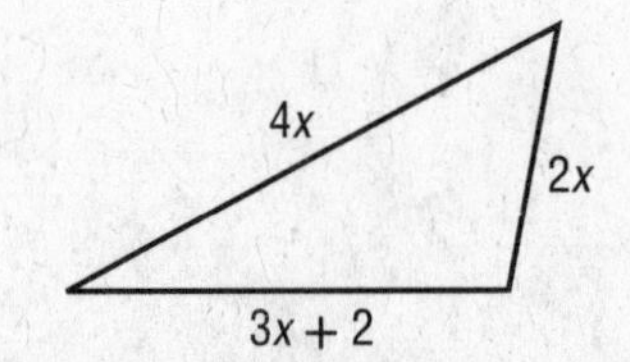

NAME ______________________ DATE ____________ PERIOD ________

9. Minnie lives 200 miles away from her grandmother. It took her 5 hours to drive this distance. Which equation can be used to find r, the rate of travel?

A. $200 = 5 + r$

B. $200r = 5$

C. $\frac{5}{r} = 200$

D. $200 = 5r$

10. What is the solution of the equation $-2 = -4 + t$?

F. -6

G. -2

H. 2

I. 6

11. THINK SOLVE EXPLAIN **SHORT RESPONSE** Solve the equation $-3x - 5 = -14$.

12. What is the solution of the equation below?

$$-2(2p - 4) = 3(p + 12)$$

A. -28

B. -4

C. 4

D. 28

13. What is the simplified form of $\frac{7x^2y^3}{28xy}$?

F. xy^2

G. $4xy^2$

H. $\frac{xy^2}{4}$

I. $\frac{yx}{4}$

14. THINK SOLVE EXPLAIN **SHORT RESPONSE** Simplify and express your answer using a positive exponent.

$$5^{-3} \cdot 5^7$$

15. Which property is illustrated by the equation below?

$$5 \cdot 12 \cdot 25 = 5 \cdot 25 \cdot 12$$

A. Associative Property of Multiplication

B. Commutative Property of Multiplication

C. Distributive Property

D. Multiplicative Identity

16. What is the solution of the equation below?

$$2p - 14 = 4(p + 5)$$

F. -17

G. -16

H. 16

I. 17

17. THINK SOLVE EXPLAIN **EXTENDED RESPONSE** The polygon below has the side lengths shown.

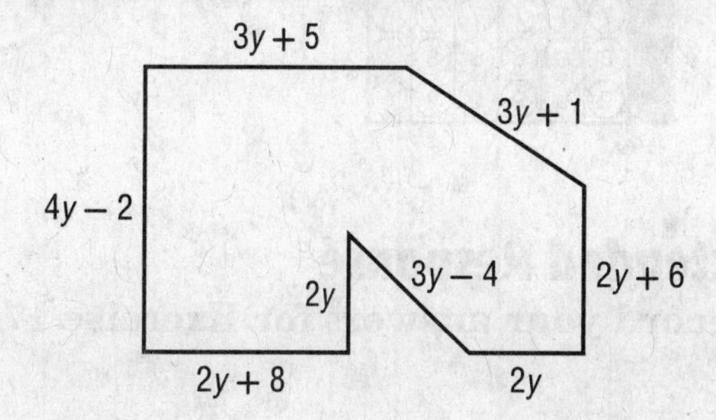

Part A Write an expression in simplest form for the perimeter of the figure.

Part B If the perimeter is 77 units, what is the value of y?

NAME ______________________________ DATE ______________ PERIOD ________

Student Recording Sheet

SCORE ________

Use this recording sheet with the Standardized Test Practice pages.

Fill in the correct answer. For gridded-response questions, write your answers in the boxes on the answer grid and fill in the bubbles to match your answers.

1. Ⓐ Ⓑ Ⓒ Ⓓ

2. ____________

3. Ⓕ Ⓖ Ⓗ Ⓘ

4. Ⓐ Ⓑ Ⓒ Ⓓ

5. ____________

6.

⊖		⊘	⊘	⊘	⊘
	⊙	⊙	⊙	⊙	⊙
	0	0	0	0	0
	1	1	1	1	1
	2	2	2	2	2
	3	3	3	3	3
	4	4	4	4	4
	5	5	5	5	5
	6	6	6	6	6
	7	7	7	7	7
	8	8	8	8	8
	9	9	9	9	9

7. Ⓕ Ⓖ Ⓗ Ⓘ

8. ____________

9. Ⓐ Ⓑ Ⓒ Ⓓ

10. Ⓕ Ⓖ Ⓗ Ⓘ

11. ____________

12. ____________

13. Ⓕ Ⓖ Ⓗ Ⓘ

14. ____________

15. Ⓐ Ⓑ Ⓒ Ⓓ

16. Ⓕ Ⓖ Ⓗ Ⓘ

Extended Response

Record your answers for Exercise 17 on the back of this paper.

NAME ______________________ DATE ____________ PERIOD ________

Extended-Response Test

SCORE ________

Below is an example of an architect's floor plan for a house she is designing for Marlien. Demonstrate your knowledge by giving a clear, concise solution to each problem. Be sure to include all relevant drawings and justify your answers. You may show your solution in more than one way or investigate beyond the requirements of the problem. If necessary, record your answers on another piece of paper.

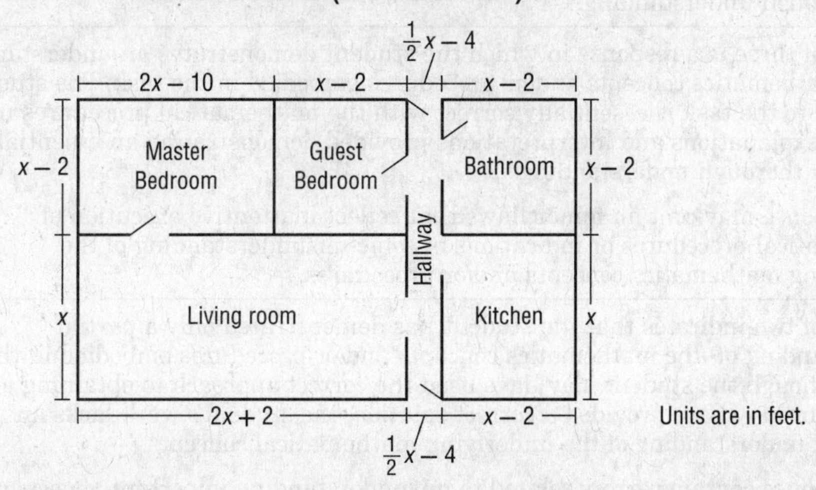

1. **a.** Write the algebraic expressions representing the length and width of Marlien's house.

   b. Write the algebraic expression representing the perimeter of Marlien's house in simplest form.

   c. If the perimeter is 142 feet, find the value of x.

2. **a.** Show how the Distributive Property can be used to find the area of Marlien's living room.

   b. Write an algebraic expression for the area of the living room.

3. **a.** Using the algebraic expressions for the length and width of Marlien's living room found in the diagram and the value of x found in Exercise 1c, find the actual length and width of her living room.

   b. Find the actual area of Marlien's living room.

4. Using the algebraic expression for area found in Exercise 2b, find the area of the living room.

5. Are your answers to Exercise 3b and Exercise 4 the same? What can you conclude?

NAME ______________________ DATE ____________ PERIOD ________

Extended-Response Rubric

SCORE ________

Score	Description
4	A score of four is a response in which the student demonstrates a thorough understanding of the mathematics concepts and/or procedures embodied in the task. The student has responded correctly to the task, used mathematically sound procedures, and provided clear and complete explanations and interpretations. The response may contain minor flaws that do not detract from the demonstration of a thorough understanding.
3	A score of three is a response in which the student demonstrates an understanding of the mathematics concepts and/or procedures embodied in the task. The student's response to the task is essentially correct with the mathematical procedures used and the explanations and interpretations provided demonstrating an essential but less than thorough understanding. The response may contain minor flaws that reflect inattentive execution of mathematical procedures or indications of some misunderstanding of the underlying mathematics concepts and/or procedures.
2	A score of two indicates that the student has demonstrated only a partial understanding of the mathematics concepts and/or procedures embodied in the task. Although the student may have used the correct approach to obtaining a solution or may have provided a correct solution, the student's work lacks an essential understanding of the underlying mathematical concepts. The response contains errors related to misunderstanding important aspects of the task, misuse of mathematical procedures, or faulty interpretations of results.
1	A score of one indicates that the student has demonstrated a very limited understanding of the mathematics concepts and/or procedures embodied in the task. The student's response is incomplete and exhibits many flaws. Although the student's response has addressed some of the conditions of the task, the student reached an inadequate conclusion and/or provided reasoning that was faulty or incomplete. The response exhibits many flaws or may be incomplete.
0	A score of zero indicates that the student has provided no response at all, or a completely incorrect or uninterpretable response, or demonstrated insufficient understanding of the mathematics concepts and/or procedures embodied in the task. For example, a student may provide some work that is mathematically correct, but the work does not demonstrate even a rudimentary understanding of the primary focus of the task.

NAME ______________________ DATE ____________ PERIOD ________

Test, Form 1A

SCORE ________

Write the letter for the correct answer in the blank at the right of each question.

Translate each sentence into an equation.

1. The sum of five times a number and −6 is −2.

A. $-6n + 5 = -2$ **C.** $5n - (-6) = -2$

B. $\frac{n}{5} - 6 = -2$ **D.** $5n + (-6) = -2$ **1.** ________

2. Three less than one-half a number is −71.

F. $-\frac{1}{2}n + 2 = -71$ **H.** $\frac{1}{2}n - 3 = -71$

G. $2n - \frac{1}{2} = -71$ **I.** $3 - \frac{1}{2}n = -71$ **2.** ________

Solve each equation.

3. $10 + \frac{1}{3}y = 1$

A. −30 **B.** −27 **C.** 27 **D.** 30 **3.** ________

4. $-0.4w = 4.2$

F. 105 **G.** −10.5 **H.** −105 **I.** 10.5 **4.** ________

5. $\frac{x}{2} - 5 = -3$

A. 4 **B.** 1 **C.** −4 **D.** −16 **5.** ________

6. $-5 - 3w = 7w$

F. 4 **G.** 2 **H.** −4 **I.** −2 **6.** ________

7. $\frac{4}{7}w = 16$

A. 4 **B.** 14 **C.** 28 **D.** 112 **7.** ________

8. Marianna wants to buy a new tennis racket that costs $57.50. She has $8 and plans to save $4.50 each week. How many weeks will it take her to save the money?

F. 24 weeks **H.** 11 weeks

G. 15 weeks **I.** 10 weeks **8.** ________

NAME ______________________ DATE ____________ PERIOD ________

Test, Form 1A *(continued)*

SCORE ________

9. In a contest, each finalist must answer 5 questions correctly. Each question is worth twice as much as the question before it. The fifth question is worth \$1,600. How much is the first question worth?

A. \$800 **C.** \$200

B. \$400 **D.** \$100 **9.** ________

Solve each equation.

10. $4x - 2 = 22 - 8x$

F. -6 **G.** -2 **H.** 2 **I.** 6 **10.** ________

11. $5n - 12 = -3n + 4$

A. 2 **C.** all real numbers

B. 1 **D.** -2 **11.** ________

12. $49 - 3m = 4m + 14$

F. all real numbers **H.** 3

G. 5 **I.** 1 **12.** ________

13. $-2y - 3y + 8 = 8 - 5y - 12$

A. -11 **B.** 2 **C.** null set **D.** 11 **13.** ________

14. $-3(p + 2) = -30$

F. $-\frac{32}{3}$ **G.** 8 **H.** null set **I.** $\frac{-32}{3}$ **14.** ________

15. $0.4(2 - q) = 0.2(q + 7)$

A. -3 **C.** 3

B. -1 **D.** all real numbers **15.** ________

16. The Hazell family has 4 children. Murphy is 1 year younger than his older brother Michael. Keira is 2 years younger than Murphy. Isabelle and Keira are twins. If Michael is 8, how old is Isabelle?

F. 8 **H.** 5

G. 7 **I.** 4 **16.** ________

17. Sarah and Bryan went shopping and spent a total of \$47.50. Bryan spent \$15.50 less than what Sarah spent. How much did Bryan spend?

A. \$31.50 **C.** \$16

B. \$31 **D.** \$15.50 **17.** ________

NAME ______________________ DATE ____________ PERIOD ________

Test, Form 1B

SCORE ________

Write the letter for the correct answer in the blank at the right of each question.

Translate each sentence into an equation.

1. Four times a number increased by 3 is −89.

A. $-3n + 4 = -89$ **C.** $4n + 3 = -89$

B. $4n - 3 = -89$ **D.** $4 + 3n = -89$ **1.** ________

2. Five more than three-fourths a number is −19.

F. $-\frac{3}{4}n + 5 = -19$ **H.** $\frac{3}{4}n + 5 = -19$

G. $5n - \frac{3}{4} = -19$ **I.** $5 - \frac{3}{4}n = -19$ **2.** ________

Solve each equation.

3. $15 + \frac{1}{4}p = 2$

A. 60 **B.** −52 **C.** 52 **D.** −60 **3.** ________

4. $-0.5x = 3.6$

F. −72 **G.** −7.2 **H.** 7.2 **I.** 72 **4.** ________

5. $\frac{d}{3} - 10 = -2$

A. 36 **B.** 24 **C.** −24 **D.** −36 **5.** ________

6. $8 - 3m = 26$

F. 18 **G.** 6 **H.** −6 **I.** −18 **6.** ________

7. $\frac{7}{9}w = 56$

A. 8 **B.** 9 **C.** 72 **D.** 504 **7.** ________

8. Guadalupe wants to buy new goggles that cost \$31.50. She has \$4.50 and plans to save \$2.25 each week. How many weeks will it take her to save the money?

F. 14 weeks **H.** 11 weeks

G. 12 weeks **I.** 10 weeks **8.** ________

Test, Form 1B *(continued)*

SCORE ________

9. In a contest, each finalist must answer 4 questions correctly. Each question is worth twice as much as the question before it. The fourth question is worth \$2,000. How much is the first question worth?

A. \$1,000 **C.** \$250
B. \$500 **D.** \$125

9. ________

Solve each equation.

10. $3x - 4 = 18 + 5x$

F. 22 **G.** 11 **H.** -11 **I.** -22

10. ________

11. $4u - 2 = -6u + 28$

A. -15 **C.** all real numbers
B. 3 **D.** 15

11. ________

12. $-3x + 3 = -15 + 6x$

F. null set **G.** 2 **H.** 4 **I.** -2

12. ________

13. $-6x - x + 10 = 15 - 7x - 5$

A. all real numbers **C.** 10
B. -12 **D.** 12

13. ________

14. $-2(p - 1) = 15$

F. $\frac{13}{2}$ **H.** 8
G. all real numbers **I.** $-\frac{13}{2}$

14. ________

15. $0.3(r + 2) = -0.1(-2r - 4)$

A. -22 **B.** -1 **C.** -2 **D.** null set

15. ________

16. The Walsh family has 4 children. Ryan is 2 years younger than his older brother Patrick. Kelly is 2 years younger than Ryan. Caroline and Kelly are twins. If Patrick is 12, how old is Caroline?

F. 8 **H.** 10
G. 9 **I.** 11

16. ________

17. Chris and Lisa went shopping and spent a total of \$25.50. Lisa spent \$13.50 more than what Chris spent. How much did Lisa spend?

A. \$12 **C.** \$19
B. \$19.50 **D.** \$6.50

17. ________

NAME ______________________ DATE ____________ PERIOD ________

Test, Form 2A

SCORE ________

Write the letter for the correct answer in the blank at the right of each question.

Translate each sentence into an equation.

1. 12 birds are 3 more than twice the number of birds Rhonda saw yesterday.

A. $12 = 3b + 2$ **C.** $12 = 2b + 3$

B. $12 = 3 - 2b$ **D.** $12 = \frac{b}{3} + 2$ **1.** ________

2. The difference between two-thirds of a number and 4 is −92.

F. $\frac{2}{3}n - 4 = -92$ **H.** $4n - \frac{2}{3} = -92$

G. $\frac{2}{3} - 4n = -92$ **I.** $-\frac{2}{3}n + 4 = -92$ **2.** ________

3. Negative 6 times the sum of a number and 4 is 2.

A. $-6n + 4 = 2$ **C.** $-6 + 4n = 2$

B. $-6(n + 4) = 2$ **D.** $-6n - 4 = 2$ **3.** ________

Solve each equation.

4. $-2.17 = 0.35r$

F. 6.2 **H.** −7.6

G. all real numbers **I.** −6.2 **4.** ________

5. $2\frac{2}{5}w = 21\frac{3}{5}$

A. −9 **B.** 5 **C.** null set **D.** 9 **5.** ________

6. $-25 = \frac{1}{3}n - 10$

F. null set **G.** 45 **H.** −15 **I.** −45 **6.** ________

7. $4 - 5y = -16$

A. −5 **C.** 5

B. 4 **D.** all real numbers **7.** ________

8. $-17 = -7c + 4$

F. $\frac{7}{13}$ **H.** 3

G. $\frac{13}{7}$ **I.** all real numbers **8.** ________

9. $\frac{x + 5}{4} = -4$

A. null set **B.** 21 **C.** −21 **D.** −36 **9.** ________

NAME ______________________ DATE ____________ PERIOD ________

Test, Form 2A *(continued)*

SCORE ________

10. Elyse wants to buy a new softball glove that costs $46.50. She has $15 and plans to save $5.25 each week. How many weeks will it take her to save the money?

F. 9 weeks **G.** 8 weeks **H.** 7 weeks **I.** 6 weeks **10.** ________

11. To catch an 8:30 A.M. bus, Kendra needs 45 minutes to shower and dress, 20 minutes for breakfast, and 10 minutes to walk to the bus stop. To catch the bus, what is the latest time she should wake up?

A. 6:45 A.M. **B.** 7:05 A.M. **C.** 7:15 A.M. **D.** 7:25 A.M. **11.** ________

Solve each equation.

12. $-5x = -40 + 3x$

F. 20 **H.** -5

G. 5 **I.** all real numbers **12.** ________

13. $\frac{3}{4}(x - 16) = -2(x - 3) + 4$

A. null set **B.** 8 **C.** 4 **D.** -4 **13.** ________

14. $-7b - 3 = -3b + 5$

F. -2 **G.** 2 **H.** -5 **I.** null set **14.** ________

15. $-2(y - 4) = 20 - 2y - 12$

A. 4 **C.** -4

B. all real numbers **D.** -12 **15.** ________

16. $2(v - 4) - 10 = -2(-1 + 4v)$ **16.** ________

17. Mabel scored 19 points more on her pre-algebra test than Nancy. Phoebe scored 10 points less on her pre-algebra test than Nancy. If Phoebe scored 23 points, how many points did Mabel score? **17.** ________

18. The figures below show sketches of Earl's and Dylan's flower gardens. If the perimeter of each of their gardens is the same, what is the length and width of Earl's garden? **18.** ________

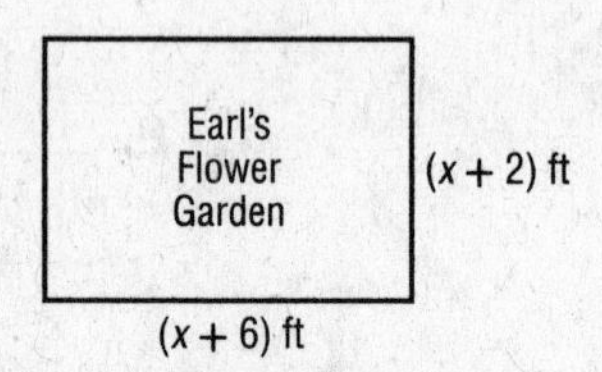

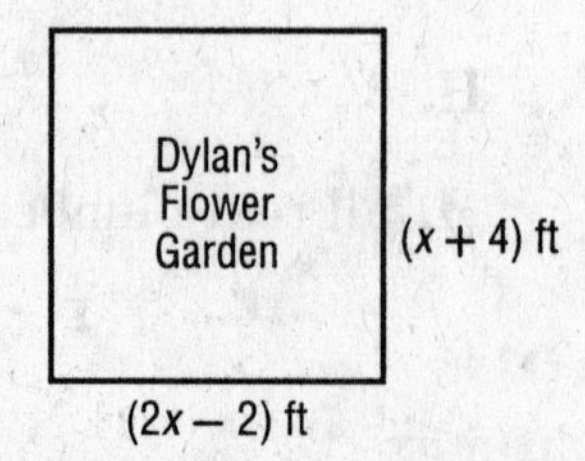

NAME ______________________ DATE ____________ PERIOD ________

Test, Form 2B

SCORE ________

Write the letter for the correct answer in the blank at the right of each question.

Translate each sentence into an equation.

1. 7 berries are 5 less than twice the number of berries Mickey had for lunch.

A. $7 = 5 - 2m$ **C.** $7 = 2m - 5$

B. $5 = 2m - 7$ **D.** $12 = \frac{m}{2}$ **1.** ________

2. The difference between three-fifths of a number and 7 is -36.

F. $-\frac{3}{5}n + 7 = -36$ **H.** $7n - \frac{3}{5} = -36$

G. $\frac{3}{5} - 7n = -36$ **I.** $\frac{3}{5}n - 7 = -36$ **2.** ________

3. Negative 4 times the difference of a number and 7 is 12.

A. $-4 + 7n = 12$ **C.** $-4n + 7 = 12$

B. $-4n - 7 = 12$ **D.** $-4(n - 7) = 12$ **3.** ________

Solve each equation.

4. $-2.73 = -0.42r$

F. 6.5 **G.** 7.2 **H.** null set **I.** -6.5 **4.** ________

5. $1\frac{7}{8}w = -11\frac{1}{4}$

A. -6 **B.** 6 **C.** null set **D.** 10 **5.** ________

6. $-17 = \frac{1}{5}n - 20$

F. 18 **H.** 15

G. all real numbers **I.** -15 **6.** ________

7. $18 - 4d = 34$

A. -5 **C.** 5

B. -4 **D.** all real numbers **7.** ________

8. $47 = 3 - 6y$

F. $-\frac{3}{22}$ **G.** $-\frac{22}{3}$ **H.** null set **I.** $-\frac{25}{3}$ **8.** ________

9. $\frac{x + 7}{2} = -10$

A. all real numbers **C.** 13

B. 27 **D.** -27 **9.** ________

NAME ______________________ DATE ____________ PERIOD ________

Test, Form 2B *(continued)*

SCORE ________

10. Cameron wants to buy new lacrosse equipment that costs \$75.25. She has \$20 and plans to save \$4.25 each week. How many weeks will it take her to save the money?

F. 14 weeks **G.** 13 weeks **H.** 12 weeks **I.** 11 weeks

10. ________

11. To go to dance class at 6:45 P.M. bus, Kelly needs 35 minutes to walk home from a friend's house, 30 minutes for dinner, and 20 minutes to drive to the class. To make class on time, what is the latest time she should leave her friend's house?

A. 4:20 P.M. **B.** 4:55 P.M. **C.** 5:20 P.M. **D.** 5:45 P.M.

11. ________

Solve each equation.

12. $-2y = 45 + 7y$

F. 20 **H.** -5
G. 5 **I.** all real numbers

12. ________

13. $\frac{2}{3}(p - 12) = -(2p - 1) + 7$

A. all real numbers **C.** -4
B. -6 **D.** 6

13. ________

14. $5g - 7 = -3g + 1$

F. -4 **G.** -1 **H.** 1 **I.** 4

14. ________

15. $-5(c - 2) = 20 - 5c + 10$

A. 4 **B.** null set **C.** 1 **D.** -1

15. ________

16. $5(m + 4) = -2(-4 - m) + 3$

16. ________

17. Gus has skydived 4 more times than Nico. Emma has skydived twice as many times as Nico. If Emma has skydived 16 times, how many times has Gus skydived?

17. ________

18. The sketches below show Cat's and Fred's driveways. If the perimeter of each of their driveways is the same, what is the length and width of Cat's driveway?

18. ________

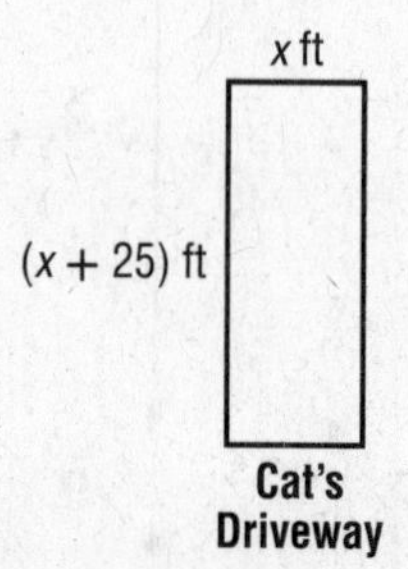

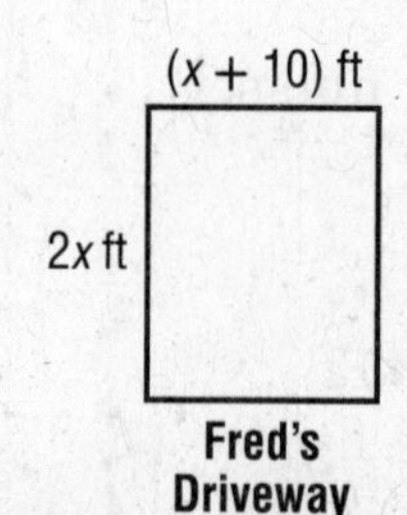

NAME ______________________ DATE ____________ PERIOD ________

Test, Form 3A

SCORE ________

1. Susan is 5 years older than her sister. The sum of their ages is 51. Define a variable. Then write an equation that could be used to find their ages.

1. ________

2. Two beakers plus their contents have a mass of 180.4 grams. The total mass of the contents is 56.8 grams. Write and solve an equation to find the mass of one beaker.

2. ________

3. At a concert, you purchase 3 T-shirts and a concert program for a total cost of $90. The program cost $15 and the T-shirts all cost the same. Write and solve an equation to find the cost of one T-shirt.

3. ________

Solve each equation.

4. $-1.4d = 0.7$

4. ________

5. $1\frac{2}{3}m + 2 = 2\frac{1}{6}$

5. ________

6. $-14.2 = -4.2g + 6.8$

6. ________

7. $-w = -10 + 4w$

7. ________

8. $\frac{3}{4}n = -1\frac{3}{4}n - 18$

8. ________

9. $-3.6b - 7.2 = -12.7 - 6.1b$

9. ________

10. An online movie streaming plan charges an annual fee of $45 plus $2.50 per movie watched. Another plan has no annual fee but charges $3.75 per movie watched. For how many movies is the cost of the plans the same?

10. ________

Test, Form 3A *(continued)*

SCORE ________

11. Find the value of x so that the polygons have the same perimeter

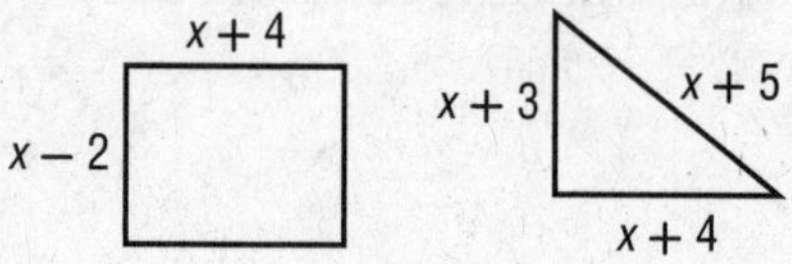

11. ________

Solve each equation.

12. $-50 = -2(a + 3)$ **12.** ________

13. $4(x - 2) = 2(x - 4) + 2x$ **13.** ________

14. $5(y - 2) - 2 = 2(y + 1) - 5$ **14.** ________

15. $-4(p + 1) = -2(8 - 2p)$ **15.** ________

16. The table shows the number of points scored by three players in last night's basketball game. If Gil and Darby scored the same number of points, how many points did Josiah score?

Player	Points
Josiah	x
Darby	$2x + 8$
Gil	$3x - 4$

16. ________

17. The table shows the number of tulip bulbs Chloe and Grady planted. If they each planted the same number of bulbs, how many did each plant?

Name	Number of Bulbs Planted
Chloe	$3(t + 1)$
Grady	$3(2t - 3)$

17. ________

18. Tony and some friends went to the movies. They bought 4 drinks and 2 tubs of popcorn and spent a total of $32.50 on the food. Each drink costs $3.50 less than a tub of popcorn.

a. Define a variable. Write an equation that can be used to find the cost of one tub of popcorn. **18a.** ________

b. Solve the equation to find the cost of a tub of popcorn. **18b.** ________

NAME ______________________ DATE ____________ PERIOD ________

Test, Form 3B

SCORE ________

1. Kenny has 9 more comic books than Bobbie. Together they have 95 comic books. Define a variable. Then write an equation that could be used to find the number of comic books they each have. **1.** ________

2. You and 3 friends pay \$26.55 for a pizza and 4 of the same kind of drinks. The pizza cost \$18.75. Write and solve an equation to find the cost of one drink. **2.** ________

3. Crystal bowled two games for a total score of 202. Her score for the second game was 30 points less than the score of her first game. Write and solve an equation to find her score for the second game. **3.** ________

Solve each equation.

4. $-0.7y = 9.1$ **4.** ________

5. $2\frac{1}{4}m + 3 = 4\frac{1}{8}$ **5.** ________

6. $-19.2 = -3.6x + 2.4$ **6.** ________

7. $-2a = 12 - 4a$ **7.** ________

8. $-2\frac{2}{3}n + 21 = \frac{-1}{3}n$ **8.** ________

9. $-2.3c - 6.6 = -12.2 - 3.9c$ **9.** ________

10. An online movie streaming plan has no annual fee but charges \$4.25 per movie watched. Another plan charges an annual fee of \$36 plus \$3.50 per movie watched. For how many movies is the cost of the plans the same? **10.** ________

Test, Form 3B *(continued)*

SCORE ________

11. Find the value of x so that the polygons have the same perimeter.

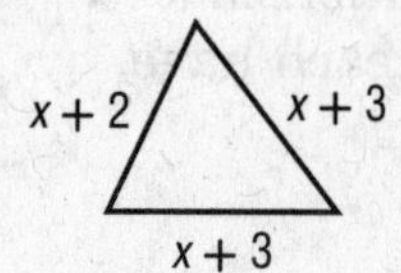

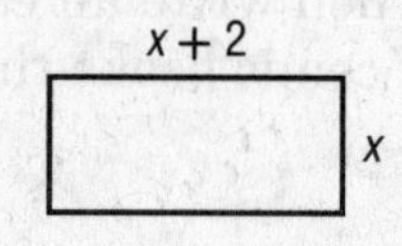

11. ________

Solve each equation.

12. $-30 = -2(-n + 3)$ **12.** ________

13. $7(1 - p) = 2(1 - 3p)$ **13.** ________

14. $-3(q - 4) + 15 = -5(q - 7) - 10$ **14.** ________

15. $3(y - 2) + 15 = -3(y - 3) + 6y$ **15.** ________

16. The table shows the number of hits made by three players in yesterday's softball game. If Mercedes and Kiaya had the same number of hits, how many hits did Evelyn have?

Player	Points
Evelyn	x
Mercedes	$3x - 1$
Kiaya	$4x - 2$

16. ________

17. The table shows the number of fish Callie and Jada each caught. If they caught the same number of fish, how many did each catch?

Name	Number of Fish Caught
Callie	$2(3t + 1)$
Jada	$4(2t - 1)$

17. ________

18. Bonnie and some friends went to an amusement park. They bought five of the same lunches and 3 desserts and spent a total of $60.25 on the food. Each dessert costs $5.25 less than one of the lunches.

a. Define a variable. Write an equation that can be used to find the cost of lunch. **18a.** ________

b. Solve the equation to find the cost of a lunch. **18b.** ________

NAME ______________________ DATE ____________ PERIOD ________

Are You Ready?

Review

To subtract integers, add the opposite of the second integer to the first integer.

Example 1

Find 5 – 11.

$5 + (-11)$ Add the opposite of 11.

$= -6$

Example 2

Find –8 – (–14).

$-8 + 14$ Add the opposite of –14.

$= 6$

Exercises

Subtract.

1. $-9 - 11$ **1.** ____________
2. $6 - 5$ **2.** ____________
3. $7 - 13$ **3.** ____________
4. $-7 - 5$ **4.** ____________

Subtract.

5. $8 - (-5)$ **5.** ____________
6. $-4 - (-1)$ **6.** ____________
7. $6 - (-6)$ **7.** ____________
8. $-9 - (-9)$ **8.** ____________
9. **WEATHER** At 5:00 A.M. the temperature in Phoenix, Arizona, was already 80°F. By 3:00 P.M. the temperature had risen to 97°F. What is the difference in the temperatures? **9.** ____________

NAME ______________________ DATE ____________ PERIOD ________

Are You Ready?

Practice

Subtract.

1. $12 - 9$ 1. ________
2. $-8 - (-3)$ 2. ________
3. $-6 - 5$ 3. ________
4. $17 - (-9)$ 4. ________
5. **WATER** The boiling point of water is 212°F, and its freezing point is 32°F. Find the difference between the two extremes. 5. ________
6. **FOOTBALL** Ogden and Sylvio are both running backs for their respective football teams. Sylvio ran for a total of 82 yards while Ogden ran for a total of −21 yards. What is the difference in the number of yards they ran? 6. ________

Evaluate each expression.

7. $\frac{7 - 2}{12 - 2}$ 7. ________
8. $\frac{6 - 7}{4 - 9}$ 8. ________
9. $\frac{13 - 5}{6 - 10}$ 9. ________
10. $\frac{1 - 6}{8 - 3}$ 10. ________
11. $\frac{6 - 3}{7 - 2}$ 11. ________
12. Evaluate $\frac{b + a}{c - a}$ if $a = -2$, $b = 6$, and $c = -4$. 12. ________

NAME ______________________ DATE ____________ PERIOD ________

Are You Ready?

Apply

1. **DEBT** Clare had $110 in her savings. She wanted to buy an MP3 player for $118. What is the difference between how much money Clare had and how much she needed?

2. **TEMPERATURE** How much warmer is it when the temperature is 17°F than when it is −11°F?

3. **KITES** Dexter is flying his kite. Initially he flies his kite at an altitude of 157 feet. The kite then descends 35 feet. What is the height of the kite?

4. **HIKING** Neddy and Louisa were hiking when they came across the following sign along the trail. What is the difference in the elevations of White's Peak and Rock Falls?

Mountain	Elevation
White's Peak	720 feet
Little Meadow	612 feet
Rock Falls	510 feet

5. **SCUBA DIVING** Guadalupe is scuba diving. Four minutes ago, she was at −3 feet, but now is at −15 feet. What is the difference in the two depths?

6. **RABBITS** Clint and his brother raise rabbits. Clint has 11 rabbits and his brother has 24. What is the difference between the number of rabbits Clint has and the number his brother has?

NAME ______________________ DATE ____________ PERIOD ________

Diagnostic Test

Subtract.

1. $7 - (-2)$ **1.** ____________

2. $-8 - 5$ **2.** ____________

3. $14 - 16$ **3.** ____________

4. $-5 - (-2)$ **4.** ____________

5. **AIRPLANES** A plane ascends to 2,010 feet and then descends 605 feet. How high is the plane? **5.** ____________

6. **GRILLING** Marty is cooking steaks on his grill. The temperature of a steak cooked to medium is 160°F. Currently, the temperature of the steaks is 144°F. What is the difference in the two temperatures? **6.** ____________

Evaluate each expression.

7. $\frac{-9 - 3}{18 - 6}$ **7.** ____________

8. $\frac{13 - 3}{-2 - 3}$ **8.** ____________

9. $\frac{5 + 8}{16 - 3}$ **9.** ____________

10. $\frac{4 - 7}{6 - 3}$ **10.** ____________

11. **ALGEBRA** Evaluate the expression $\frac{a + b}{c - b}$ if $a = 8$, $b = -2$, and $c = -1$. **11.** ____________

NAME ______________________ DATE ______________ PERIOD ________

Pretest

1. Determine whether the relationship between the two quantities described in the table is linear. If so, find the constant rate of change. If not, explain your reasoning.

x	1	2	3	4
y	5	7	9	11

1. ______________

2. Find the slope of the line that passes through the points $(-2, 6)$ and $(9, -5)$.

2. ______________

3. **PEANUTS** The cost of peanuts varies directly with the number of pounds bought. If 5 pounds cost \$7.50, find the cost of 3 pounds.

3. ______________

4. State the slope and y-intercept for the graph of the equation $y = -2x + 7$.

4. ______________

5. State the x- and y-intercepts of $-2x + 5y = -10$. Then graph the function.

5. ______________

6. Solve the given system by graphing.

$x + y = -6$
$x - y = 2$

6.

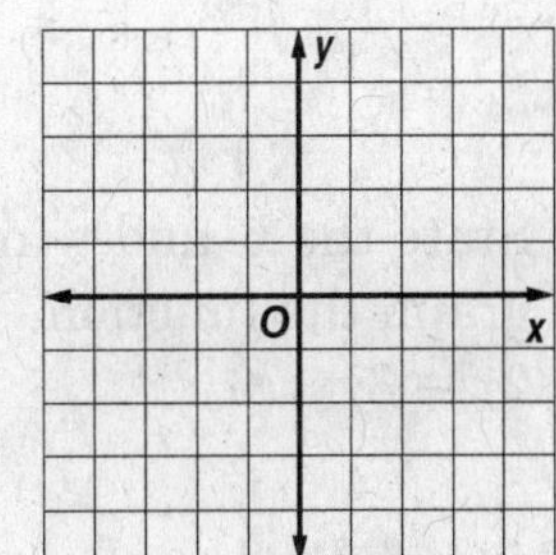

7. Solve the given system by substitution.

$y = -2x - 2$
$y = -3x$

7. ______________

NAME _______________ DATE _______________ PERIOD _______________

Chapter Quiz

1. Determine whether the relationship between the two quantities described in the table is linear. If so, find the constant rate of change. If not, explain your reasoning.

Hours Rented (h)	Cost ($)
2	50
4	100
6	150
8	200

1. _______________

2. Find the slope of the line that passes through the points $A(0, 2)$ and $B(4, -1)$.

2. _______________

3. **SHOPPING** A supermarket sells 2 cans of ground coffee for $18.50. The cost of coffee varies directly with the number of cans. How much do 5 cans of coffee cost?

3. _______________

State the slope and the *y*-intercept for the graph of each equation.

4. $y = -3x - 2$

4. _______________

5. $y + 5x = 7$

5. _______________

6. State the x- and y-intercepts for the given function. Then graph the function.
$2x - 3y = 6$

6. _______________

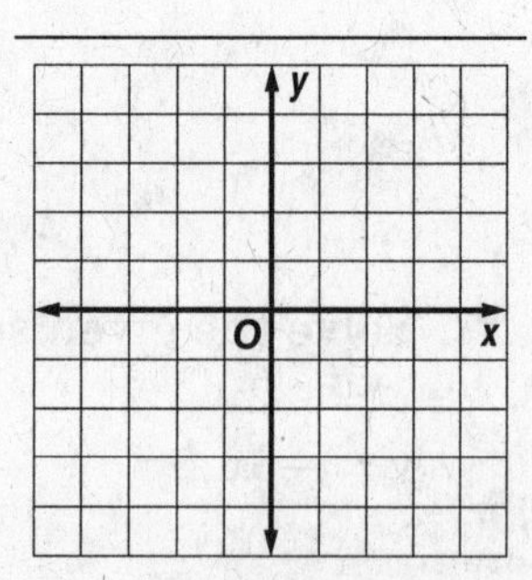

NAME ______________________ DATE ______________ PERIOD ________

Vocabulary Test

SCORE ________

constant of proportionality	point-slope form	standard form
constant of variation	rise	substitution
constant rate of change	run	system of equations
direct variation	slope	*x*-intercept
linear relationship	slope-intercept form	*y*-intercept

State whether each statement is *true* or *false*.

1. A linear equation in two different unknowns, x and y, is called a system of equations. **1.** ____________

2. The constant ratio in a direct variation is called the constant of variation. **2.** ____________

3. Slope is the ratio of the run, or horizontal change, to the rise, or vertical change. **3.** ____________

4. Linear functions written in the form $Ax + By = C$ are written in slope-intercept form. **4.** ____________

5. Substitution is an algebraic model that can be used to find the exact solution of a system of equations. **5.** ____________

6. Rise is the vertical change between any two points. **6.** ____________

7. The x-intercept of a function is the x-coordinate of the point where the graph crosses the x-axis. **7.** ____________

8. When the ratio of two variables' quantities varies, the relationship is called a direct variation. **8.** ____________

Define in your own words.

9. run **9.** ____________

10. y-intercept **10.** ____________

NAME ______________________ DATE ______________ PERIOD ________

Standardized Test Practice

Read each question. Then fill in the correct answer on the answer document provided by your teacher or on a sheet of paper.

1. Which statement is true about the slope of line RT?

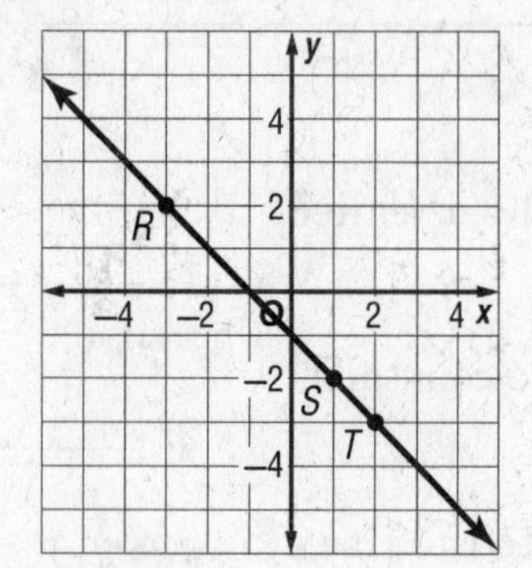

   A. The slope is the same between any two points.
   B. The slope between point R and point S is greater than the slope between point S and point T.
   C. The slope between point R and point T is greater than the slope between point S and point T.
   D. The slope is positive.

2. A truck used 6.3 gallons of gasoline to travel 107 miles. How many gallons of gasoline would it need to travel an additional 250 miles?

   F. 8.4 gallons
   G. 14.7 gallons
   H. 18.9 gallons
   I. 21.0 gallons

3. **GRIDDED RESPONSE** The cost of a pair of inline skates is \$63. If the inline skates are on sale for 35% off, what is the sale price of the inline skates in dollars?

4. Which of the following is the graph of $y = \frac{3}{4}x + 2$?

A.
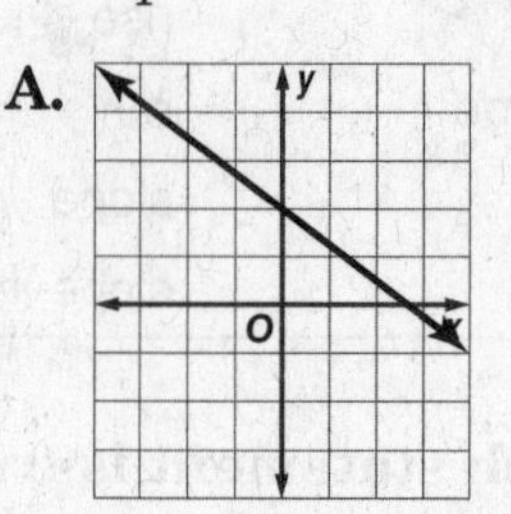

C.
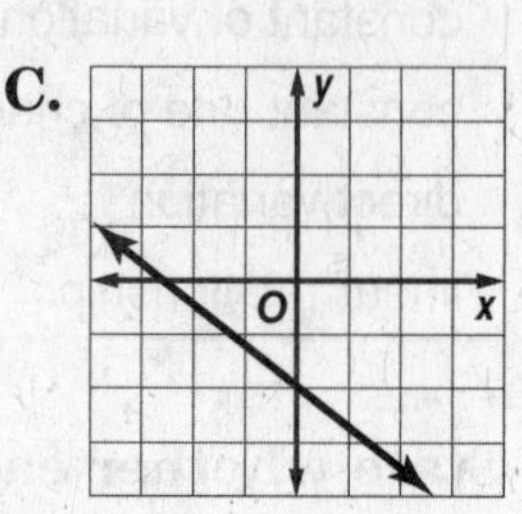

B.
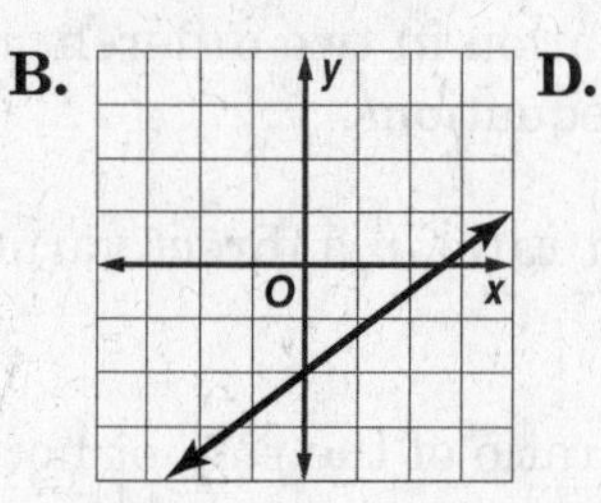

D.
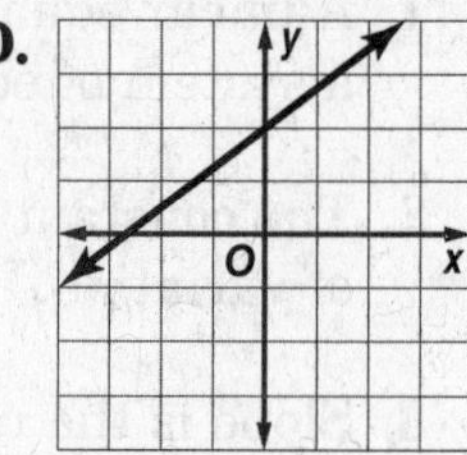

5. THINK SOLVE EXPLAIN **SHORT RESPONSE** What are the slope and x- and y-intercepts of the equation below?

$$3x - 6y = 18$$

6. The table below shows how much Heather pays to rent DVDs.

Number of DVDs	Cost (\$)
2	5
4	10
6	15
8	20

   Which of the following expressions can be used to find the total cost of renting any number n of DVDs?

   F. $5n$
   G. $2.5n$
   H. $5n - 2$
   I. $2.5n - 2$

7. **GRIDDED RESPONSE** What is the product of the fractions below?

$$\frac{3}{5} \cdot \frac{10}{12}$$

NAME ______________________ DATE ____________ PERIOD ________

8. The equation $c = 0.8t$ represents c, the cost of t tickets on a ferry. Which table contains values that satisfy this equation?

A.

Cost of Ferry Tickets				
t	1	2	3	4
c	\$0.80	\$1.00	\$1.20	\$1.40

B.

Cost of Ferry Tickets				
t	1	2	3	4
c	\$0.80	\$1.60	\$2.40	\$3.20

C.

Cost of Ferry Tickets				
t	1	2	3	4
c	\$0.75	\$1.50	\$2.25	\$3.00

D.

Cost of Ferry Tickets				
t	1	2	3	4
c	\$1.80	\$2.60	\$3.40	\$4.20

9. **GRIDDED RESPONSE** The slope of the line shown below is $\frac{4}{5}$.

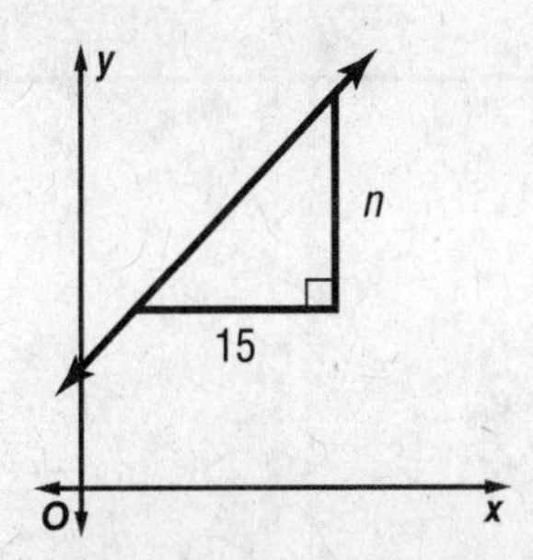

What is the value of n?

10. What is the solution of the system of equations below?

$$y = x - 4$$
$$y = 3x$$

F. (3, 4) **H.** (−2, −6)
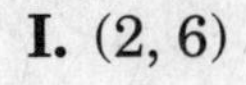
G. (−3, 4) **I.** (2, 6)

11. What does the slope of the line in the graph below represent?

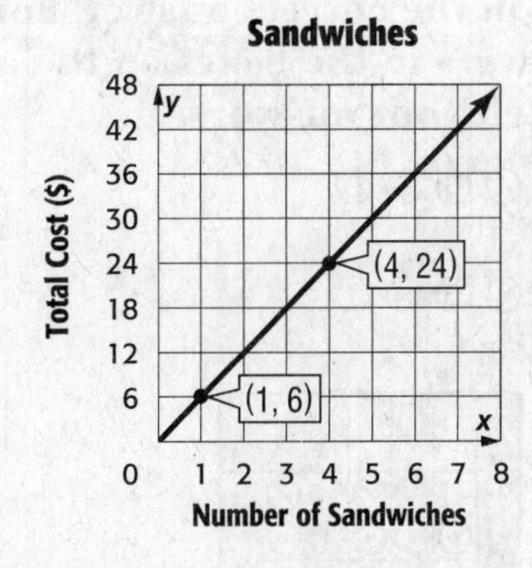

A. total number of sandwiches purchased
B. total cost of one sandwich
C. total cost of any number of sandwiches
D. None of the above

12. THINK SOLVE EXPLAIN **EXTENDED RESPONSE** The table below shows the admission rates at the local aquarium.

Ticket	Price (\$)
Adult	15
Child	9

Part A Mr. Reilly spent \$90 on tickets to the aquarium. Write an equation in standard form to represent the situation.

Part B What are the x- and y-intercepts of the function? What do they represent?

Part C Write the equation from Part **A** in slope-intercept form.

Part D Graph the equation on a coordinate plane.

Part E What is the slope of the line? What does the slope represent?

NAME ______________________ DATE ____________ PERIOD ________

Student Recording Sheet

SCORE ________

Use this recording sheet with the Standardized Test Practice pages.

Fill in the correct answer. For gridded-response questions, write your answers in the boxes on the answer grid and fill in the bubbles to match your answers.

1. Ⓐ Ⓑ Ⓒ Ⓓ

2. Ⓕ Ⓖ Ⓗ Ⓘ

3.

4. Ⓐ Ⓑ Ⓒ Ⓓ

5. ______________________

6. Ⓕ Ⓖ Ⓗ Ⓘ

7.

8. Ⓐ Ⓑ Ⓒ Ⓓ

9.

10. Ⓕ Ⓖ Ⓗ Ⓘ

11. Ⓐ Ⓑ Ⓒ Ⓓ

Extended Response

Record your answers for Exercise 12 on the back of this paper.

NAME ______________________ DATE ____________ PERIOD ________

Extended-Response Test

SCORE ________

Car Depreciation

The value of a new car depreciates (loses value) in many ways. First, a new car can depreciate in value up to 25% once it is purchased and driven off of the car lot. After that the value of a car depreciates each year because it is one year older. Let's consider two different new cars.

	Car *A*	Car *B*
Purchase Price	$30,000	$25,000
Percent of Initial Depreciation	20%	10%
Yearly Depreciation	$2,000	$1,500

1. What is the value of Car *A* and Car *B* immediately after they are purchased and driven off of the car lot?

2. Complete the table and find the value of each vehicle at the end of the indicated year.

Car Value at the End of the Indicated Year	Car *A*	Car *B*
1st		
2nd		
3rd		
4th		
5th		

3. Find the constant rate of change in the value of Car *A* and in the value of Car *B* after the initial depreciation.

4. Based on the data in the table and the constant rate of change for each vehicle, write an equation to predict the value of Car *A* and Car *B* at the end of *t* years.

5. Graph both equations on the graph shown. Label both the x- and y-intercepts and the point of intersection.

6. Interpret the meaning of both the x- and y-intercepts for each line and the point of intersection.

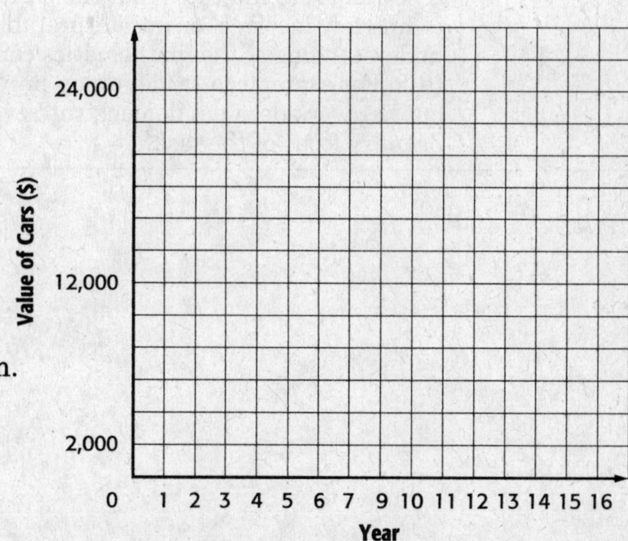

NAME ______________________ DATE ____________ PERIOD ________

Extended-Response Rubric

SCORE ________

Score	Description
4	A score of four is a response in which the student demonstrates a thorough understanding of the mathematics concepts and/or procedures embodied in the task. The student has responded correctly to the task, used mathematically sound procedures, and provided clear and complete explanations and interpretations. The response may contain minor flaws that do not detract from the demonstration of a thorough understanding.
3	A score of three is a response in which the student demonstrates an understanding of the mathematics concepts and/or procedures embodied in the task. The student's response to the task is essentially correct with the mathematical procedures used and the explanations and interpretations provided demonstrating an essential but less than thorough understanding. The response may contain minor flaws that reflect inattentive execution of mathematical procedures or indications of some misunderstanding of the underlying mathematics concepts and/or procedures.
2	A score of two indicates that the student has demonstrated only a partial understanding of the mathematics concepts and/or procedures embodied in the task. Although the student may have used the correct approach to obtaining a solution or may have provided a correct solution, the student's work lacks an essential understanding of the underlying mathematical concepts. The response contains errors related to misunderstanding important aspects of the task, misuse of mathematical procedures, or faulty interpretations of results.
1	A score of one indicates that the student has demonstrated a very limited understanding of the mathematics concepts and/or procedures embodied in the task. The student's response is incomplete and exhibits many flaws. Although the student's response has addressed some of the conditions of the task, the student reached an inadequate conclusion and/or provided reasoning that was faulty or incomplete. The response exhibits many flaws or may be incomplete.
0	A score of zero indicates that the student has provided no response at all, or a completely incorrect or uninterpretable response, or demonstrated insufficient understanding of the mathematics concepts and/or procedures embodied in the task. For example, a student may provide some work that is mathematically correct, but the work does not demonstrate even a rudimentary understanding of the primary focus of the task.

NAME _______________ DATE _______________ PERIOD _______________

Test, Form 1A

SCORE _______________

Write the letter for the correct answer in the blank at the right of each question.

1. What is the constant rate of change between the values of x and y in the table?

x	1	5	9	13
y	−6	−3	0	3

A. $-\frac{4}{3}$ **B.** $-\frac{3}{4}$ **C.** $\frac{3}{4}$ **D.** $\frac{4}{3}$ **1.** _______________

2. What is the slope of the line that passes through the points $A(-2, -1)$ and $D(3, 5)$?

F. $\frac{6}{5}$ **G.** $\frac{5}{6}$ **H.** $-\frac{5}{6}$ **I.** $-\frac{6}{5}$ **2.** _______________

3. What are three numbers that have a sum of 35 if the greatest number is 14 more than the least number?

A. 6, 7, 20 **B.** 5, 11, 19 **C.** 10, 11, 24 **D.** 1, 15, 15 **3.** _______________

4. The costs of cookies at store A are shown in the graph. The cost y for x cookies at store B is represented by the equation $y = 0.30x$. Which of the following statements is true?

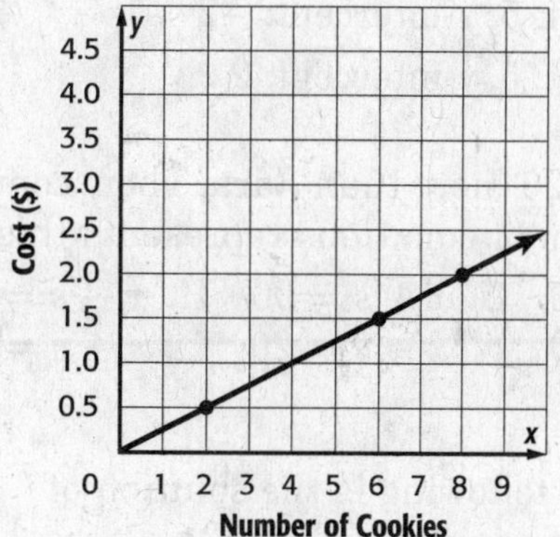

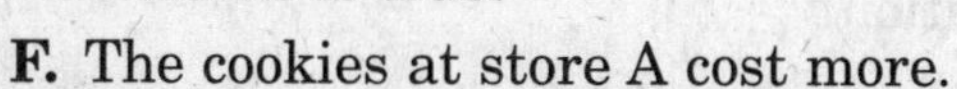

F. The cookies at store A cost more.

G. The cookies at store A cost $0.50 each.

H. The cookies at store B cost $0.15 each

I. The cookies at store B cost more. **4.** _______________

5. What are the slope and y-intercept for the graph of $y - 7x = 10$?

A. slope: 7, y-intercept: 10 **C.** slope: −7, y-intercept: 10

B. slope: 7, y-intercept: −10 **D.** slope: −7, y-intercept: −10 **5.** _______________

6. Which is the equation in slope-intercept form for the graph of the line shown?

F. $y = -3x - 2$ **H.** $y = 3x - 2$

G. $y = -3x + 2$ **I.** $y = 3x + 2$ **6.** _______________

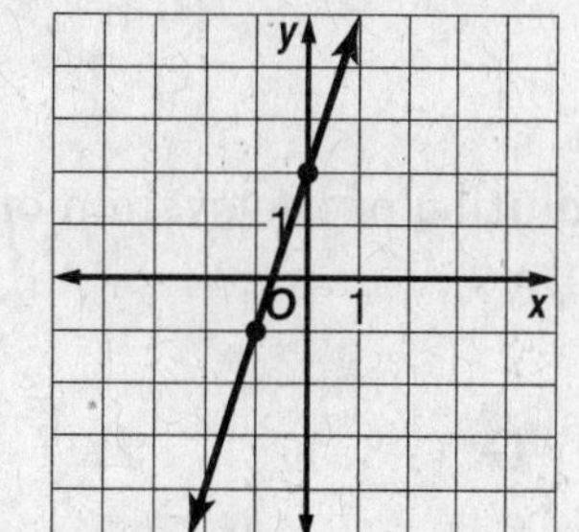

Test, Form 1A *(continued)*

SCORE ______

7. David is having his birthday party at a water park. The park charges $150 plus $10 per guest. The total cost of the party y can be represented by the equation $y = 10x + 150$. What does the slope represent?

A. the number of guests

B. the cost to rent the water park

C. the cost per guest

D. David's age

7. ______

8. Which equation, in point-slope form, passes through (3, −1) and has a slope of 2?

F. $y + 1 = 2(x - 3)$ **H.** $y + 1 = 2(x + 3)$

G. $y - 1 = 2(x + 3)$ **I.** $y - 1 = 2(x - 3)$

8. ______

9. What are the x- and y-intercepts for the graph of $2x - 5y = 10$?

A. x-intercept: −5, y-intercept: 2

B. x-intercept: −5, y-intercept: −2

C. x-intercept: 5, y-intercept: −2

D. x-intercept: 5, y-intercept: 2

9. ______

10. Xavier has $20 more than Sara. Their combined money totals $90. Which system of equations represents this situation?

F. $x + s = 90$, $s + x = 20$ **G.** $x + s = 90$, $x - s = 20$ **H.** $x - s = 90$, $s + s = 20$ **I.** $s - x = 90$, $x - s = 20$

10. ______

11. Which of the following is the solution of the system of equations shown?

A. (2, 2) **C.** (2, −2)

B. (−2, 2) **D.** (−2, −2)

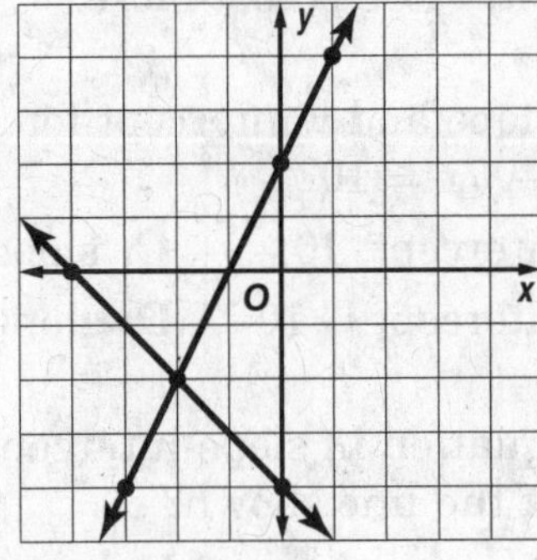

11. ______

12. What is the solution of the system of equations?

$y = x - 4$

$y = -3x$

F. (3, −1) **G.** (−3, 1) **H.** (−1, 3) **I.** (1, −3)

12. ______

13. What is the solution of the system of equations?

$y = x - 10$

$y = 2x + 5$

A. (15, 25) **B.** (15, −25) **C.** (−15, −25) **D.** (−15, 25)

13. ______

NAME ______________________ DATE ______________ PERIOD ________

Test, Form 1B

SCORE ________

Write the letter for the correct answer in the blank at the right of each question.

1. What is the constant rate of change between the values of x and y in the table?

x	−3	−1	1	3
y	7	4	1	−2

A. $\frac{3}{2}$ **B.** $\frac{2}{3}$ **C.** $-\frac{2}{3}$ **D.** $-\frac{3}{2}$ **1.** ________

2. What is the slope of the line that passes through the points $C(-2, 4)$ and $D(1, -1)$?

F. $-\frac{5}{3}$ **G.** $-\frac{3}{5}$ **H.** $\frac{3}{5}$ **I.** $\frac{5}{3}$ **2.** ________

3. What are three numbers that have a sum of 44 if the greatest number is 11 more than the least?

A. 1, 15, 12 **B.** 9, 14, 20 **C.** 8, 17, 19 **D.** 11, 16, 22 **3.** ________

4. The profits from selling T-shirts at store A are shown in the graph. The profit y for selling x T-shirts at store B is represented by the equation $y = 3.75x$. Which of the following statements is true?

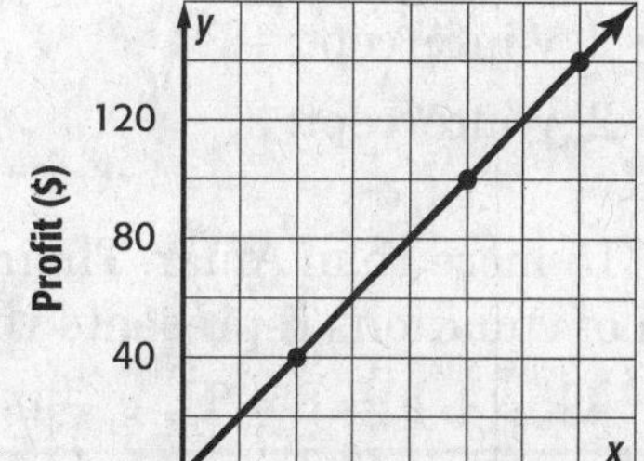

F. Store A made a greater profit per T-shirt.
G. Store B made a greater profit per T-shirt.
H. Store A made a profit of $3.50 per T-shirt.
I. Store B made a profit of $4 per T-shirt. **4.** ________

5. What are the slope and y-intercept for the graph of $y + 9x = -6$?

A. slope: 9, y-intercept: −6 **C.** slope: −9, y-intercept: −6
B. slope: −6, y-intercept: 9 **D.** slope: −6, y-intercept: −9 **5.** ________

6. What is the equation in slope-intercept form for the graph of the line shown?

F. $y = -2x - 1$ **H.** $y = 2x - 1$
G. $y = -2x + 1$ **I.** $y = 2x + 1$ **6.** ________

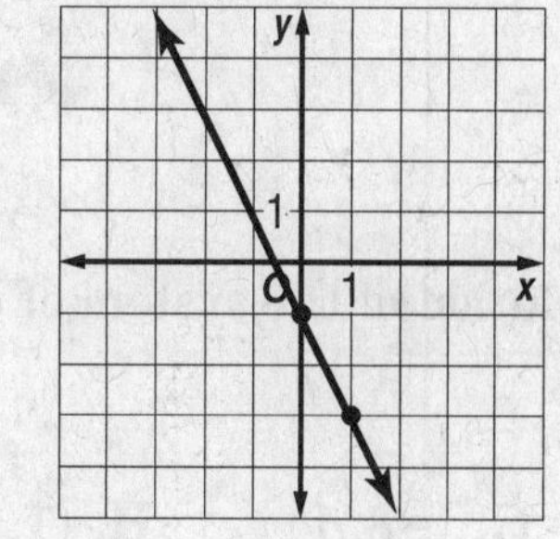

Test, Form 1B *(continued)*

SCORE ________

7. Alice is having her birthday party at a game center. The center charges $100 plus $20 per guest. The total cost of the party y can be represented by the equation $y = 20x + 100$. What does the y-intercept represent?

A. the number of guests

B. the cost to rent the game center

C. the cost per guest

D. Alice's age

7. ________

8. Which equation, in point-slope form, passes through $(-2, 4)$ and has a slope of 3?

F. $y - 4 = 3(x - 2)$ **H.** $y + 4 = 3(x - 2)$

G. $y - 4 = 3(x + 2)$ **I.** $y + 4 = 3(x + 2)$

8. ________

9. What are the x- and y-intercepts for the graph of $3x - 2y = 6$?

A. x-intercept: -2, y-intercept: 3

B. x-intercept: -2, y-intercept: -3

C. x-intercept: 2, y-intercept: -3

D. x-intercept: 2, y-intercept: 3

9. ________

10. Candace has $15 more than Amar. Their combined money totals $85. Which system of equations represents this situation?

F. $c + a = 85$, $a + c = 15$ **G.** $c + a = 85$, $c - a = 15$ **H.** $c - a = 85$, $c + a = 15$ **I.** $a - c = 85$, $c - a = 15$

10. ________

11. Which of the following is the solution of the system of equations shown?

A. $(-3, -4)$ **C.** $(3, 4)$

B. $(-3, 4)$ **D.** $(3, -4)$

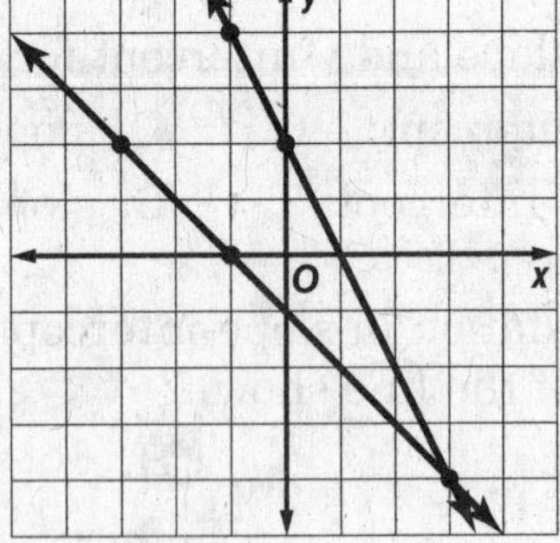

11. ________

12. What is the solution of the system of equations below?

$y = x + 2$

$y = 3x$

F. $(3, -1)$ **G.** $(-3, 1)$ **H.** $(1, 3)$ **I.** $(1, -3)$

12. ________

13. What is the solution of the system of equations below?

$y = 2x + 2$

$y = 4x - 2$

A. $(2, -6)$ **B.** $(-2, -6)$ **C.** $(-2, 6)$ **D.** $(2, 6)$

13. ________

NAME ______________________ DATE ____________ PERIOD ________

Test, Form 2A

SCORE ________

Write the letter for the correct answer in the blank at the right of each question.

1. What is the slope (grade) of a road that rises 6 feet for every horizontal change of 100 feet?

A. $\frac{1}{100}$ **B.** $\frac{1}{6}$ **C.** $\frac{3}{50}$ **D.** $\frac{50}{3}$ **1.** ________

2. What is the constant rate of change between the two quantities in the table?

Time (minutes) (x)	15	30	45	60
Number of Pages Read (y)	10	20	30	40

F. $\frac{30}{15}$ **G.** $\frac{15}{1}$ **H.** $\frac{2}{3}$ **I.** $\frac{1}{3}$ **2.** ________

3. What is the slope of the line that passes through the points $E(-1, 4)$ and $F(2, 6)$?

A. $-\frac{3}{2}$ **B.** $-\frac{2}{3}$ **C.** $\frac{2}{3}$ **D.** $\frac{3}{2}$ **3.** ________

4. The cost of nails varies directly with the number of pounds bought. If 4 pounds of nails cost $11.60, what is the cost of 3.5 pounds?

F. $5.80 **G.** $10.15 **H.** $11.60 **I.** $13.05 **4.** ________

5. What are the slope and y-intercept for the graph of $y - 4x = -2$.

A. slope: -4, y-intercept: -2

B. slope: 4, y-intercept: -2

C. slope: -4, y-intercept: 2

D. slope: 4, y-intercept: 2 **5.** ________

6. What is the equation in slope-intercept form for the graph shown?

F. $y + x = -3$ **H.** $y - 3x = 1$

G. $y = 3x + 1$ **I.** $y = -3x + 1$

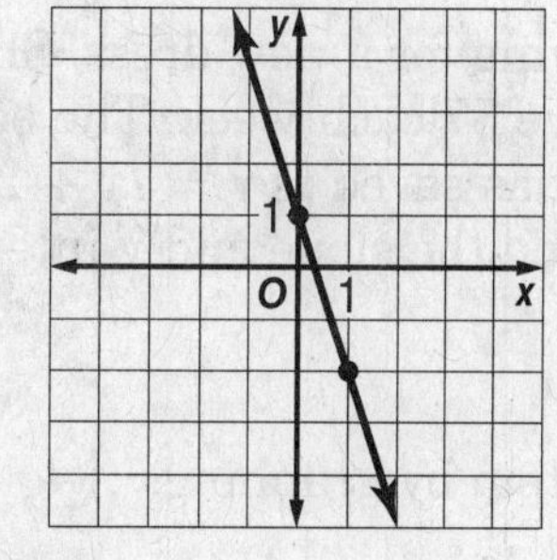

6. ________

7. What are the x- and y-intercepts for the graph of $-3x + 5y = -15$?

A. x-intercept: -5, y-intercept: -3

B. x-intercept: -5, y-intercept: 3

C. x-intercept: 5, y-intercept: 3

D. x-intercept: 5, y-intercept: -3 **7.** ________

Test, Form 2A *(continued)*

SCORE ________

8. At store A, pencils are sold individually. The cost y of x pencils is represented by the equation $y = 0.55x$. The costs of pencils at store B are shown in the table.

Number of Pencils (x)	6	12	18	24
Cost (y)	\$3.06	\$6.12	\$9.18	\$12.24

Which of the following statements is true?

F. The pencils at store A cost more.

G. The pencils at store A cost \$0.27 each.

H. The pencils at store B cost \$0.30 each.

I. The pencils at store B cost more.

8. ________

9. What is the equation in slope-intercept form for the line that passes through the points $(-2, -1)$ and $(1, 5)$?

A. $y = 2x - 3$ **C.** $y = -2x - 3$

B. $y = 2x + 3$ **D.** $y = -2x + 3$

9. ________

10. What is the solution of the system of equations?

$y - 2x = -6$

$y - 4x = 0$

F. $(-3, -12)$ **G.** $(-3, 12)$ **H.** $(3, -12)$ **I.** $(3, 12)$

10. ________

11. Theo is renting two kinds of tables for his party. One type of table seats 4 people and the other seats 6 people. If 36 people will be at his party and he rents 7 tables, how many of each type of table does he rent?

11. ________

12. Geneva is saving for a new dress. She already has \$20 saved and intends to save \$7 each week. The equation for the amount of money y she has saved is $y = 7x + 20$, where x is the number of weeks. What do the slope and y-intercept represent?

12. ________

13. Solve the system by graphing.

$y = -2x + 3$

$y = -x - 1$

13. ________

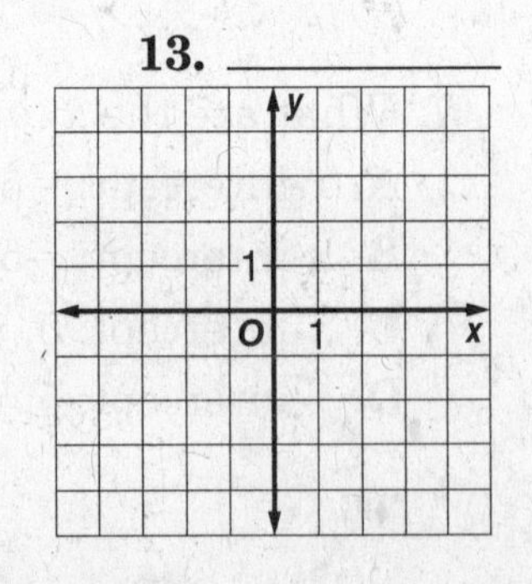

NAME ______________________ DATE ____________ PERIOD ________

Test, Form 2B

SCORE ________

Write the letter for the correct answer in the blank at the right of each question.

1. What is the slope of a ski run that rises 5 feet for every horizontal change of 20 feet?
   A. $\frac{1}{20}$ B. $\frac{1}{5}$ C. $\frac{1}{4}$ D. $\frac{20}{5}$ 1. ________

2. What is the constant rate of change between the two quantities in the table?

Number of Hours (x)	2	4	6	8
Snowfall (inches) (y)	3	6	9	12

   F. $\frac{3}{2}$ G. $\frac{2}{3}$ H. $\frac{-2}{3}$ I. $\frac{-3}{2}$ 2. ________

3. What is the slope of the line that passes through the points $E(5, 1)$ and $F(2, -7)$?
   A. $\frac{8}{3}$ B. $\frac{3}{8}$ C. $-\frac{3}{8}$ D. $-\frac{8}{3}$ 3. ________

4. The cost of peanuts varies directly with the number of pounds bought. If 3 pounds of peanuts cost \$6.30, what is the cost of 4.5 pounds?
   F. \$7.35 G. \$8.40 H. \$9.45 I. \$10.05 4. ________

5. What are the slope and y-intercept for the graph of $y - 3x = -1$.
   A. slope: 1, y-intercept: 3
   B. slope: 3, y-intercept: -1
   C. slope: -3, y-intercept: 1
   D. slope: -1, y-intercept: -3 5. ________

6. What is the equation in slope-intercept form for graph shown?
   F. $y + x = -2$
   G. $y - 2x = 1$
   H. $y = 2x + 1$
   I. $y = -2x + 1$

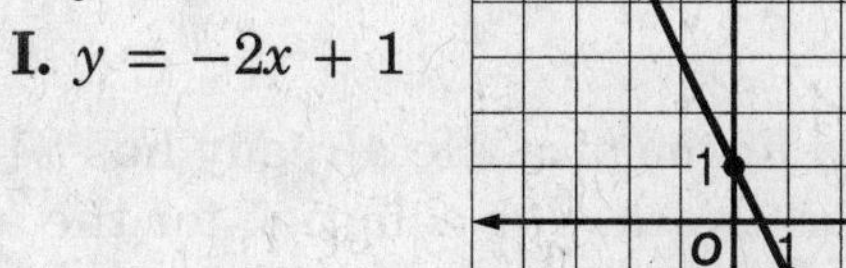

6. ________

7. What are the x- and y-intercepts for the graph of $4x - 3y = -12$?
   A. x-intercept: -3, y-intercept: -4
   B. x-intercept: 3, y-intercept: 4
   C. x-intercept: 3, y-intercept: -4
   D. x-intercept: -3, y-intercept: 4 7. ________

Test, Form 2B *(continued)*

SCORE ________

8. At store A, rulers are sold individually. The cost y of x rulers is represented by the equation $y = 0.95x$. The costs of rulers at store B are shown in the table.

Number of Rulers (x)	5	10	15	20
Cost (y)	\$4.60	\$9.20	\$13.80	\$18.40

Which of the following statements is true?

F. The rulers at store A cost more.

G. The rulers at store A cost \$0.90 each.

H. The rulers at store B cost \$0.90 each.

I. The rulers at store B cost more.

8. ________

9. What is the equation in slope-intercept form for the line that passes through the points $(-1, 3)$ and $(-2, -3)$?

A. $y = 6x - 9$ **C.** $y = -6x - 9$

B. $y = 6x + 9$ **D.** $y = -6x + 9$

9. ________

10. What is the solution of the system of equations below?

$y + 2x = 2$

$y + 4x = 0$

F. $(1, -4)$ **G.** $(-1, -4)$ **H.** $(-1, 4)$ **I.** $(1, 4)$

10. ________

11. Georgia is renting two kinds of rowboats for the campout. One type of rowboat seats 3 people and the other seats 5 people. If 53 people will be at the campout and she rents 13 boats, how many of each type of boat does she rent?

11. ________

12. Homer is saving for a harmonica. He already has \$15 saved and intends to save \$4 each week. The equation for the amount of money y he has saved is $y = 4x + 15$, where x is the number of weeks. What do the slope and y-intercept represent?

12. ________

13. Solve the system by graphing.

$y = 3x + 4$

$y = x + 2$

13. ________

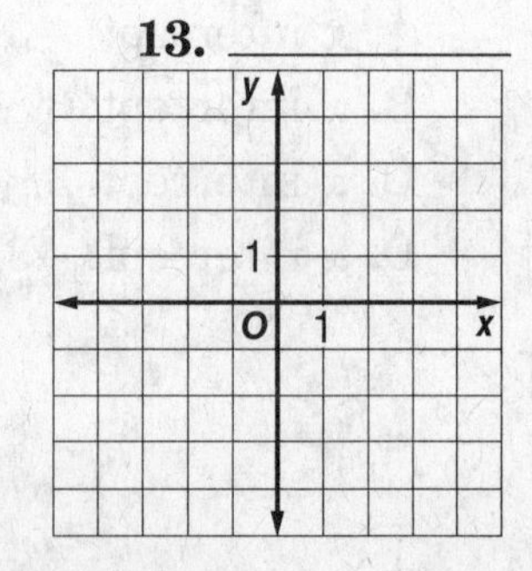

NAME ______________________ DATE ______________ PERIOD ________

Test, Form 3A

SCORE ________

1. Juanita is bringing the snacks for her daughter's soccer team. Each girl on the team will eat $\frac{1}{3}$ of an orange and drink one serving of juice or $\frac{1}{9}$ of the amount in a bottle. How many oranges and how many juice bottles will she need for all 18 girls?

1. ____________

2. The top of Angie's ladder is resting against the side of her house 22 feet above the ground. If the base of the ladder is 5 feet from the house, what is the slope of the ladder?

2. ____________

3. The framing gallery can frame 4 pictures per hour. Write and solve a direct variation equation to find how many pictures they can expect to frame in a $6\frac{1}{2}$ hour shift.

3. ____________

4. Store A is offering four bottles of nail polish for $15. The costs for nail polish at Store B are shown in the table. Assume the cost for the nail polish varies directly with the number of bottles. At which store does the nail polish cost more? Explain.

4. ____________

Number of Bottles	2	4	6
Cost ($)	7	14	21

5. State the slope and y-intercept for the graph of $-8x + y = -12$.

5. ____________

6. Write an equation in slope-intercept form for the graph of the line shown.

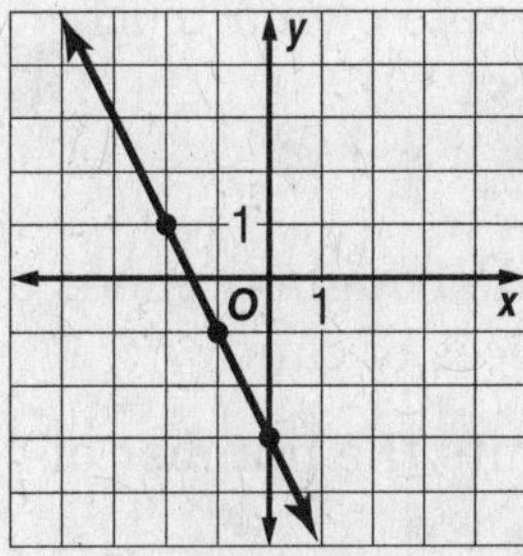

6. ____________

7. An albatross is flying at a height of 300 feet and slowly descending at a rate of 73 feet per second. The equation for the height of the bird y is $y = 300 - 73x$, where x is the number of seconds in descent. What do the slope and y-intercept represent?

7. ____________

NAME ______________________ DATE ____________ PERIOD ________

Test, Form 3A *(continued)*

SCORE ________

8. State the x- and y-intercepts for the graph of $-2y - 5x = -20$. **8.** ________

9. The table shows the items and their individual prices that Lakasha brought to donate for a charity. Altogether, she spent \$420. This is represented by the function $20x + 70y = 420$.

9a.

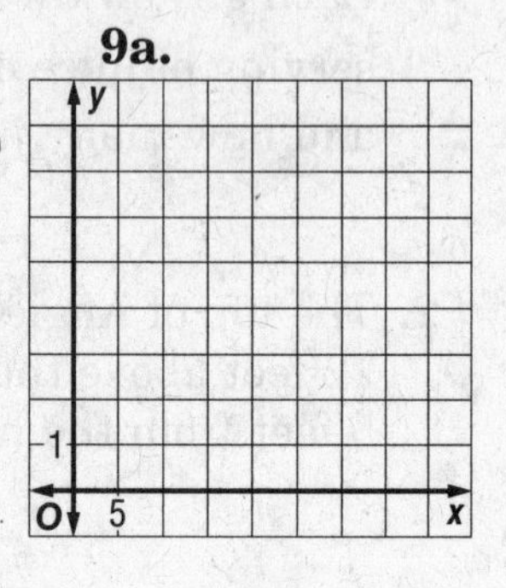

	Hats	Coats
Cost (\$)	\$20	\$70
Amount Bought	x	y

a. Graph the function.

b. Interpret the x- and y-intercepts. **9b.** ________

10. Solve the system of equations by graphing. **10.** ________

$y = 3x - 2$

$x + y = 6$

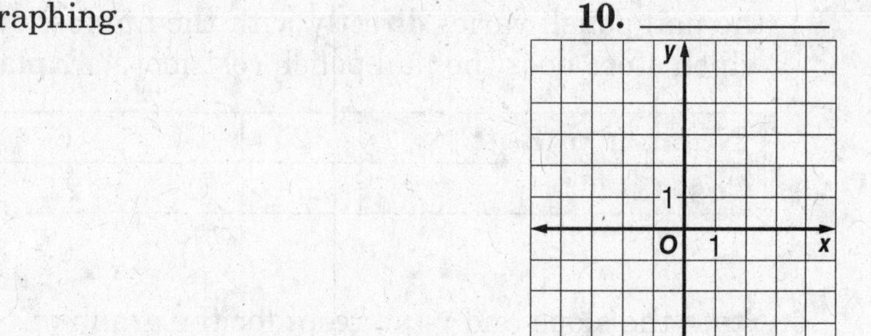

11. Logan asked his 20 coworkers whether they own a car or a truck. There were 6 more car owners than truck owners.

a. Write a system of equations that can be used to find out how many people own a car and how many people own a truck. **11a.** ________

b. Solve the system. **11b.** ________

12. Isaiah bought a total of 32 pieces of candy. He bought 3 times as many soft pieces of candy as he did hard pieces of candy.

a. Write a system of equations that represents the number of pieces of candy Isaiah bought. **12a.** ________

b. Solve the system. **12b.** ________

c. Interpret the solution. **12c.** ________

NAME ______________________ DATE ______________ PERIOD ________

Test, Form 3B

SCORE ________

1. Tyrell wants to buy bagels and cream cheese for his 16 coworkers at the office. He expects that each worker will eat $1\frac{1}{2}$ bagels and 2 servings of cream cheese. The cream cheese comes in 4-serving containers. How many bagels and containers of cream cheese will he need?

1. ________

2. To get into her tree house, Annabeth rests a ladder against the tree. The top of the ladder is 13 feet above the ground. The base of the ladder is 3 from the tree. What is the slope of the ladder?

2. ________

3. Mrs. Potts can make 5 dozen ravioli in 1 hour. Write and solve a direct variation equation to find how many she can make in $2\frac{1}{2}$ hours.

3. ________

4. Store A is offering two tubes of lip gloss for $7. The costs for lip gloss at Store B are shown in the table. Assume the cost for the lip gloss varies directly with the number of tubes. At which store does the lip gloss cost more? Explain.

Number of Tubes	3	5	7
Cost ($)	12	20	28

4. ________

5. State the slope and y-intercept for the graph of $7x + y = 3$.

5. ________

6. Write an equation in slope-intercept form for the graph of the line shown.

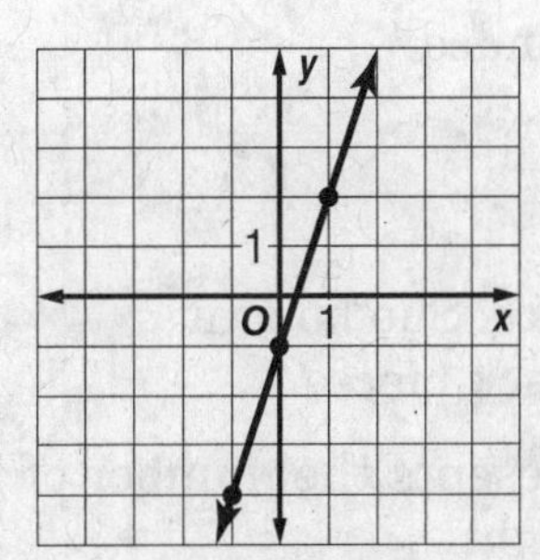

6. ________

7. An eagle is flying at a height of 275 feet and climbing at a rate of 65 feet per second. The equation for the height of the bird y is $y = 275 + 65x$, where x is the number of seconds in flight. What do the slope and y-intercept represent?

7. ________

NAME ______________________ DATE ____________ PERIOD ________

Test, Form 3B *(continued)*

SCORE ________

8. State the x- and y-intercepts for the graph of $2y + 3x = -18$. **8.** ________

9. The table shows the items and their individual prices that Amy bought for her party. Altogether, she spent $18. This is represented by the function $2x + 3y = 18$.

	Streamers	Balloons
Cost ($)	$2	$3
Amount Bought	x	y

9a.

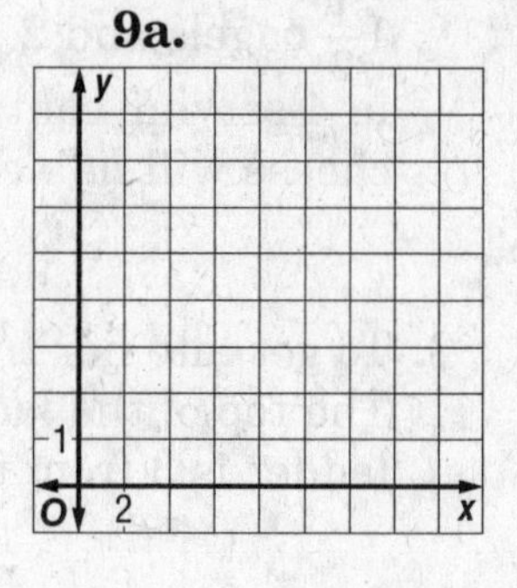

a. Graph the function.

b. Interpret the x- and y-intercepts. **9b.** ________

10. Solve the system of equations by graphing. **10.** ________

$y = -4x + 3$

$-x + y = -2$

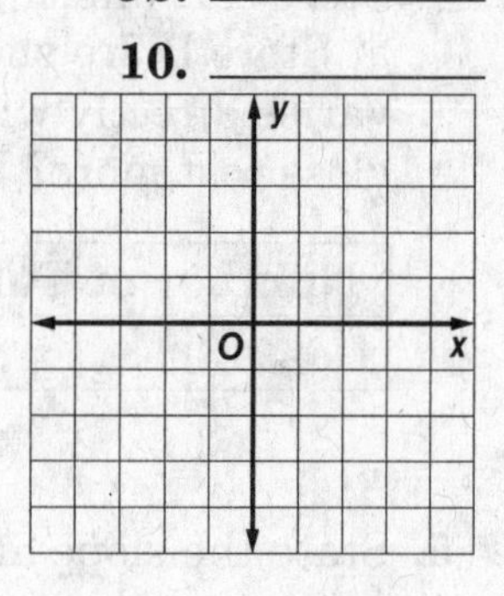

11. Cooghan asked his 19 classmates whether they were right or left-handed. There were 5 more right-handed classmates than left-handed classmates.

a. Write a system of equations that can be used to find out how many classmates were right or left handed. **11a.** ________

b. Solve the system. **11b.** ________

12. Gwen bought a total of 35 pieces of licorice. She bought 4 times as many red pieces as she did black pieces.

a. Write a system of equations that represents the number of pieces of each kind of licorice that Gwen bought. **12a.** ________

b. Solve the system by substitution. **12b.** ________

c. Interpret the solution. **12c.** ________

NAME ______________________ DATE ____________ PERIOD ________

Are You Ready?

Review

To evaluate an algebraic expression, first replace the variable or variables with the known values to produce a numerical expression. Then find the value of the numerical expression using the order of operations.

Example 1

Evaluate $3x + 4$ if $x = 6$.

$$3x + 4 = 3(6) + 4$$
$$= 18 + 4$$
$$= 22$$

Example 2

Evaluate $5x - 3$ if $x = 7$.

$$5x - 3 = 5(7) - 3$$
$$= 35 - 3$$
$$= 32$$

Exercises

Evaluate each expression if $x = 5$.

1. $6x - 8$ **1.** ________

2. $24 - 3x$ **2.** ________

3. $\frac{8x}{2}$ **3.** ________

4. $12 + x$ **4.** ________

5. $\frac{12x}{3}$ **5.** ________

6. $7x$ **6.** ________

Evaluate.

7. $5m - 4$ if $m = 6$ **7.** ________

8. $6t + 8$ if $t = 4$ **8.** ________

9. $\frac{8m}{3}$ if $m = 12$ **9.** ________

10. $39 - 15x$ if $x = 2$ **10.** ________

Are You Ready?

Practice

Name the ordered pair for each point.

1. W
2. X
3. Y
4. Z

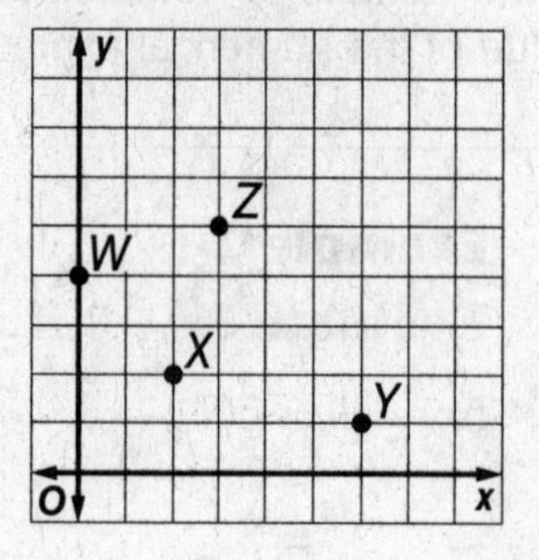

1. ____________
2. ____________
3. ____________
4. ____________

Graph each point on a coordinate grid.

5. $A(0, 4)$
6. $B(4, 1)$
7. $C(3, 6)$
8. $D(2, 6)$
9. $E(5, 1)$
10. $F(3, 0)$

5–10.

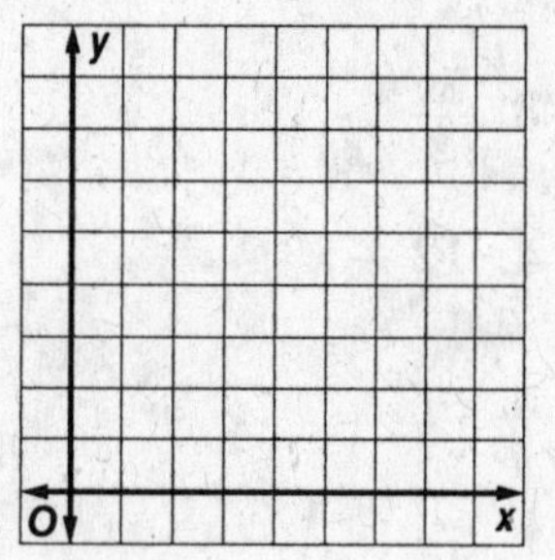

Evaluate each expression if $x = 6$.

11. $12x$
12. $4x + 2$
13. $8 + x$
14. $\frac{7x}{3}$
15. $\frac{x}{2} + 9$

11. ____________
12. ____________
13. ____________
14. ____________
15. ____________

NAME ______________________ DATE ____________ PERIOD ________

Are You Ready?

Apply

1. **WALKING** From his cabin, Rodney walked 2 miles north and 4 miles east, where he rested. If the origin represents the cabin, graph the point on the coordinate grid representing Rodney's resting point.

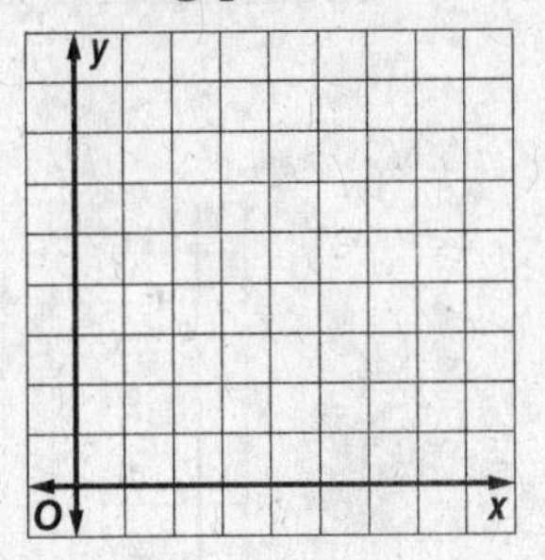

2. **PROFIT** The monthly profit of a flute store is $150x - 1{,}180$, where x represents the number of units sold. Find the monthly profit if the store sells 1,120 flutes.

3. **DVDs** Shonda received 2 DVDs free when she joined an online movie store. She must buy 3 DVDs per month after that. Write an expression to find the total number of DVDs she will have after m months. How many DVDs will she have after 5 months?

4. **DEBT** Kayla owes \$200 more dollars on her student loan than her brother Darin does on his student loan. Write an expression to find Kayla's debt if z is the amount Darin owes. How much does Kayla owe if Darin owes \$450?

5. **CELL PHONE BILL** Jonathan's cell phone bill before taxes is $\$49.99 + 0.10m$, where m is the number of minutes over 400. What is Jonathan's bill if he used 550 minutes?

6. **CAR SALES** Samara's salary rate is $\$17h$, where h is the number of hours. In addition, she earns \$100 commission for each car she sells. How much did she earn last week if she worked 40 hours and sold three cars?

NAME ______________________ DATE ____________ PERIOD ________

Diagnostic Test

Name the ordered pair for each point.

1. J

2. K

3. L

4. M

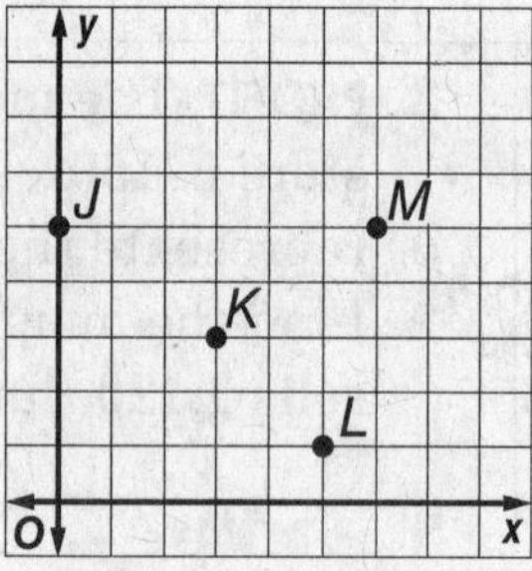

1. ____________
2. ____________
3. ____________
4. ____________

5. WALKING From her cabin, Rosalind walked 4 miles north and 6 miles east, where she rested. If the origin represents the cabin, graph the point representing Rosalind's resting point.

5.

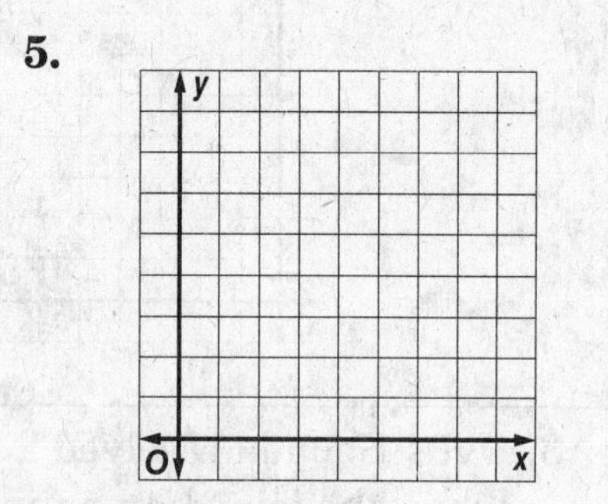

Evaluate each expression if $x = 4$.

6. $6x$

7. $5x + 7$

8. $\frac{8x}{3}$

9. $6x - 3$

10. $12 + x$

11. PROFIT The monthly profit of a skateboard company is $120x - 2,240$, where x represents the number of units sold. Find the monthly profit if the company sells 42 skateboards.

6. ____________
7. ____________
8. ____________
9. ____________
10. ____________
11. ____________

NAME ______________________ DATE ____________ PERIOD ________

Pretest

1. Find $f(3)$ if $f(x) = 2x + 4$. **1.** ____________

2. State the domain and range for the relation.
$\{(3, -2), (4, 1), (-1, -3), (5, 0)\}$ **2.** ____________

3. **PARKING** The table shows the total cost of parking in a garage. Write an expression that can be used to find the nth term of the sequence. Then use the expression to find the next three terms. **3.** ____________

Number of Hours	1	2	3	4
Cost ($)	3	5	7	9

4. **RUNNING** The table below shows how many miles Jarred ran each week. Graph the ordered pairs.

Week	1	2	3	4
Miles	2	5	8	11

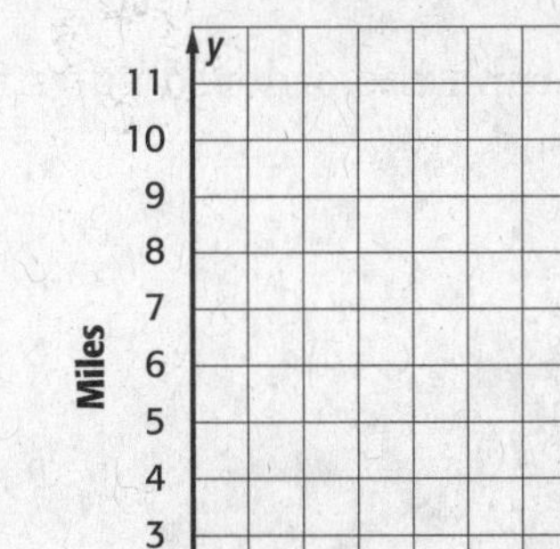

4. ____________

5. **TEXT MESSAGES** Each month, Daneen pays $8.00 plus $0.10 for each text message she sends.

a. Write a function to represent the situation. **5a.** ____________

b. Last month, she sent 46 text messages. How much did she pay? **5b.** ____________

c. Make a table of values for 1, 2, 3, and 4 text messages sent. **5c.** ____________

NAME ____________________ DATE ____________ PERIOD ________

Chapter Quiz

1. **MONEY** Diana earns \$8.50 an hour working as a lifeguard. Write an equation to find Linda's money earned m for any number of hours h.

1. 

2. Make a table to find Diana's earnings if she works 4, 5, 6, or 7 hours.

2.

Hours, h	Money, m

3. Graph the ordered pairs from Exercise 3.

3. 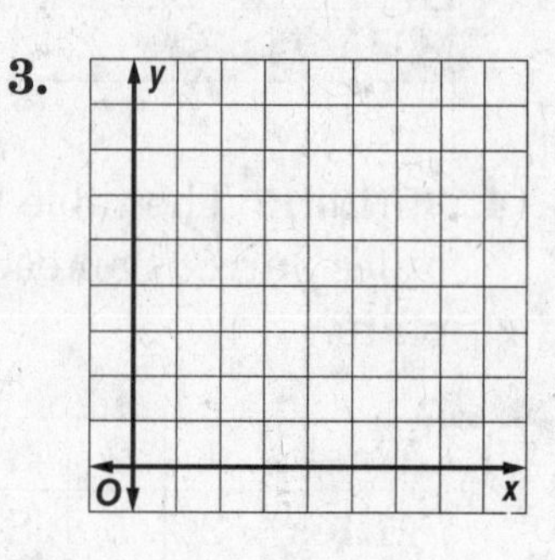

4. Find $f(2)$ if $f(x) = 2x + 1$.

4. ________

5. Find $f(-4)$ if $f(x) = -x - 5$.

5. ________

6. Complete the function table in the answer column at the right if $f(x) = x + 4$.

6.

x	$x + 4$	$f(x)$
−3		
−1		
2		
11		

7. Graph $y = -\frac{1}{2}x$.

8. Graph $y = 2x$.

7–8.

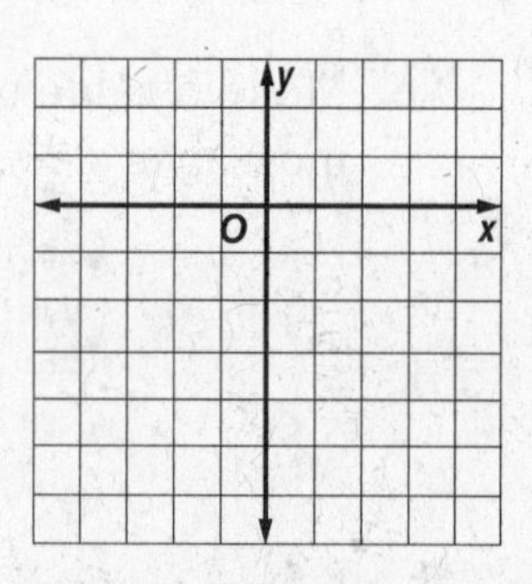

NAME ______________________ DATE ______________ PERIOD ________

Vocabulary Test

SCORE ________

continuous data	function table	quadratic function
dependent variable	independent variable	qualitative graph
discrete data	linear equation	range
domain	linear function	relation
function	nonlinear function	

Choose from the terms above to complete each sentence.

1. A(n) ______________ can be used to organize function values. 1. ________
2. A set of ordered pairs is a ______________. 2. ________
3. The ______________ of a relation is the set of *x*-coordinates. 3. ________
4. A(n) ______________ has a rate of change that is not constant. 4. ________
5. A(n) ______________ is used to represent situations that may not have numerical values. 5. ________
6. The ______________ of a relation is the set of *y*-coordinates. 6. ________
7. A graph of ______________ has space between possible data values. 7. ________
8. A relation in which each member of the domain is paired with exactly one member of the range is called a(n) ______________. 8. ________

Define each term in your own words.

9. dependent variable 9. ________

10. quadratic function 10. ________

NAME ______________________ DATE ____________ PERIOD ________

Standardized Test Practice

Read each question. Then fill in the correct answer on the answer document provided by your teacher or on a sheet of paper.

1. Beth's monthly charge for Internet access c is represented by the function $c = 12 + 2.50h$, where h represents the number of hours of usage during a month. What is the total charge for a month in which Beth used the Internet for 9 hours?

   A. $39.95
   B. $34.50
   C. $27.00
   D. $22.50

2. The graph of the line $y = -2x + 1$ is shown below. Which table of ordered pairs contains only points on this line?

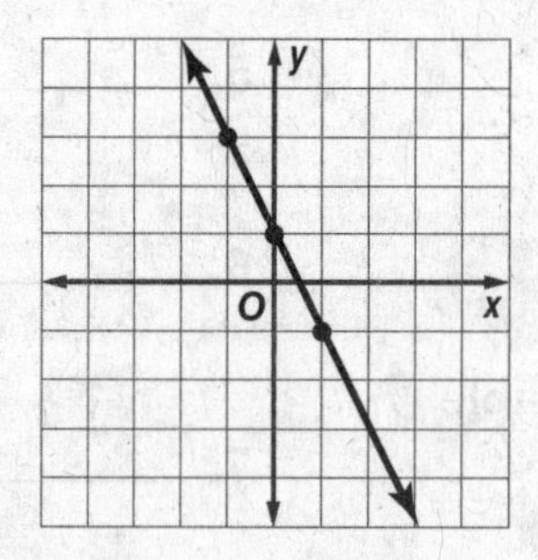

F.

x	−2	−1	0
y	5	3	−1

G.

x	−2	−1	0
y	3	1	−1

H.

x	−1	0	1
y	−3	−1	1

I.

x	−1	0	1
y	3	1	−1

3. THINK SOLVE EXPLAIN **SHORT RESPONSE** What is the domain of the relation {(−1, 4), (4, 6), (−3, −7), (2, −1)}?

4. The table shows the number of shoppers in a store on each of the first four days after its grand opening.

Day	Shoppers
1	250
2	310
3	370
4	430

Suppose the pattern continues. Which expression can be used to find the number of shoppers on any day?

   A. $250d$
   B. $250d + 60$
   C. $60d$
   D. $60d + 190$

5. Which of the following is the best estimate for the square root of 82?

   F. 7
   G. 8
   H. 9
   I. 10

6. The graph shows the speed of a car.

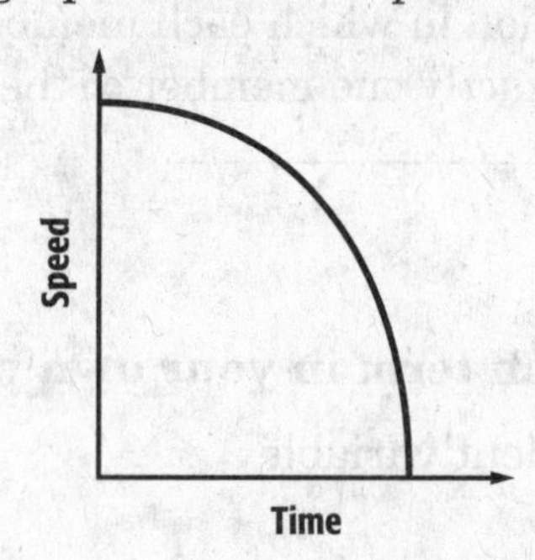

Which statement is true?

   A. The speed is increasing.
   B. The speed is decreasing.
   C. The speed is constant.
   D. The speed is increasing then decreasing.

7. What is the range of the relation {(0, 2), (1, 3), (2, 4), (1, 4)}

F. {0, 1, 2, 3}

G. {1, 2, 3, 4}

H. {0, 1, 2}

I. {2, 3, 4}

8. THINK SOLVE EXPLAIN **SHORT RESPONSE** Landon is charged for each text message that he sends. The graph shows the total cost of sending text messages.

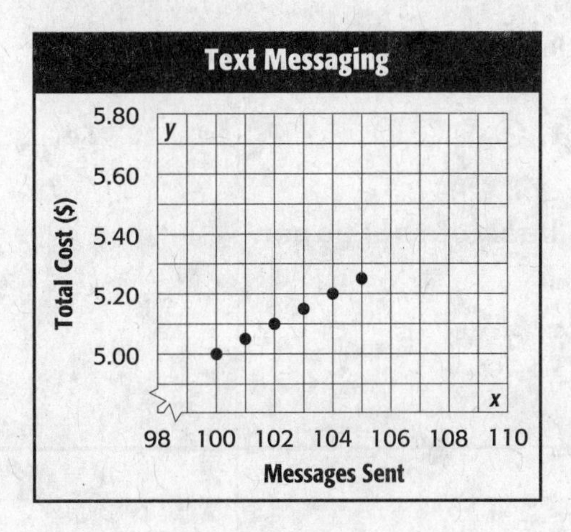

Write an expression that can be used to find the total cost of sending any number of text messages.

9. Which equation represents a linear function?

A. $y = -x^2 - 4$

B. $-3x^2 + 1 = y$

C. $y = x^2$

D. $x + 2 = y$

10. Which of the following is the value of $f(-4)$ in the function below?

$$f(x) = -2x - 3$$

F. -11

G. -5

H. 5

I. 11

11. There are four children in the Velez family. Alano is $1\frac{1}{2}$ times as tall as Lupe, and 6 inches taller than Olivia. Nelia is 56 inches tall, which is 2 inches taller than Olivia. How tall is Lupe?

A. 40 in.

B. 50 in.

C. 54 in.

D. 56 in.

12. THINK SOLVE EXPLAIN **EXTENDED RESPONSE** Video Mania is having a sale on DVDs. If you buy one DVD for $5, the cost of any additional DVDs is $2. The table shows the total cost c of any number of DVDs d.

Number of DVDs	Total Cost ($)
1	5
2	7
3	9
⋮	⋮
10	■

Part A Copy and complete the table.

Part B Graph the ordered pairs (DVDs, total cost) on a coordinate plane.

Part C Write an equation to find the total cost of any number of DVDs.

Part D How much would it cost to purchase 15 DVDs?

NAME ______________________________ DATE ______________ PERIOD ________

Student Recording Sheet

SCORE ________

Use this recording sheet with the Standardized Test Practice pages.

Fill in the correct answer. For gridded-response questions, write your answers in the boxes on the answer grid and fill in the bubbles to match your answers.

1. Ⓐ Ⓑ Ⓒ Ⓓ

2. Ⓕ Ⓖ Ⓗ Ⓘ

3. ______________

4. Ⓐ Ⓑ Ⓒ Ⓓ

5. Ⓕ Ⓖ Ⓗ Ⓘ

6. Ⓐ Ⓑ Ⓒ Ⓓ

7. Ⓕ Ⓖ Ⓗ Ⓘ

8. ______________

9. Ⓐ Ⓑ Ⓒ Ⓓ

10. Ⓕ Ⓖ Ⓗ Ⓘ

11. Ⓐ Ⓑ Ⓒ Ⓓ

Extended Response

Record your answers for Exercise 12 on the back of this paper.

NAME ______________________________ DATE ______________ PERIOD __________

Extended-Response Test

SCORE __________

1. Consider the function $y = 2x - 3$.

a. Complete the following function table.

x	$2x - 3$	y
-2		
-1		
0		
1		

b. Is the function *linear* or *nonlinear*? How can you determine this using the table?

c. Graph the function. Connect the points and describe the graph.

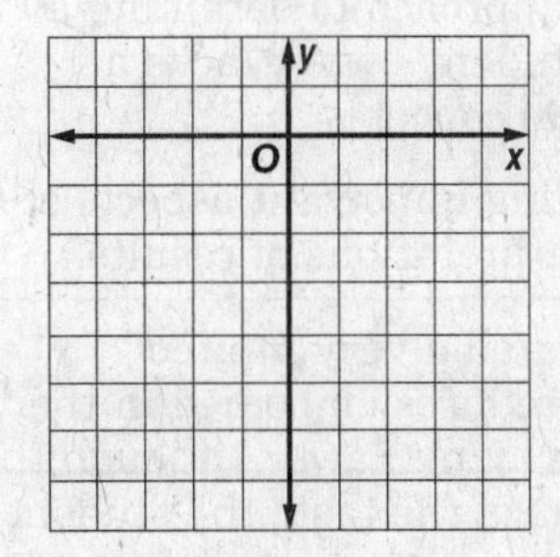

2. Mr. James fills his swimming pool with water. Once the pool is filled he turns off the water. Sketch a qualitative graph of the situation.

NAME ______________________ DATE ____________ PERIOD ________

Extended-Response Rubric

SCORE ________

Score	Description
4	A score of four is a response in which the student demonstrates a thorough understanding of the mathematics concepts and/or procedures embodied in the task. The student has responded correctly to the task, used mathematically sound procedures, and provided clear and complete explanations and interpretations. The response may contain minor flaws that do not detract from the demonstration of a thorough understanding.
3	A score of three is a response in which the student demonstrates an understanding of the mathematics concepts and/or procedures embodied in the task. The student's response to the task is essentially correct with the mathematical procedures used and the explanations and interpretations provided demonstrating an essential but less than thorough understanding. The response may contain minor flaws that reflect inattentive execution of mathematical procedures or indications of some misunderstanding of the underlying mathematics concepts and/or procedures.
2	A score of two indicates that the student has demonstrated only a partial understanding of the mathematics concepts and/or procedures embodied in the task. Although the student may have used the correct approach to obtaining a solution or may have provided a correct solution, the student's work lacks an essential understanding of the underlying mathematical concepts. The response contains errors related to misunderstanding important aspects of the task, misuse of mathematical procedures, or faulty interpretations of results.
1	A score of one indicates that the student has demonstrated a very limited understanding of the mathematics concepts and/or procedures embodied in the task. The student's response is incomplete and exhibits many flaws. Although the student's response has addressed some of the conditions of the task, the student reached an inadequate conclusion and/or provided reasoning that was faulty or incomplete. The response exhibits many flaws or may be incomplete.
0	A score of zero indicates that the student has provided no response at all, or a completely incorrect or uninterpretable response, or demonstrated insufficient understanding of the mathematics concepts and/or procedures embodied in the task. For example, a student may provide some work that is mathematically correct, but the work does not demonstrate even a rudimentary understanding of the primary focus of the task.

NAME ______________________ DATE ____________ PERIOD ________

Test, Form 1A

SCORE ________

Write the letter for the correct answer in the blank at the right of each question.

1. Miguel and Molly are cyclists. The graph shows the distance Miguel biked one day. Molly biked at a rate of 0.15 mile per minute. Which statement about their speeds is true?

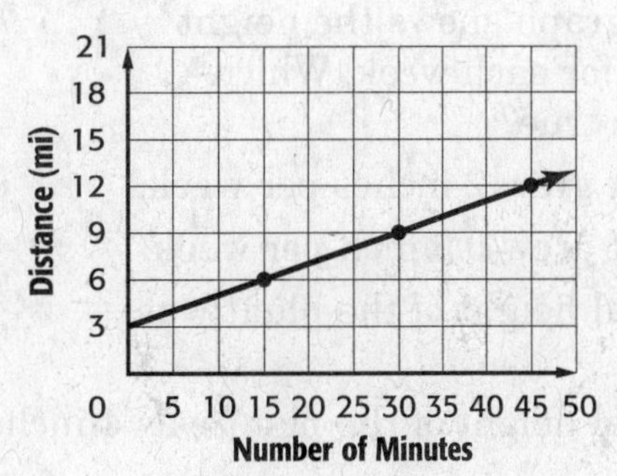

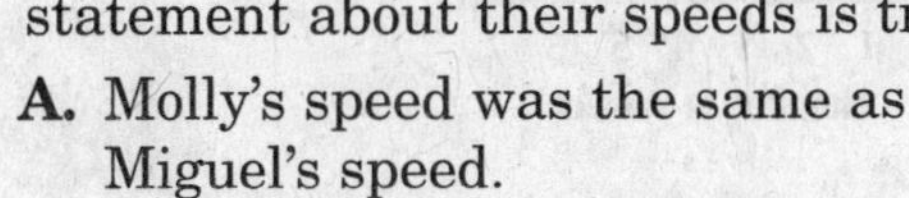

A. Molly's speed was the same as Miguel's speed.

B. Molly's speed was greater than Miguel's speed.

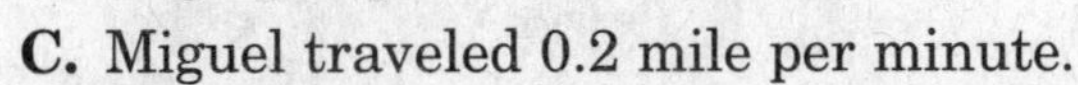

C. Miguel traveled 0.2 mile per minute.

D. Miguel traveled 5 miles per minute. **1.** ________

2. What is $f(3)$ if $f(x) = -4x + 1$?

F. -44 **G.** -11 **H.** $\frac{1}{2}$ **I.** 13 **2.** ________

3. Which table represents a nonlinear function?

A.

x	-1	0	1	2
y	5	7	9	11

C.

x	-5	0	5	10
y	1	3	7	15

B.

x	5	9	13	17
y	-6	-4	-2	0

D.

x	6	4	2	0
y	1	5	9	13

3. ________

4. Which function is graphed at the right?

F. $y = \frac{1}{2}x + 2$

G. $y = x + 2$

H. $y = -\frac{1}{2}x + 2$

I. $y = -2x + 2$

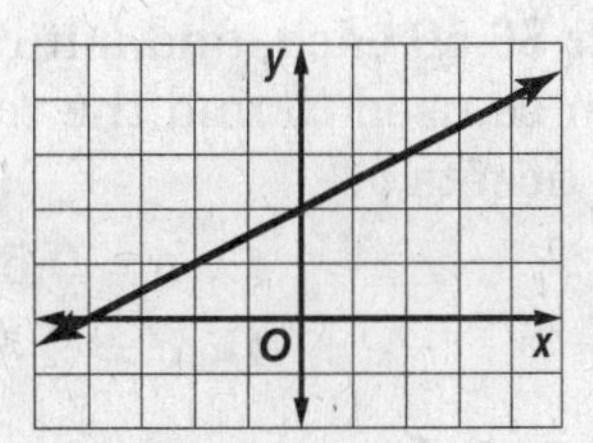

4. ________

5. Which function matches the function table at the right?

A. $f(x) = x + 3$

B. $f(x) = 2x$

C. $f(x) = 4x - 1$

D. $f(x) = x + 2$

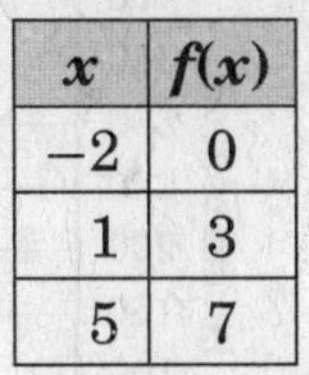

x	$f(x)$
-2	0
1	3
5	7

5. ________

6. Graphs that represent situations that may not have numerical values are called?

F. linear **G.** nonlinear **H.** qualitative **I.** quadratic **6.** ________

NAME ______________________ DATE ______________ PERIOD ________

Test, Form 1A *(continued)*

SCORE ________

7. A plant is a certain height. The height of the plant is measured for several weeks. The graph shows the height of the plant for each week. Which statement is true?

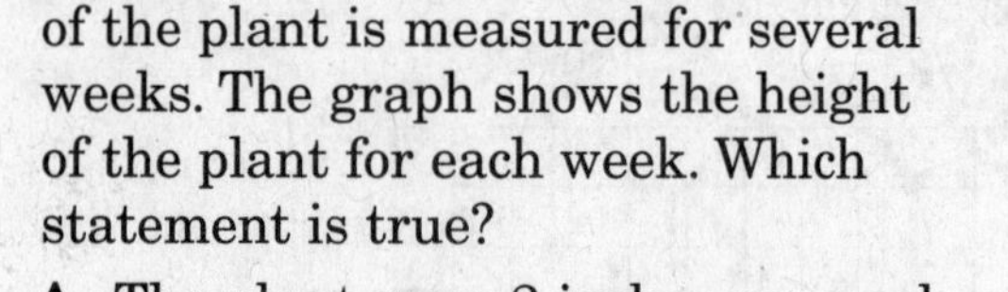

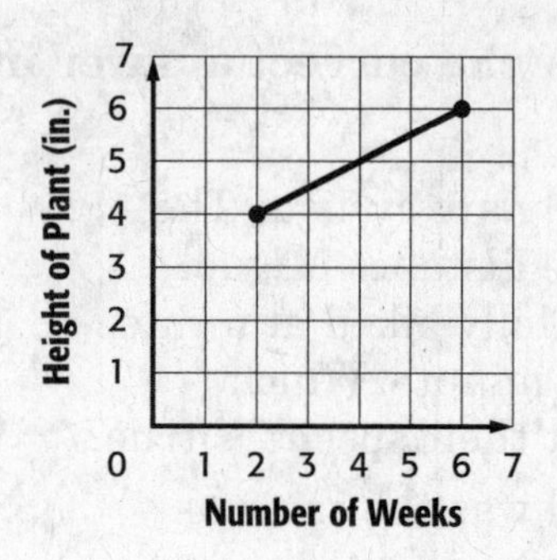

A. The plant grew 2 inches per week.
B. The plant grew 0.5 inch per week.
C. The initial height of the plant was 2 inches.
D. The initial height of the plant was 4 inches.

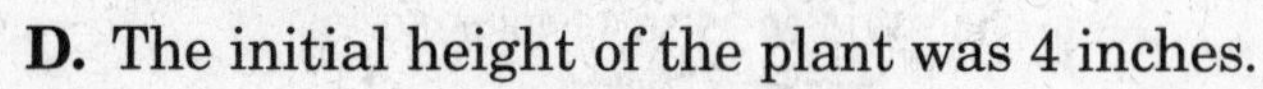

7. ________

8. What is $f(3)$ if $f(x) = 2x + 1$?

F. 4 **G.** 7 **H.** 8 **I.** 11

8. ________

9. What is the domain of the relation $\{(-2, 4), (1, 3), (0, -4), (3, 2)\}$?

A. $\{-2, 0, 1, 3\}$ **C.** $\{0, 1, 2, 4\}$
B. $\{-4, -2, 2, 3\}$ **D.** $\{-4, 2, 3, 4\}$

9. ________

10. Which equation represents the graph at the right?

F. $y = 2x^2 - 2$ **H.** $y = x^2 - 2$
G. $y = -2x^2$ **I.** $y = -x^2$

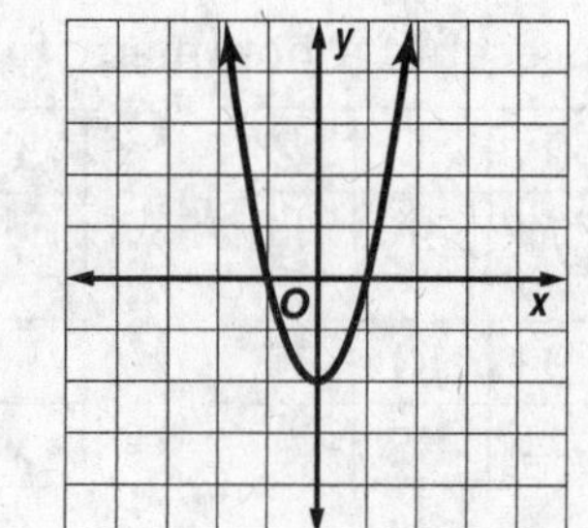

10. ________

11. Student tickets cost $6.50 each, and adult tickets cost $9.50 each. Which equation can be used to find the total cost of c of any number of student tickets t?

A. $t = 6.5c$ **C.** $t = 9.5c$
B. $c = 6.5t$ **D.** $c = 9.5t$

11. ________

12. The graph shows the amount of food Dan's bobwhite quails eat each week. Which equation can be used to find the number of pounds y eaten after any number of weeks x?

F. $y = 150x$
G. $y = 112x$
H. $y = 37.5x$
I. $y = 75x$

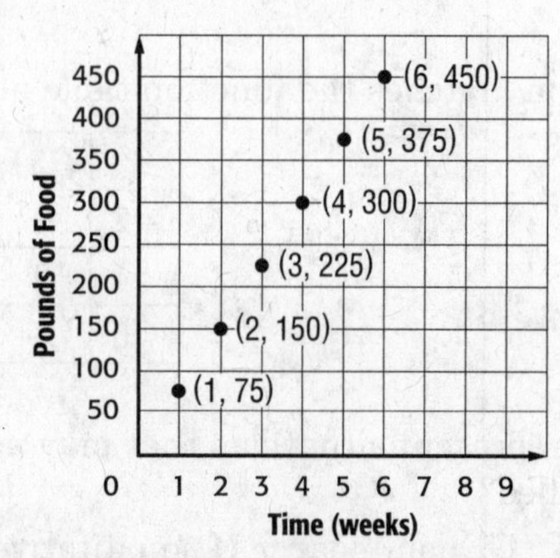

12. ________

NAME ______________________ DATE ______________ PERIOD ________

Test, Form 1B

SCORE ________

Write the letter for the correct answer in the blank at the right of each question.

1. The graph shows the distance a cheetah ran. A giraffe ran at a rate of 0.25 mile per minute. Which statement about their speeds is true?

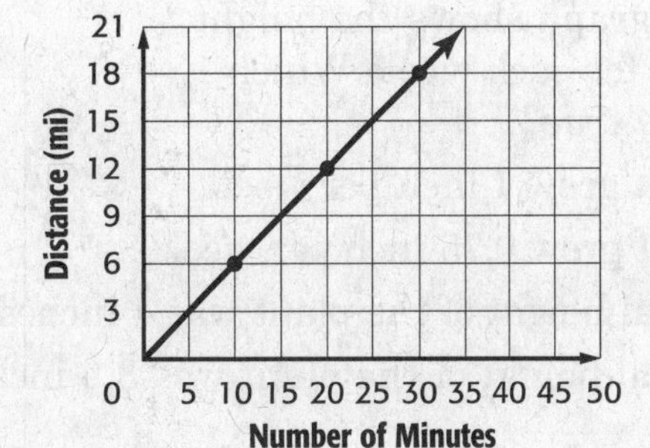

A. The cheetah traveled 0.6 mile per minute.

B. The cheetah traveled 3 miles per minute.

C. The cheetah was twice as fast as the giraffe.

D. The cheetah and the giraffe traveled at the same rate.

1. ________

2. What is $f(7)$ if $f(x) = -4x + 9$?

F. -19 **G.** -4 **H.** 4 **I.** 37

2. ________

3. Which table represents a linear function?

A.

x	1	2	3	4
y	0	2	6	12

C.

x	-3	-2	-1	0
y	5	3	1	-1

B.

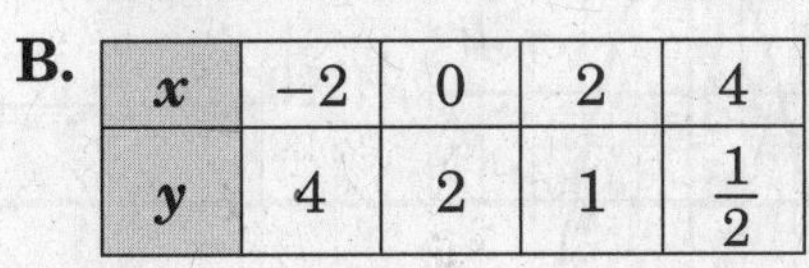

x	-2	0	2	4
y	4	2	1	$\frac{1}{2}$

D.

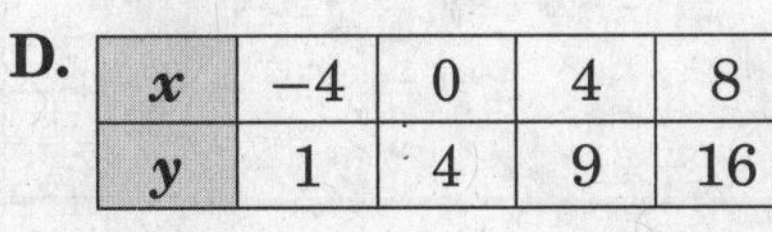

x	-4	0	4	8
y	1	4	9	16

3. ________

4. Which function is graphed at the right?

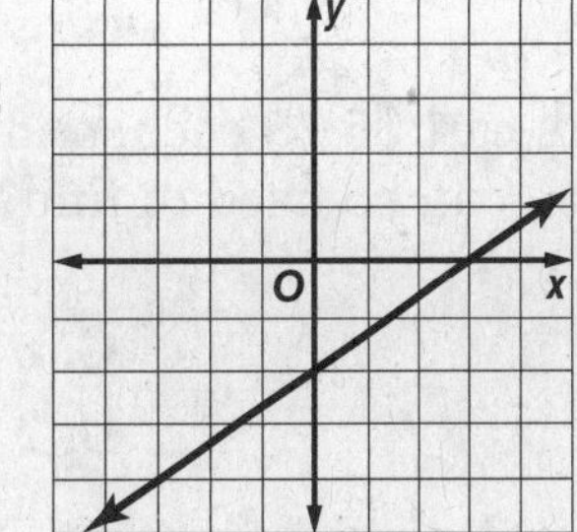

F. $y = -\frac{3}{2}x - 2$ **H.** $y = -\frac{2}{3}x - 2$

G. $y = \frac{3}{2}x - 2$ **I.** $y = \frac{2}{3}x - 2$

4. ________

5. Which function matches the function table at the right?

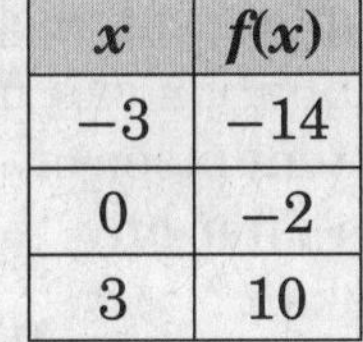

x	$f(x)$
-3	-14
0	-2
3	10

A. $f(x) = 4x - 2$ **C.** $f(x) = 2x + 4$

B. $f(x) = 5x + 1$ **D.** $f(x) = 4x + 2$

5. ________

6. Graphs that represent situations that may not have numerical values are called?

F. linear **G.** nonlinear **H.** quadratic **I.** qualitative

6. ________

NAME ____________________ DATE __________ PERIOD ______

Test, Form 1B *(continued)*

SCORE ______

7. A plant is a certain height. The height of the plant is measured for several weeks. The graph shows the height of the plant for each week. Which statement is true?

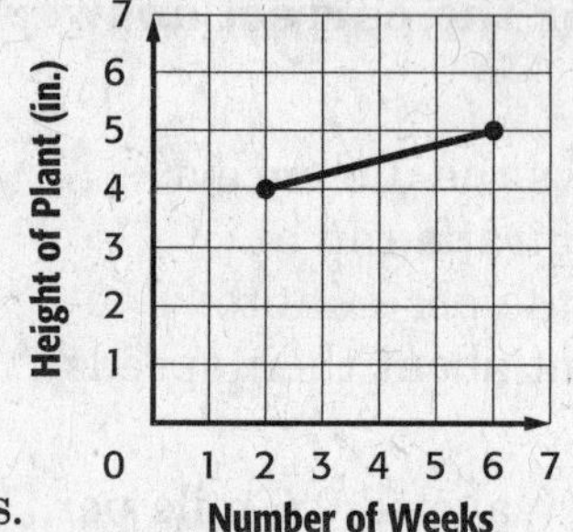

A. The plant grew 1 inch per week.

B. The plant grew 0.75 inch per week.

C. The initial height of the plant was 4 inches.

D. The initial height of the plant was 3.5 inches.

7. ______

8. What is $f(4)$ if $f(x) = 2x - 2$?

F. 6 **G.** 10 **H.** 12 **I.** 14

8. ______

9. What is the domain of the relation $\{(-2, 4), (1, 3), (0, -4), (3, 2)\}$?

A. $\{0, 1, 2, 4\}$ **C.** $\{-2, 0, 1, 3\}$

B. $\{-4, -2, 2, 3\}$ **D.** $\{-4, 2, 3, 4\}$

9. ______

10. Which equation represents the graph at the right?

F. $y = x^2 + 3$ **H.** $y = -3x^2$

G. $y = -x^2$ **I.** $y = -x^2 + 3$

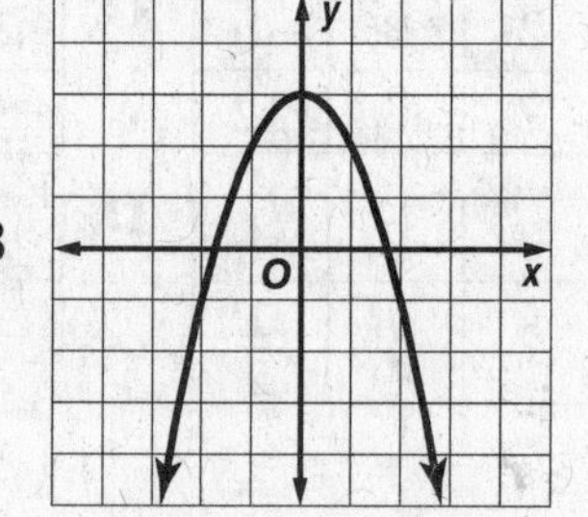

10. ______

11. Student tickets cost \$5.75 each, and adult tickets cost \$8.50 each. Which equation can be used to find the total cost c of any number of adult tickets t?

A. $c = 8.5t$ **C.** $c = 5.75t$

B. $t = 8.5c$ **D.** $t = 5.75c$

11. ______

12. The graph shows the amount of food Ian's rabbits eat each week. Which equation can be used to find the number of pounds y eaten after any number of weeks x?

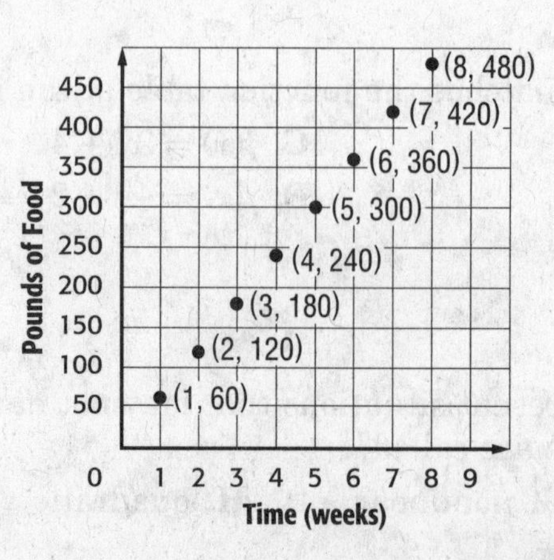

F. $y = 120x$

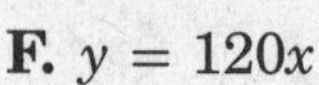

G. $y = 60x$

H. $y = 30x$

I. $y = 15x$

12. ______

NAME ______________________ DATE ______________ PERIOD __________

Test, Form 2A

SCORE __________

Write the letter for the correct answer in the blank at the right of each question.

1. Which ordered pair is *not* a point on the graph of $y = \frac{1}{2}x - 7$?

A. $\left(1, -6\frac{1}{2}\right)$ **B.** $(-2, -8)$ **C.** $(0, -7)$ **D.** $(2, 8)$ **1.** ________

2. What is $f(-2)$ if $f(x) = \frac{1}{2}x$?

F. -2 **G.** -1 **H.** 0 **I.** 1 **2.** ________

3. The graph at the right shows Jeremy's distance from home each hour he is on a car trip. How many miles will he be from home after 10 hours?

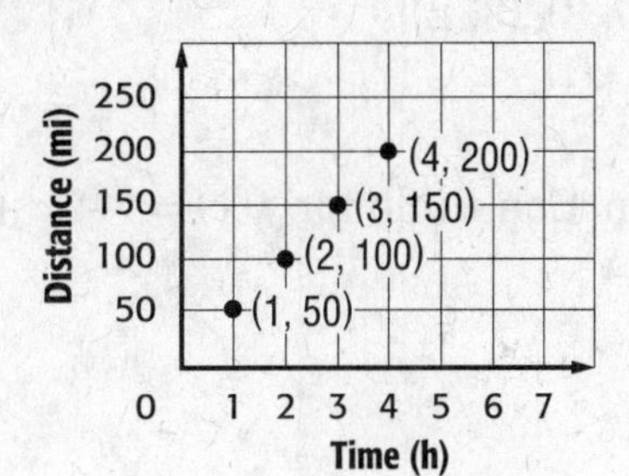

A. 350 miles **C.** 500 miles

B. 400 miles **D.** 550 miles **3.** ________

4. Which table represents a linear function?

F.

x	5	3	1	-1
y	6	8	10	12

H.

x	-2	0	2	4
y	0	1	3	6

G.

x	-3	-1	1	3
y	1	4	9	16

I.

x	7	4	1	-2
y	-1	-3	-6	-9

4. ________

5. Juana's monthly cost of sending text messages can be represented by the function $y = 0.05x$, where y represents the total cost and x represents the number of text messages. The table shows Tanya's monthly cost of sending text messages. Which statement is *not* true?

Messages	Cost ($)
20	10
30	11
40	12
50	13

A. Tanya's initial cost is greater than Juana's initial cost.

B. Tanya pays more per text than Juana.

C. Juana pays $7.50 for sending 150 text messages.

D. Tanya pays $20 for sending 150 text messages. **5.** ________

6. Which of the following represents a nonlinear function?

F. $y = 5x + 7$ **G.** $y = x^2$ **H.** $y = -2x$ **I.** $y = x$ **6.** ________

Test, Form 2A *(continued)*

SCORE ________

7. Nate has a certain number of songs on his MP3 player. Each week, he plans to add 2 more songs. After 5 weeks, he had 25 songs on his MP3 player. Which statement is true?

A. Nate adds 5 songs on his MP3 player per week.

B. Nate adds 10 songs on his MP3 player per week.

C. The initial number of songs on Nate's MP3 player is 15.

D. The initial number of songs on Nate's MP3 player is 2.

7. ____________

8. State the domain and range for the following relation.
{(−4, 4), (1, 2), (0, 3), (3, 2)}

8. ____________

9. Complete the function table for $f(x) = -2x + 1$.

9.

x	$f(x)$
−2	
0	
1	
2	

For Exercises 10 and 11, consider the following situation.

The grocery store sells cantaloupes for $4.50 per pound.

10. Write a function to represent the situation.

10. ____________

11. Is the function continuous or discrete? Explain.

11. ____________

12. Graph $y = x^2 - 2$.

12. (coordinate grid with axes labeled y, x, and origin O)

13. The value of a painting has increased steadily over time. Sketch a qualitative graph to represent this situation.

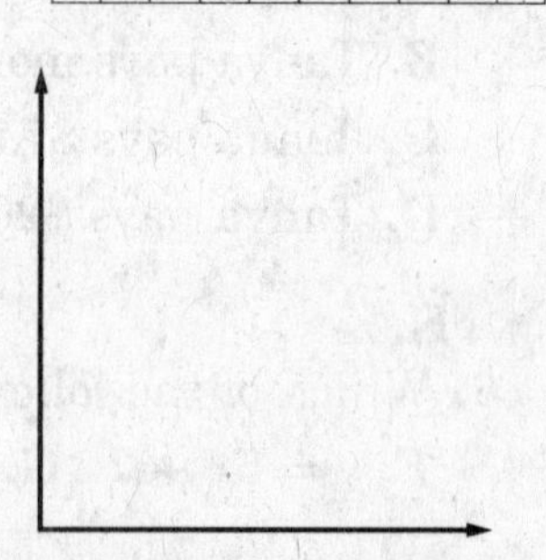

13. ____________

NAME ______________________ DATE ____________ PERIOD ________

Test, Form 2B

SCORE ________

Write the letter for the correct answer in the blank at the right of each question.

1. Which ordered pair is *not* a point on the graph of $y = -5x + 2$?

A. $(-1, 6)$ **B.** $(0, 2)$ **C.** $(-2, 12)$ **D.** $(2, -8)$ **1.** ________

2. What is $f(-3)$ if $f(x) = \frac{1}{3}x$?

F. 3 **G.** 1 **H.** -1 **I.** -3 **2.** ________

3. The graph at the right shows Lanna's total distance in miles for each day she is training for a marathon. What is her distance on day 10?

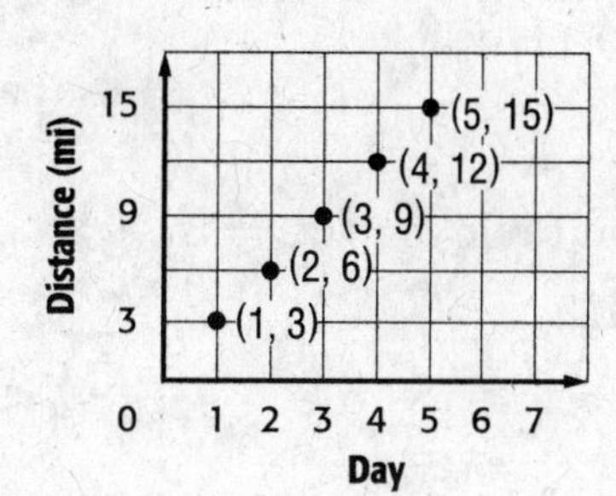

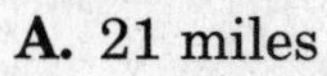

A. 21 miles **C.** 30 miles

B. 27 miles **D.** 33 miles **3.** ________

4. Which table represents a linear function?

F.

x	5	2	-1	-4
y	6	7	10	12

H.

x	4	6	9	15
y	3	4	5	6

G.

x	-2	0	2	4
y	0	1	2	3

I.

x	7	4	1	-2
y	-1	-3	-6	-10

4. ________

5. Kayla's monthly cost of sending text messages can be represented by the function $y = 0.07x$, where y represents the total cost and x represents the number of text messages. The table shows Aubrey's monthly cost of sending text messages. Which statement is *not* true?

Messages	Cost ($)
30	18
40	19
50	20
60	21

A. Kayla pays $10.50 for sending 150 text messages.

B. Aubrey pays $30 for sending 150 text messages.

C. Aubrey pays more per text than Kayla.

D. Aubrey's initial cost is greater than Kayla's initial cost. **5.** ________

6. Which of the following represents a nonlinear function?

F. $y = 4x^2$ **G.** $y = x$ **H.** $y = -9x$ **I.** $y = 8x + 10$ **6.** ________

NAME ______________________ DATE ______________ PERIOD ________

Test, Form 2B *(continued)*

SCORE ________

7. Roberto has a certain number of songs on his MP3 player. Each week, he plans to add 4 more songs. After 5 weeks, he had 40 songs on his MP3 player. Which statement is true?

A. Roberto adds 5 songs on his MP3 player per week.

B. Roberto adds 10 songs on his MP3 player per week.

C. The initial number of songs on Roberto's MP3 player is 10.

D. The initial number of songs on Roberto's MP3 player is 20.

7. ________

8. State the domain and range for the following relation.
$\{(4, -1), (3, 2), (0, -3), (1, 4)\}$

8. ________

9. Complete the function table for $f(x) = 3x + 2$.

9.

x	$f(x)$
−2	
−1	
0	
1	

For Exercises 10 and 11, consider the following situation.

The grocery store sells bacon for $5.30 per pound.

10. Write a function to represent the situation.

10. ________

11. Is the function continuous or discrete? Explain.

11. ________

12. Graph $y = -2x^2 + 4$.

12.

y

O x

13. The value of a football card has increased steadily over time. Sketch a qualitative graph of the situation.

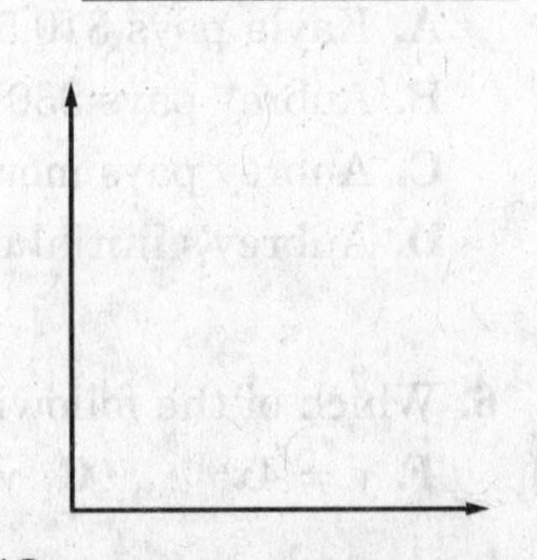

13. ________

NAME ______________________ DATE ____________ PERIOD ________

Test, Form 3A

SCORE ________

For Exercises 1–5, consider the following situation.

The deli in the grocery store gives each customer a free cup of coffee worth $1.50.

1. Write a function to represent the situation. **1.** ____________

2. Make a function table to find the total cost of the coffee if 5, 10, 15, or 20 customers come in. **2.**

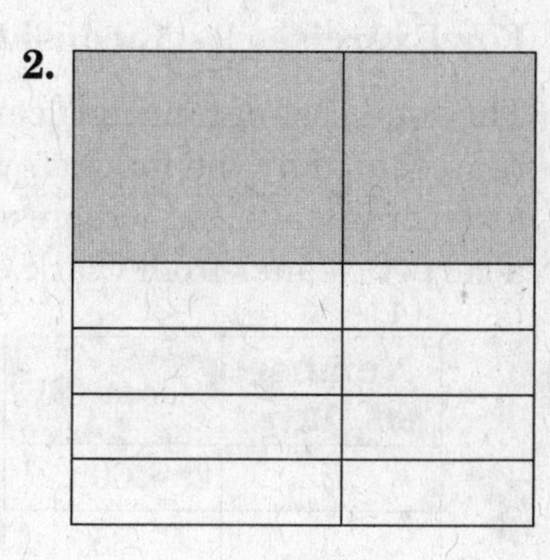

3. Graph the function. **3.**

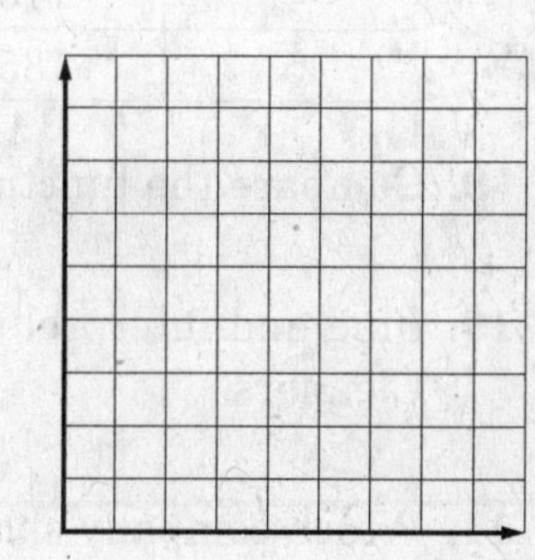

4. State the domain and range of the function. **4.** ____________

5. Is the function continuous or discrete? Explain. **5.** ____________

For Exercises 6 and 7, find each function value.

6. $f(-4)$ if $f(x) = 4x - 2$ **6.** ____________

7. $f(9)$ if $f(x) = -6x - 1$ **7.** ____________

NAME ______________________ DATE ____________ PERIOD ________

Test, Form 3A *(continued)*

SCORE ________

8. A circle has a radius of r inches. The area of a circle is represented by the expression $3.14r^2$. The area of a circle is a function of the radius. Does this situation represent a linear or nonlinear function? Explain. **8.** ____________

For Exercises 9–11, consider the following situation.

The total cost of renting a carpet cleaner from Carpets Inc. is represented by the function $y = 20x + 15$, where x represents the number of days and y represents the total cost. The cost of renting a carpet cleaner from Clark Cleaners is shown in the table.

Number of Days	Cost ($)
2	60
3	85
4	110
5	135

9. Compare the functions' rates of change. **9.** ____________

10. Find and interpret the initial value of renting from Clark Cleaners. **10.** ____________

11. Which company should you use if you rent the carpet cleaner for 6 days? **11.** ____________

12. Sketch a qualitative graph that represents a cup of hot coffee cooling down to room temperature quickly. **12.**

13. Graph $y = -x^2 + 5$. **13.**

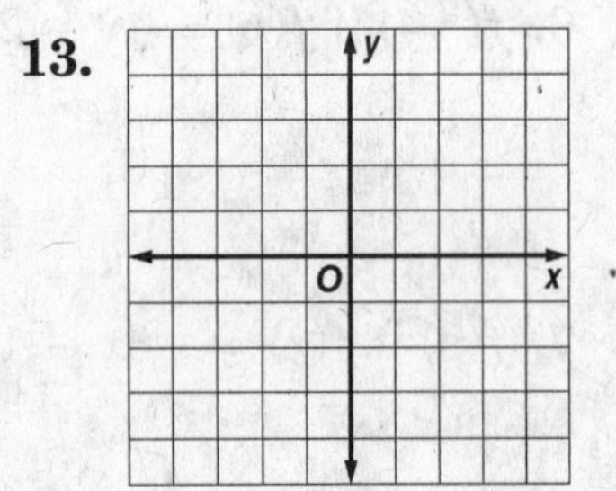

NAME ______________________ DATE ____________ PERIOD ________

Test, Form 3B

SCORE ________

For Exercises 1–5, consider the following situation.

Marylou buys bagels for a number of office staff each day. Each bagel costs \$1.75.

1. Write a function to represent the situation. **1.** ____________

2. Make a function table to find the total cost if 3, 5, 7, or 9 office workers want bagels. **2.**

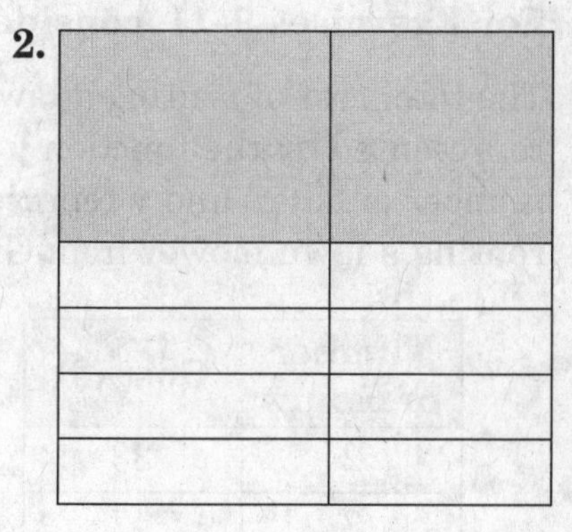

3. Graph the function. **3.**

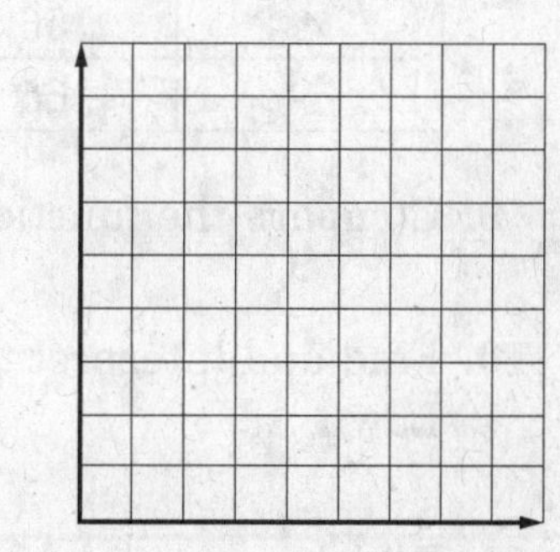

4. State the domain and range of the function. **4.** ____________

5. Is the function continuous or discrete? Explain. **5.** ____________

For Exercises 6 and 7, find each function value.

6. $f(7)$ if $f(x) = -3x + 2$ **6.** ____________

7. $f(-8)$ if $f(x) = 4x - 5$ **7.** ____________

NAME ______________________ DATE ____________ PERIOD ________

Test, Form 3B *(continued)*

SCORE ________

8. A cube has a side length of s inches. The surface area of a cube is represented by the expression $6s^2$. The surface area of a cube is a function of the side length. Does this situation represent a linear or nonlinear function? Explain. **8.** ____________

For Exercises 9–11, consider the following situation.

The total cost of renting a lawn mower from Lawns Inc. is represented by the function $y = 10x + 15$, where x represents the number of hours and y represents the total cost. The cost of renting a lawn mower from Green Lawn is shown in the table.

Number of Hours	Cost ($)
2	38
3	47
4	56
5	65

9. Compare the functions' rates of change. **9.** ____________

10. Find and interpret the initial value of renting from Green Lawn. **10.** ____________

11. Which company should you use if you rent the lawn mower for 6 hours? **11.** ____________

12. Sketch a qualitative graph that represents a cup of soup quickly cooling down. **12.**

13. Graph $y = -3x^2 + 2$. **13.**

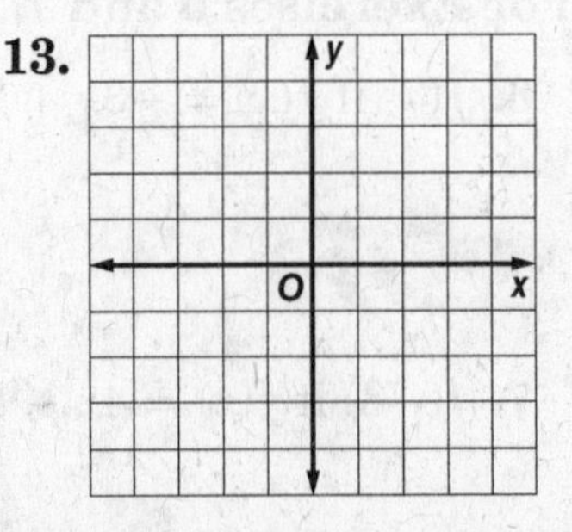

NAME ____________________ DATE ____________ PERIOD ________

Are You Ready?

Review

Example 1

Graph the point (3, –2) on the coordinate grid.

Starting at the origin, move 3 units to the right and 2 units down. Label the point (3, –2).

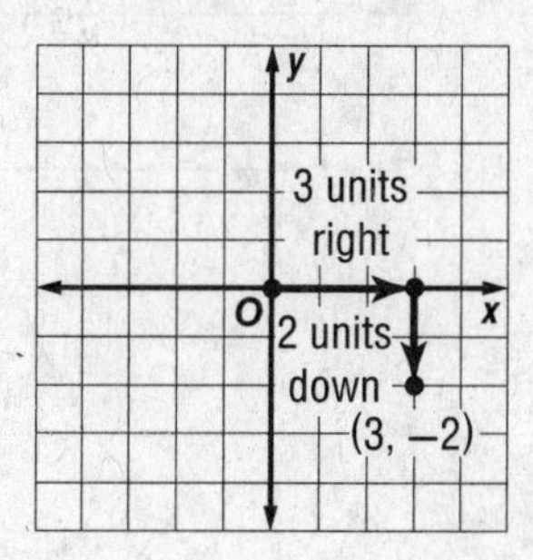

Example 2

Graph the point (–4, 3) on the coordinate grid.

Starting at the origin, move 4 units to the left and 1 unit up. Label the point (–4, 3).

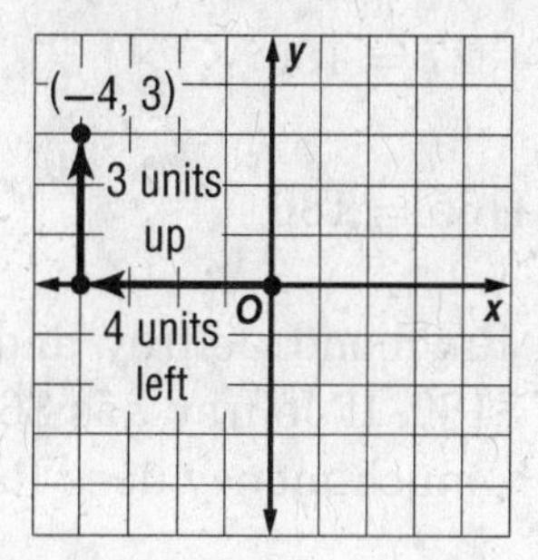

Exercises

Graph and label each point on the coordinate grid.

1. $A(2, 3)$
2. $B(-4, 1)$
3. $C(3, -2)$
4. $D(-2, -1)$

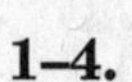

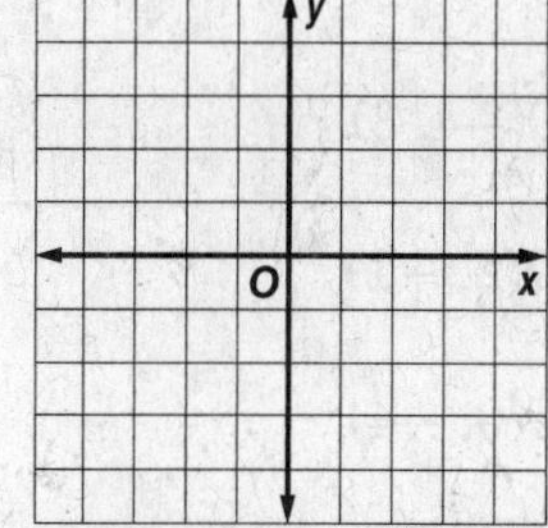

Graph and label each point on the coordinate grid.

5. $E(0, -3)$
6. $F(2, 0)$
7. $G(-4, 0)$
8. $H(0, 2)$

5–8.

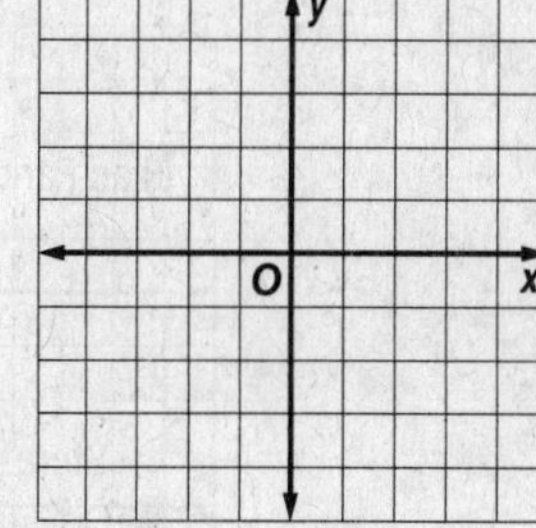

NAME ______________________ DATE ____________ PERIOD ________

Are You Ready?

Practice

Solve each equation.

1. $62 + 62 + x = 180$ **1.** ____________

2. $12 + 92 + c = 180$ **2.** ____________

3. $54 + b + 27 = 180$ **3.** ____________

4. $44 + s + 65 = 180$ **4.** ____________

5. **MONEY** Raymond, Jeffrey, and Avery have a total of \$180. If Jeffrey has \$65 and Avery has \$58, how much money does Raymond have? **5.** ____________

Graph and label each point on the coordinate grid.

6. $A(-2, -4)$

7. $B(3, 1)$

8. $C(-4, 3)$

9. $D(2, -1)$

10. $E(0, -4)$

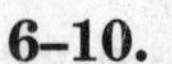
6–10.

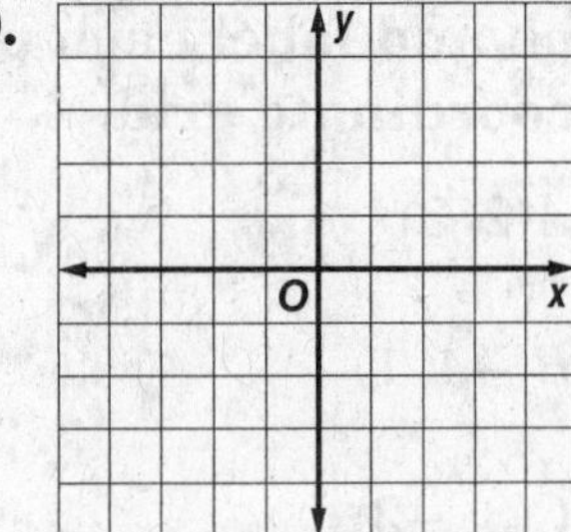

11. **MAPPING** Lilac Park is located at point $(-2, 3)$ on the grid below. Describe the location of the park with respect to the school.

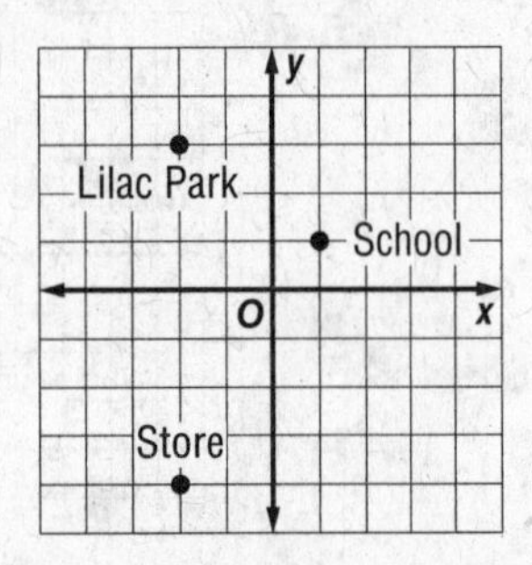

11. ____________

NAME ________________ DATE ________ PERIOD ________

Are You Ready?

Apply

1. **TRAVEL** The total distance that Fred, Amy, and Leah need to drive is 180 miles. Fred will drive 65 miles and Leah will drive 45 miles. How many miles will Amy need to drive?

2. **WEDDING** There were a total of 180 guests at a wedding. Of those guests, 37 ordered fish and 60 ordered chicken. If the remainder of the guests ordered steak, how many ordered steak?

3. **MAPPING** Conner's work place is located at the point (1, –3) on the grid below. Describe the location of his work place with respect to his home.

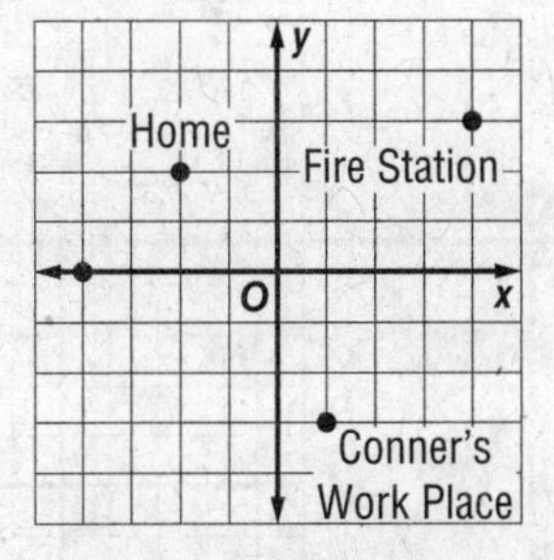

4. **MAPPING** Vikram's martial arts school is located at (–2, 4) on the grid below. Describe the location of his school with respect to his home.

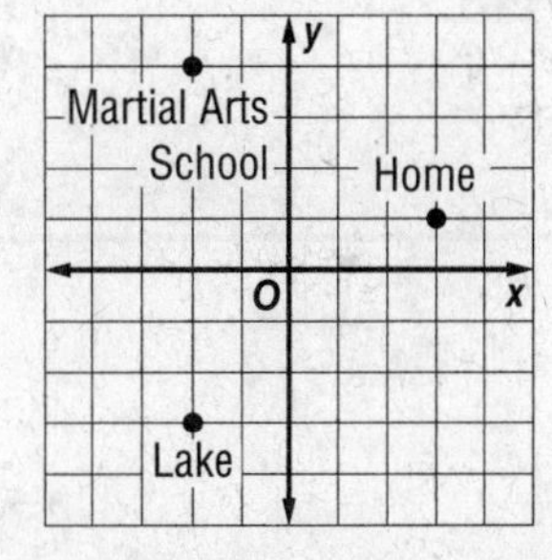

5. **MONEY** Dinner and admission tickets to the aquarium costs $180 for three adults. If two people each pay $70, how much will the third person have to pay?

6. **RUNNING** Michelle's fitness goal is to run 180 miles in 6 months. In months one and two, she ran 52 miles. In months three and four, she ran 58 miles. How many miles does Michelle need to run in months five and six to reach her fitness goal?

NAME ______________________ DATE ____________ PERIOD ________

Diagnostic Test

Solve each equation.

1. $26 + 27 + x = 180$

2. $84 + x + 12 = 180$

3. $x + 44 + 73 = 180$

4. $x + 16 + 95 = 180$

5. **WEIGHT** The Martin family has two dogs and a cat. The total weight of the three pets is 180 pounds. Use the table below to find the missing weight.

Pet	Weight (pounds)
Bud	12
Max	67
Sam	■

1. ____________

2. ____________

3. ____________

4. ____________

5. ____________

Graph and label each point on the coordinate grid.

6. $A(-2, -2)$

7. $B(4, 1)$

8. $C(-3, 2)$

9. $D(2, -4)$

10. $E(0, -2)$

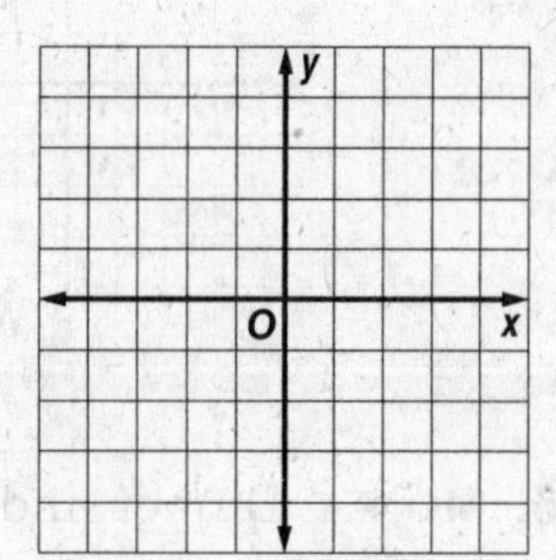

6–10.

11. **MAPPING** The bowling alley is located at $(2, -4)$ on the grid below. Describe the location of the bowling alley with respect to the police station.

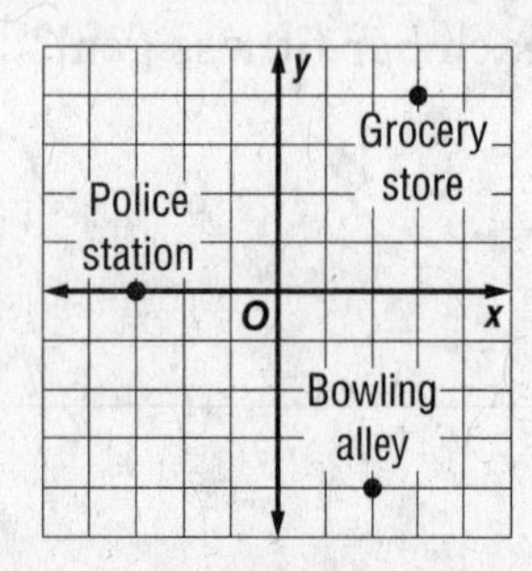

11. ____________

NAME ______________________ DATE ______________ PERIOD ________

Pretest

For Exercises 1 and 2, use the figure below.

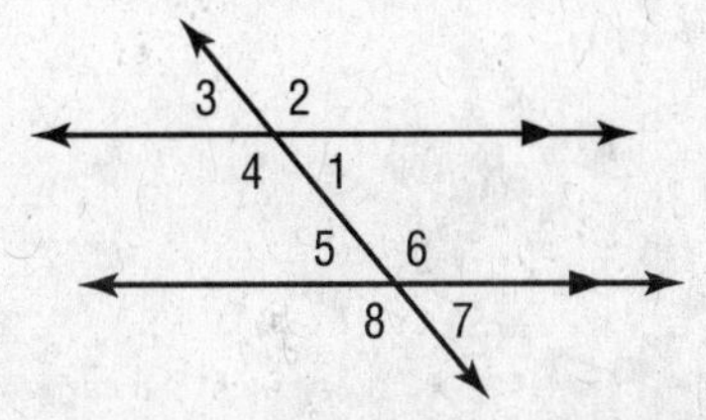

1. If $m\angle 2 = 110°$, find $m\angle 8$. **1.** ______________

2. If $m\angle 5 = 70°$, find $m\angle 7$. **2.** ______________

3. **RIGHT TRIANGLE** A right triangle has an acute angle that measures 42°. What are the measures of the other angles? **3.** ______________

4. **POLYGONS** Find the sum of the measures of the interior angle of the figure shown below.

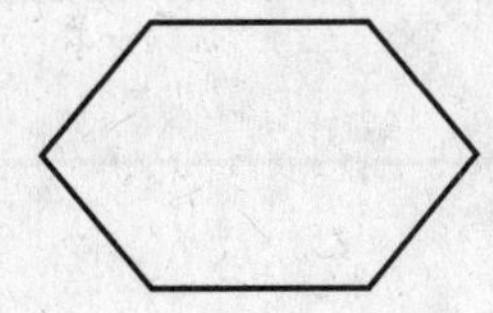

4. ______________

5. Write an equation you could use to find the length of the missing side of the triangle. Then find the missing length. Round to the nearest tenth if necessary.

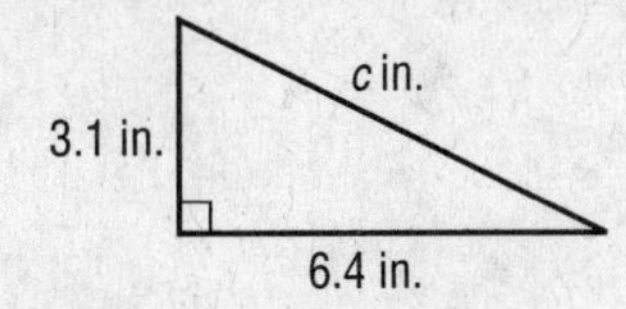

5. ______________

6. Use the Distance Formula to find the distance between the pair of points. Round to the nearest tenth if necessary.

$A(2, 11), B(-7, 9)$ **6.** ______________

NAME ______________________ DATE ____________ PERIOD ________

Chapter Quiz

For Exercises 1–3, use the figure below.

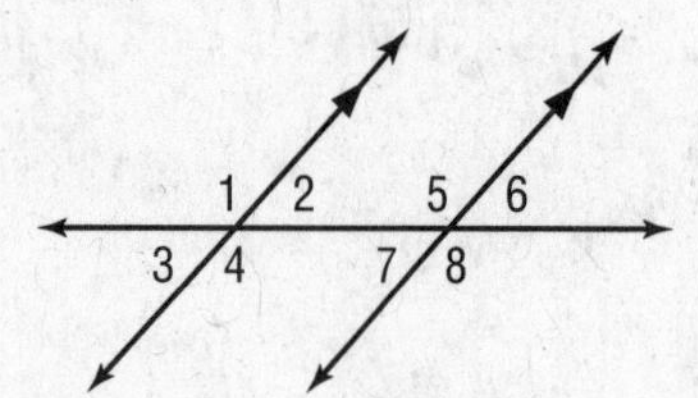

1. Look at $\angle 1$ and $\angle 2$. Classify the angle pair using all names that apply. **1.** ________

2. Find $m\angle 4$ if $m\angle 1 = 100°$. **2.** ________

3. Find $m\angle 5$ if $m\angle 1 = 100°$. **3.** ________

4. $\angle A$ and $\angle B$ are alternate exterior angles formed by two parallel lines cut by a transversal. Find $m\angle B$ if $m\angle A = 38°$. **4.** ________

5. $\angle 9$ and $\angle 10$ are corresponding and complementary angles formed by two parallel lines cut by a transversal. Find $m\angle 9$. **5.** ________

6. Find the third angle of a right triangle if the measure of one of the angles is 25°. **6.** ________

7. Find the value of x in the triangle.

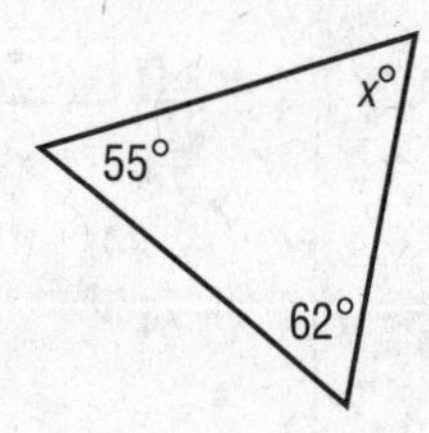

7. ________

8. Find the measure of an exterior angle of the regular polygon shown below.

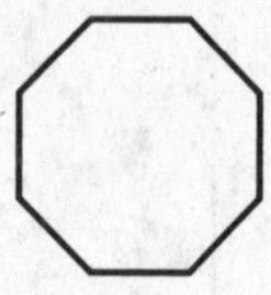

8. ________

9. The measure of the exterior angle of a regular polygon is 30 degrees. How many sides does the polygon have? **9.** ________

10. Refer to Exercise 6. Prove how you know the measure of third angle. **10.** ________

NAME _______________ DATE _______________ PERIOD _______________

Vocabulary Test

SCORE _______________

alternate exterior angles	hypotenuse	proof
alternate interior angles	inductive reasoning	Pythagorean Theorem
converse	informal proof	regular polygon
corresponding angles	interior angles	remote interior angles
deductive reasoning	legs	theorem
Distance Formula	paragraph proof	transversal
equiangular	parallel lines	triangle
exterior angles	perpendicular lines	two-column proof
formal proof	polygon	

State whether each statement is *true* or *false*.

1. The legs are the sides that form the right angle in a right triangle. **1.** _______________

2. The Pythagorean Theorem is used to find the ratio of the angle measures of two angles in a right triangle. **2.** _______________

3. A triangle is formed by three line segments that intersect only at their endpoints. **3.** _______________

4. Each exterior angle of a triangle has three remote interior angles. **4.** _______________

5. The hypotenuse is the side adjacent to the right angle. **5.** _______________

6. The converse of the Pythagorean Theorem is true. **6.** _______________

Define each term in your own words

7. regular polygon **7.** _______________

8. interior angles **8.** _______________

NAME ______________________ DATE ____________ PERIOD ________

Standardized Test Practice

Read each question. Then fill in the correct answer on the answer document provided by your teacher or on a sheet of paper.

1. Justin is flying a kite as shown below.

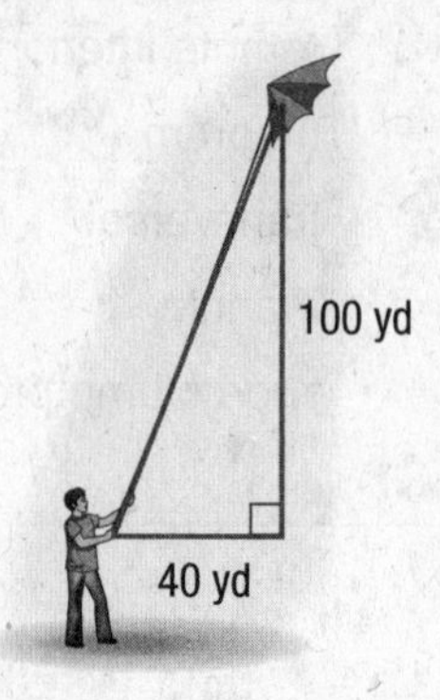

Which of the following is closest to the length of the string?

A. 70 yd
B. 92 yd
C. 108 yd
D. 146 yd

2. THINK SOLVE EXPLAIN **SHORT RESPONSE** What is the value of x in the triangle below?

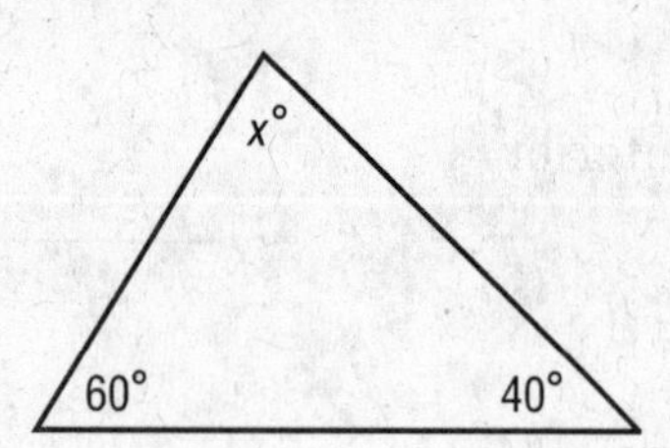

3. In 2003, a new planet was discovered. This new planet is 10^{10} miles from the Sun. Which of the following represents this number in standard form?

F. 10,000,000,000 mi
G. 10,000,000 mi
H. 10,000 mi
I. 100 mi

4. Which of the following is closest to the perimeter of the triangle below?

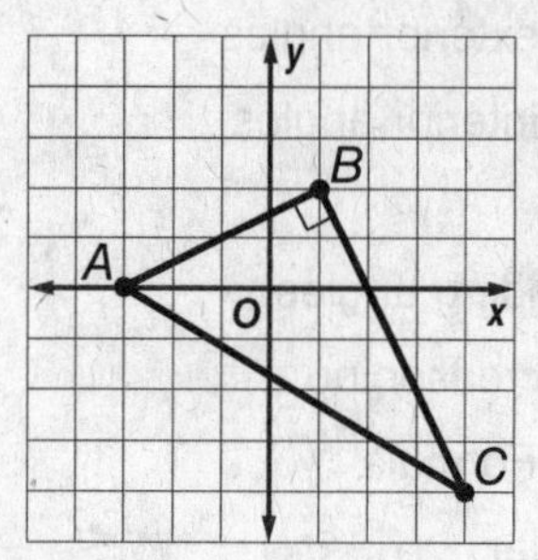

A. 17 units **C.** 21 units
B. 19 units **D.** 23 units

5. THINK SOLVE EXPLAIN **SHORT RESPONSE** Shanelle purchased a new computer for $1,099 and a computer desk for $699 including tax. She plans to pay the total amount in 24 equal monthly payments. What will be the amount of her monthly payments?

6. The area of a square is 20 square inches. Which **best** represents the length of a side of the square?

F. 4.5 inches **H.** 10 inches
G. 5 inches **I.** 11 inches

7. The proposed location of a new water tower intersects a section of an existing service road. What is x, the inside length of the section of road that is intersected by the water tower?

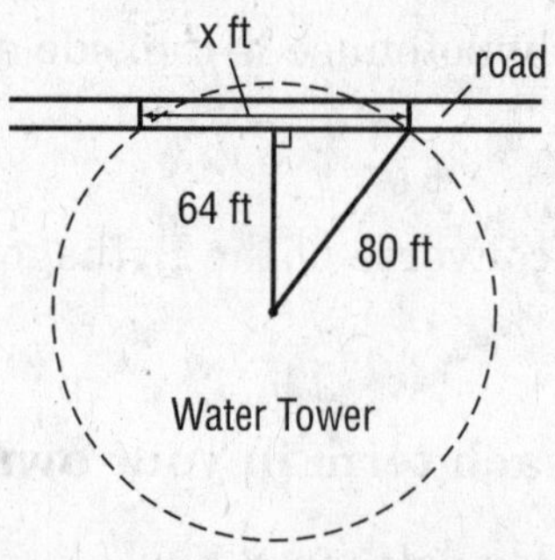

A. 36 ft **C.** 96 ft
B. 48 ft **D.** 112 ft

8. Which irrational number is closest to the number 5?

F. $\sqrt{30}$

G. $\sqrt{27}$

H. $\sqrt{20}$

I. $\sqrt{18}$

9.  **SHORT RESPONSE** Use scientific notation to find the product of 25,000,000 and 160,000.

10. Which of the following is the graph of $y = \frac{2}{3}x + 2$?

A.

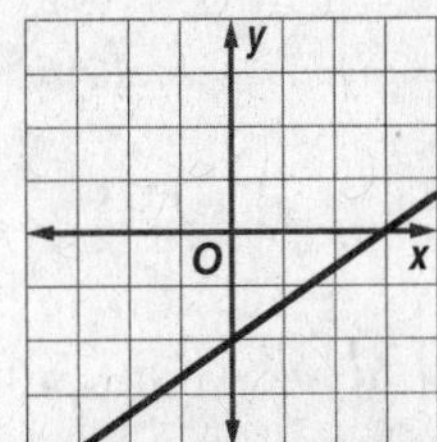

C.

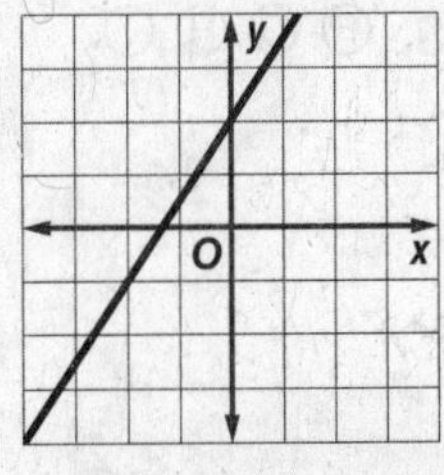

B.

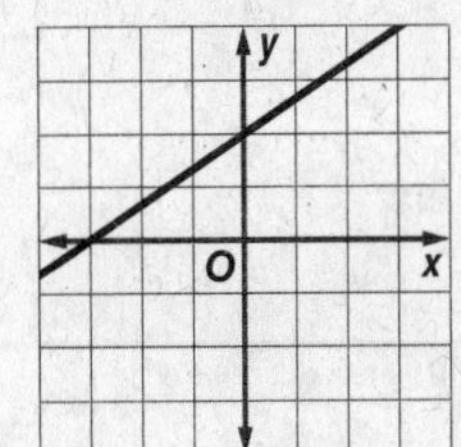

D.

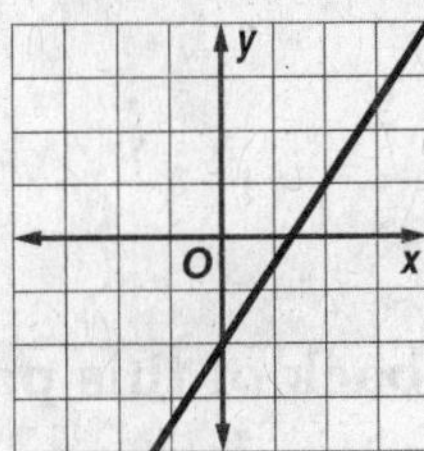

11. Mario drew a sketch of a skateboard ramp he wants to build. What is the height in feet of the skateboard ramp shown below?

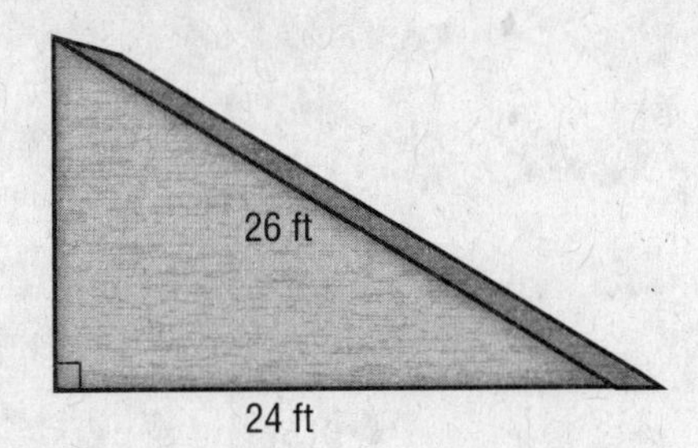

F. 10 ft

G. 22 ft

H. 25 ft

I. 34 ft

12. A stained glass window is in the shape of a regular decagon. What is the measure of one interior angle of the decagon?

A. 1,800°

B. 1,440°

C. 180°

D. 144°

13. The measure of the exterior angle of a regular polygon is 24 degrees. How many sides does the polygon have?

F. 16

G. 15

H. 14

I. 13

14. **EXTENDED RESPONSE** Refer to the park map below.

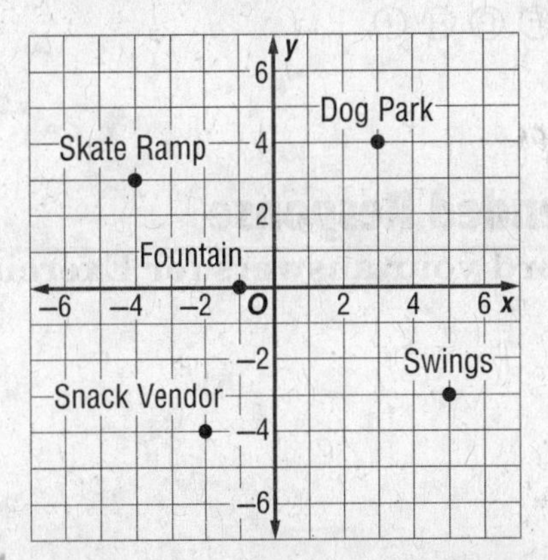

Part A At what point is the snack vendor located?

Part B At what point is the dog park?

Part C What is the distance between the dog park and the snack vendor? Round to the nearest tenth.

Part D Each unit on the map represents 15 meters. To the nearest meter, what is the approximate distance between the snack vendor and the dog park?

NAME ______________________ DATE ____________ PERIOD ________

Student Recording Sheet

SCORE ________

Use this recording sheet with the Standardized Test Practice pages.

Fill in the correct answer. For gridded-response questions, write your answers in the boxes on the answer grid and fill in the bubbles to match your answers.

1. Ⓐ Ⓑ Ⓒ Ⓓ

2. ______________

3. Ⓕ Ⓖ Ⓗ Ⓘ

4. Ⓐ Ⓑ Ⓒ Ⓓ

5. ______________

6. Ⓕ Ⓖ Ⓗ Ⓘ

7. Ⓐ Ⓑ Ⓒ Ⓓ

8. Ⓕ Ⓖ Ⓗ Ⓘ

9. ______________

10. Ⓐ Ⓑ Ⓒ Ⓓ

11. Ⓕ Ⓖ Ⓗ Ⓘ

12. Ⓐ Ⓑ Ⓒ Ⓓ

13. Ⓕ Ⓖ Ⓗ Ⓘ

Extended Response

Record your answers for Exercise 14 on the back of this paper.

NAME ____________________ DATE ____________ PERIOD ________

Extended-Response Test

SCORE ________

Demonstrate your knowledge by giving a clear, concise solution to each problem. Be sure to include all relevant drawings and justify your answers. You may show your solution in more than one way or investigate beyond the requirements of the problem. If necessary, record your answer on another piece of paper.

1. a. Explain what it means for two angles to be congruent.

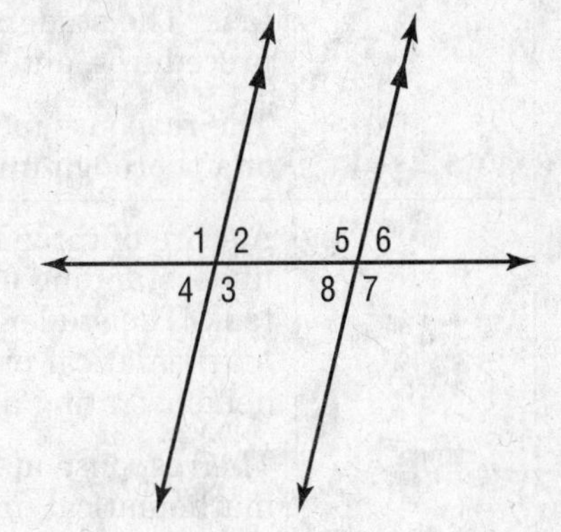

b. In the figure, which angles are congruent? Find the measure of all the angles if $m\angle 2 = 76°$

2. a. Draw a figure that is a polygon and another figure that is not a polygon. Explain how the figures are different.

b. Explain what is meant by *regular polygon*.

3. a. Explain how the diagram below demonstrates the Pythagorean Theorem for a right triangle with legs of length 2.

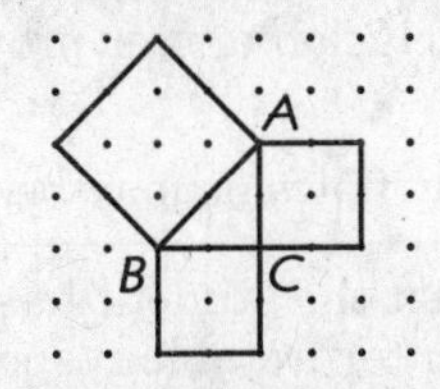

b. Write a word problem that can be solved by using the Pythagorean Theorem.

c. Solve the problem in part **b**. Explain each step.

NAME ______________________ DATE ______________ PERIOD ________

Extended-Response Rubric

SCORE ________

Score	Description
4	A score of four is a response in which the student demonstrates a thorough understanding of the mathematics concepts and/or procedures embodied in the task. The student has responded correctly to the task, used mathematically sound procedures, and provided clear and complete explanations and interpretations. The response may contain minor flaws that do not detract from the demonstration of a thorough understanding.
3	A score of three is a response in which the student demonstrates an understanding of the mathematics concepts and/or procedures embodied in the task. The student's response to the task is essentially correct with the mathematical procedures used and the explanations and interpretations provided demonstrating an essential but less than thorough understanding. The response may contain minor flaws that reflect inattentive execution of mathematical procedures or indications of some misunderstanding of the underlying mathematics concepts and/or procedures.
2	A score of two indicates that the student has demonstrated only a partial understanding of the mathematics concepts and/or procedures embodied in the task. Although the student may have used the correct approach to obtaining a solution or may have provided a correct solution, the student's work lacks an essential understanding of the underlying mathematical concepts. The response contains errors related to misunderstanding important aspects of the task, misuse of mathematical procedures, or faulty interpretations of results.
1	A score of one indicates that the student has demonstrated a very limited understanding of the mathematics concepts and/or procedures embodied in the task. The student's response is incomplete and exhibits many flaws. Although the student's response has addressed some of the conditions of the task, the student reached an inadequate conclusion and/or provided reasoning that was faulty or incomplete. The response exhibits many flaws or may be incomplete.
0	A score of zero indicates that the student has provided no response at all, or a completely incorrect or uninterpretable response, or demonstrated insufficient understanding of the mathematics concepts and/or procedures embodied in the task. For example, a student may provide some work that is mathematically correct, but the work does not demonstrate even a rudimentary understanding of the primary focus of the task.

NAME ______________________ DATE ______________ PERIOD ________

Test, Form 1A

SCORE ________

Write the letter for the correct answer in the blank at the right of each question.

For Exercises 1 and 2, use the figure at the right.

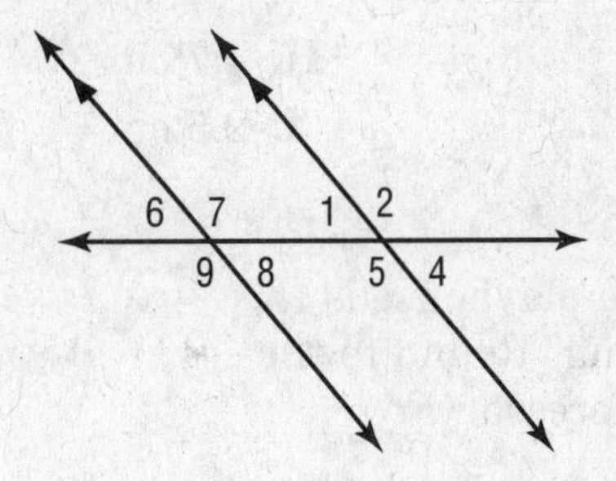

1. What is $m\angle 8$ if $m\angle 4 = 50°$?

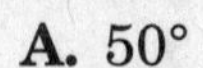

A. 50°
B. 180°
C. 70°
D. 40°

1. ________

2. If $m\angle 2 = 130°$, what is $m\angle 6$?

F. 40°
G. 130°
H. 50°
I. 180

2. ________

3. In the figure below, what is $m\angle 1$ if $m\angle 7 = 60°$?

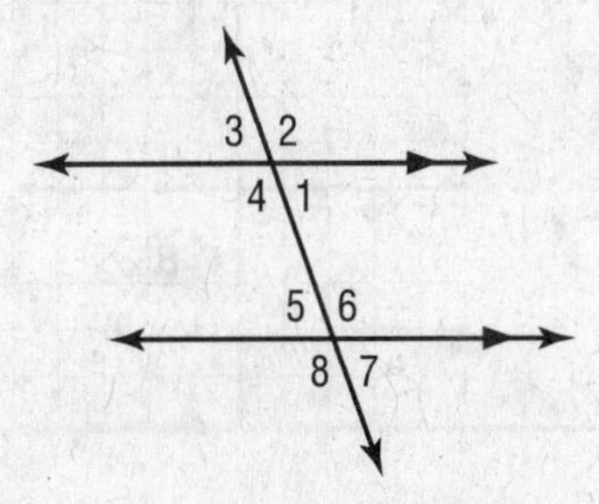

A. 60° **C.** 120°
B. 90° **D.** 150°

3. ________

4. A triangle has angles measuring 15° and 45°. What is the measure of the triangle's third angle?

F. 30° **H.** 120°
G. 90° **I.** 180°

4. ________

5. What is the value of x in the triangle at the right?

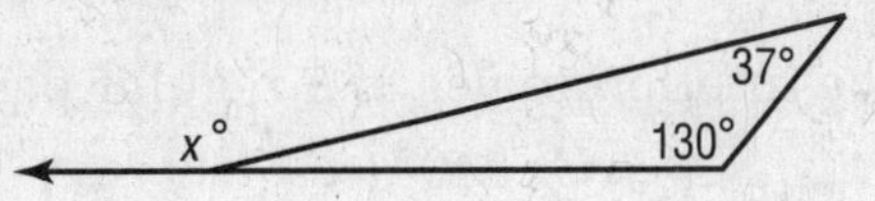

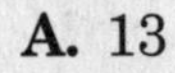

A. 13
B. 37
C. 130
D. 167

5. ________

NAME ______________________ DATE ____________ PERIOD ________

Test, Form 1A (continued)

SCORE ________

6. One leg of a right triangle is 3.2 centimeters long. The length of the second leg is 5.7 centimeters. What is the length of the hypotenuse? Round to the nearest tenth if necessary.

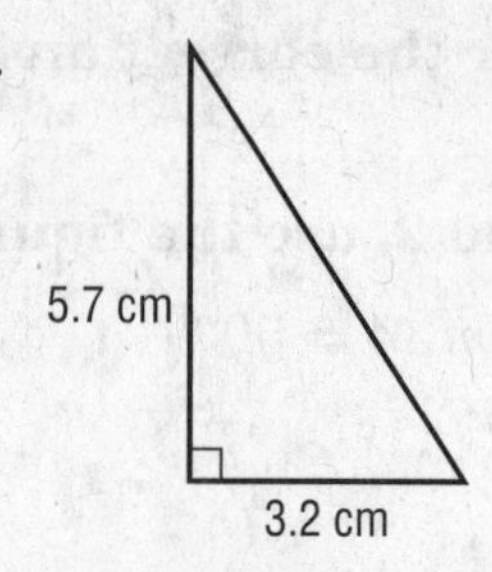

F. 6.5 cm **H.** 4.7 cm

G. 6.4 cm **I.** 2.5 cm

6. ________

7. How far up on the playhouse is the baseball bat resting? Round to the nearest tenth if necessary.

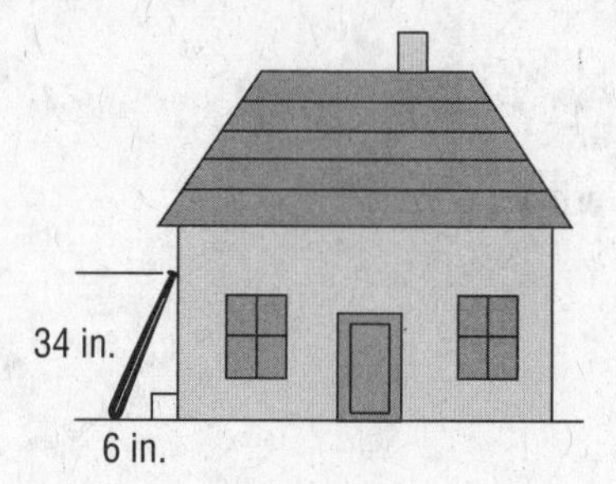

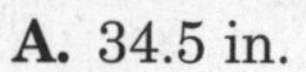

A. 34.5 in.

B. 33.5 in.

C. 28 in.

D. 5.3 in.

7. ________

8. What is the distance between points $A(-3, 4)$ and $B(1, -2)$? Round to the nearest tenth if necessary.

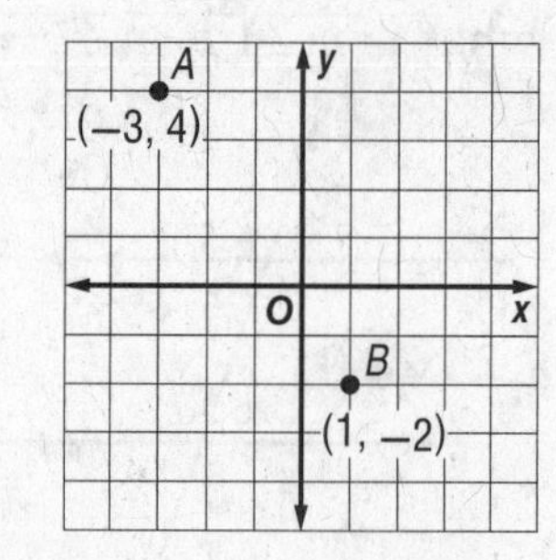

F. 6.08 units

G. 6.1 units

H. 6.5 units

I. 7.2 units

8. ________

9. Which of the following could be the side measures of a right triangle?

A. 6 ft, 5 ft, 4 ft **C.** 10 ft, 8 ft, 6 ft

B. 8 ft, 7 ft, 6 ft **D.** 12 ft, 10 ft, 8 ft

9. ________

10. Which is the measures of an exterior angle of a regular octagon?

F. 40° **H.** 50°

G. 45° **I.** 55°

10. ________

11. What is the sum of the interior angles of a regular pentagon?

A. 90° **C.** 540°

B. 180° **D.** 720°

11. ________

NAME ______________________ DATE ____________ PERIOD ________

Test, Form 1B

SCORE ________

Write the letter for the correct answer in the blank at the right of each question.

For Exercises 1 and 2, use the figure at the right.

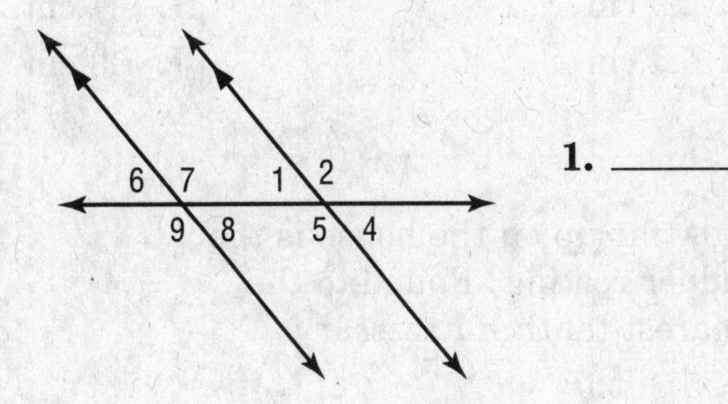

1. What is $m\angle 4$ if $m\angle 8 = 40°$?

A. 50°

B. 180°

C. 70°

D. 40°

1. ________

2. If $m\angle 9 = 130°$, what is $m\angle 4$?

F. 40° **H.** 50°

G. 130° **I.** 180

2. ________

3. In the figure below, what is $m\angle 2$ if $m\angle 6 = 120°$?

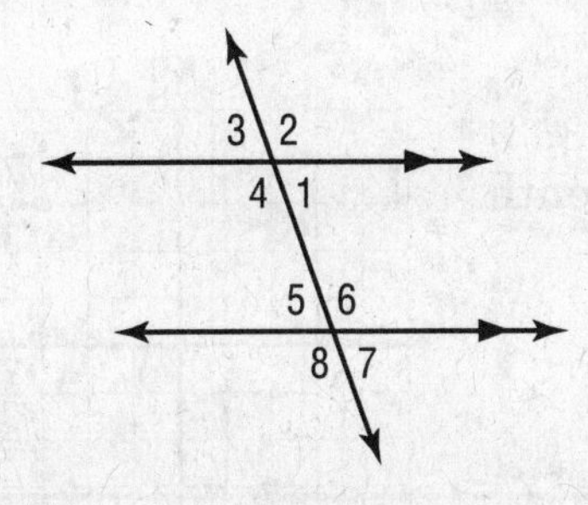

A. 60° **C.** 120°

B. 90° **D.** 150°

3. ________

4. A triangle has angles measuring 25° and 45°. What is the measure of the triangle's third angle?

F. 20° **H.** 110°

G. 90° **I.** 180°

4. ________

5. What is the value of x in the triangle at the right?

A. 23

B. 32

C. 125

D. 148

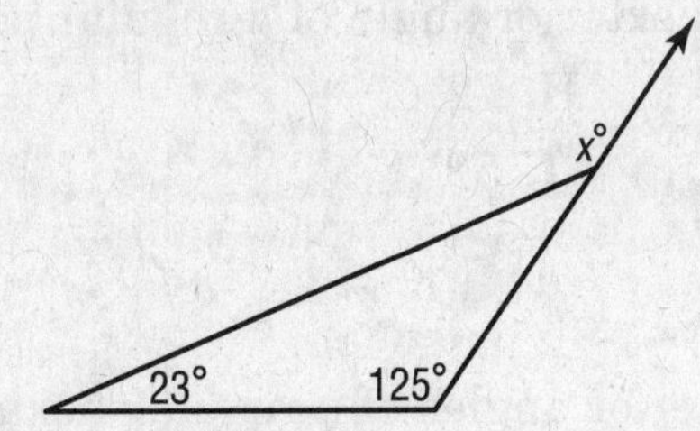

5. ________

NAME ________________ DATE ________ PERIOD ______

Test, Form 1B *(continued)*

SCORE ______

6. One leg of a right triangle is 6.4 centimeters long. The length of the second leg is 3.5 centimeters. What is the length of the hypotenuse? Round to the nearest tenth if necessary.

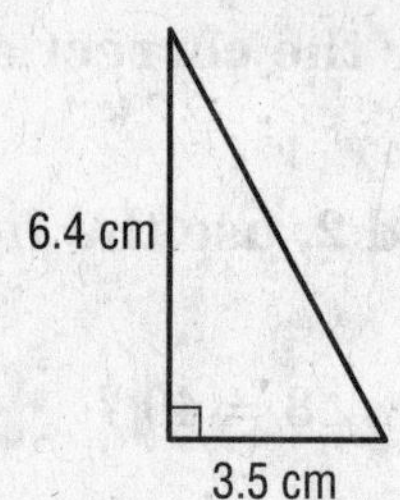

F. 2.9 cm   H. 6.1 cm
G. 5.4 cm   I. 7.3 cm

6. ________

7. How far up on the house is the ladder resting? Round to the nearest tenth if necessary.

A. 5.3 ft
B. 21 ft
C. 24.8 ft
D. 25.2 ft

7. ________

8. What is the distance between points $V(3, 3)$ and $W(-2, -3)$? Round to the nearest tenth if necessary.

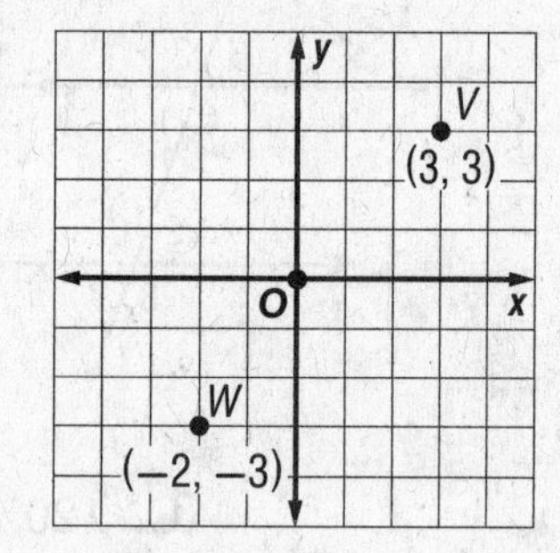

F. 12.6 units
G. 11.2 units
H. 7.8 units
I. 5.7 units

8. ________

9. Which of the following could be the side measures of a right triangle?

A. 5 ft, 4 ft, 3 ft   C. 11 ft, 10 ft, 6 ft
B. 6 ft, 4 ft, 2 ft   D. 15 ft, 10 ft, 5 ft

9. ________

10. What is the measure of an exterior angle of a regular pentagon?

F. 108°   H. 72°
G. 80°   I. 58°

10. ________

11. What is the sum of the interior angles of a regular hexagon?

A. 90°   C. 540°
B. 180°   D. 720°

11. ________

NAME ______________________ DATE ______________ PERIOD __________

Test, Form 2A

SCORE __________

Write the letter for the correct answer in the blank at the right of each question.

1. What is the value of x in the figure below?

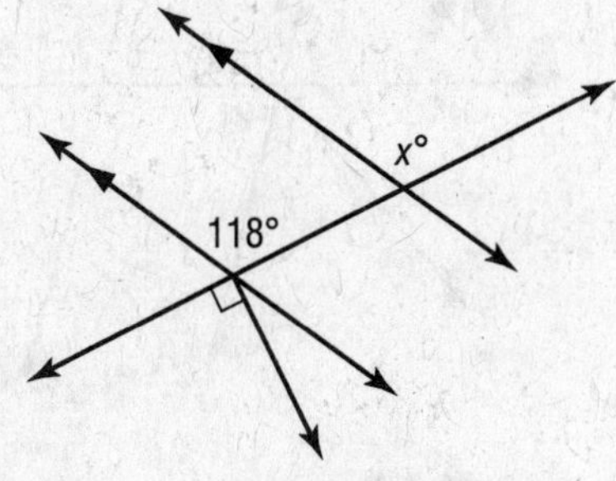

A. 28 **B.** 62 **C.** 90 **D.** 118 **1.** __________

2. Which pair of angles is *not* congruent?

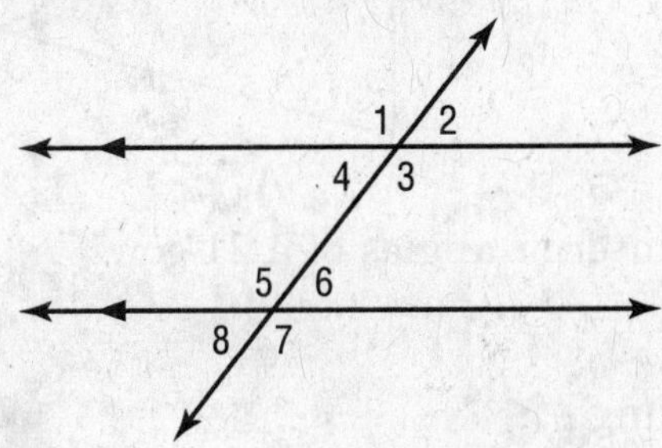

F. $\angle 1$ and $\angle 7$ **H.** $\angle 4$ and $\angle 6$
G. $\angle 3$ and $\angle 5$ **I.** $\angle 2$ and $\angle 5$ **2.** __________

3. In the figure below, $m\angle 1 = x$ and $m\angle 2 = x - 4$. Which statement could be used to prove that $x = 47$?

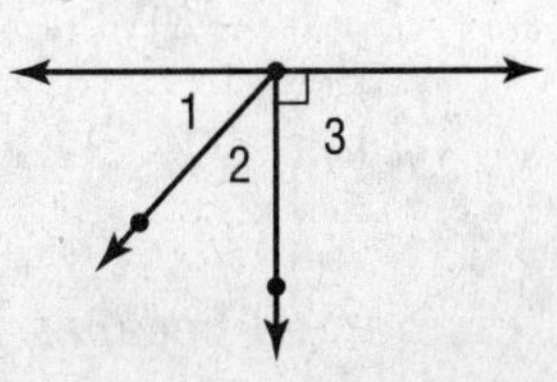

A. $m\angle 1 = m\angle 2$ **C.** $m\angle 1 + m\angle 2 = 90$
B. $m\angle 2 = 47$ **D.** $m\angle 1 + m\angle 2 = 180$ **3.** __________

4. Rini used a stick to draw a right triangle in the ground. The hypotenuse of her triangle is 24 inches and one of the legs is 12 inches. What is the length of the third side? Round to the nearest tenth if necessary.

F. 2 in. **G.** 12 in. **H.** 20.8 in. **I.** 26.8 in. **4.** __________

5. What is the measure of an exterior angle of a regular hexagon?

A. 72° **B.** 60° **C.** 45° **D.** 40° **5.** __________

Test, Form 2A *(continued)*

SCORE ________

6. In the figure at the right, find the $m\angle 6$ if $m\angle 2 = 75°$.

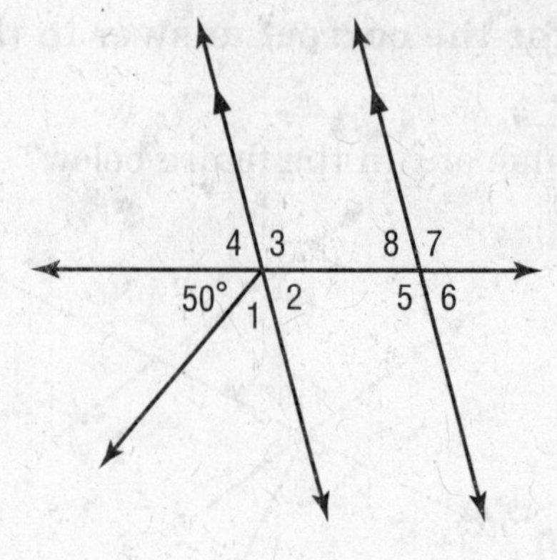

6. ____________

7. What is the value of x in the triangle at the right?

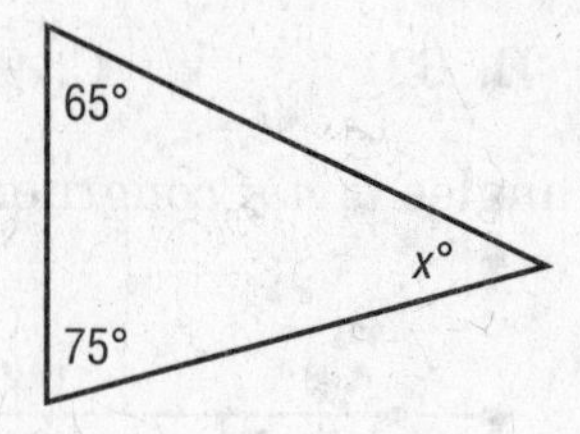

7. ____________

8. Find the sum of the measures of the interior angles of a 21-gon. **8.** ____________

9. In the diagram, Jorge let go of the string tied to his balloon. Write and solve an equation to find how far above Jorge's head the balloon is. Round your answer to the nearest tenth if necessary.

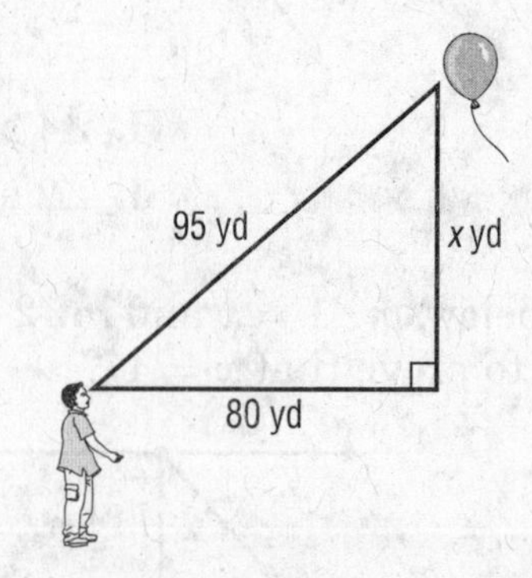

9. ____________

10. The Pentagon building in Washington, D.C., is named because it is in the shape of a regular pentagon. What is the measure of each interior angle? **10.** ____________

11. Maude's living room is in the shape of a rectangle. Its dimensions are 21 feet by 14 feet. Find the length of the diagonal of the living room. Round your answer to the nearest tenth if necessary. **11.** ____________

12. Find the distance between points $M(-4, 6)$ and $N(10, -5)$. Round to the nearest tenth if necessary. **12.** ____________

NAME ________________ DATE ____________ PERIOD ________

Test, Form 2B

SCORE ________

Write the letter for the correct answer in the blank at the right of each question.

1. What is the value of x in the figure at the right?

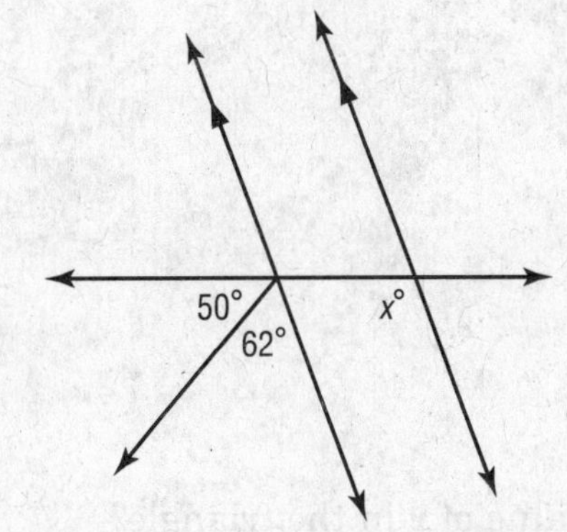

A. 50
B. 62
C. 68
D. 112

1. ________

2. Which pair of angles is *not* congruent?

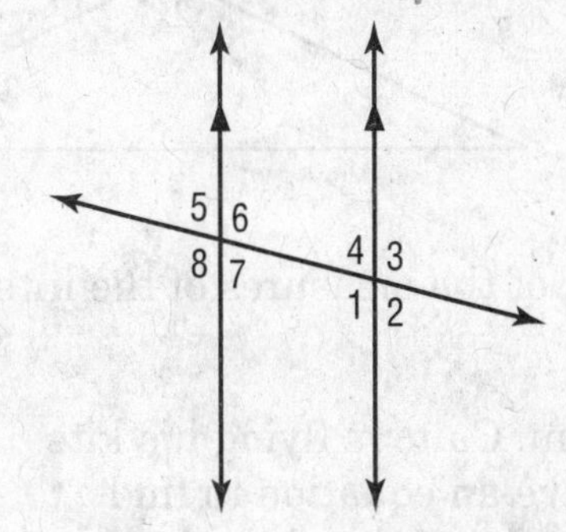

F. $\angle 1$ and $\angle 8$
G. $\angle 3$ and $\angle 5$
H. $\angle 4$ and $\angle 7$
I. $\angle 2$ and $\angle 5$

2. ________

3. In the figure below, $m\angle 1 = x$ and $m\angle 2 = x - 8$. Which statement could be used to prove that $x = 49$?

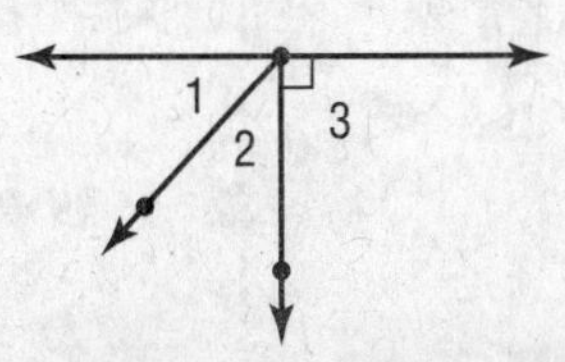

A. $m\angle 1 = m\angle 2$
B. $m\angle 2 = 49$
C. $m\angle 1 + m\angle 2 = 90$
D. $m\angle 1 + m\angle 2 = 180$

3. ________

4. Coty painted a right triangle on paper. The hypotenuse of his triangle is 18 inches and one of the legs is 7 inches. What is the length of the third side? Round to the nearest tenth if necessary.

F. 19.3 in. **G.** 16.6 in. **H.** 14.2 in. **I.** 11 in.

4. ________

5. What is the measure of an exterior angle of a regular octagon?

A. 72° **B.** 60° **C.** 45° **D.** 40°

5. ________

Test, Form 2B *(continued)*

SCORE ________

6. In the figure at the right, find the $m\angle 5$ if $m\angle 3 = 115°$.

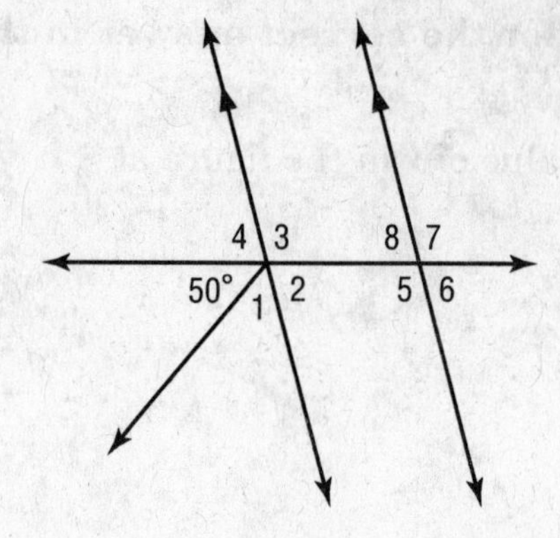

6. ____________

7. What is the value of x in the triangle?

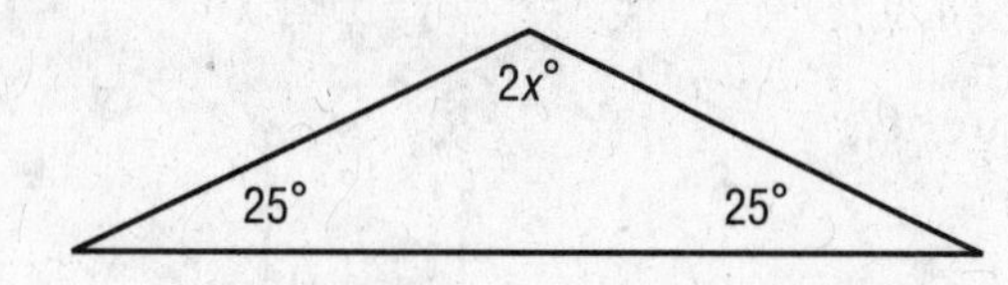

7. ____________

8. Find the sum of the measures of the interior angles of an 18-gon.

8. ____________

9. In the diagram, Caito is flying his kite. Write and solve an equation to find how far above Caito's head the kite is. Round your answer to the nearest tenth if necessary.

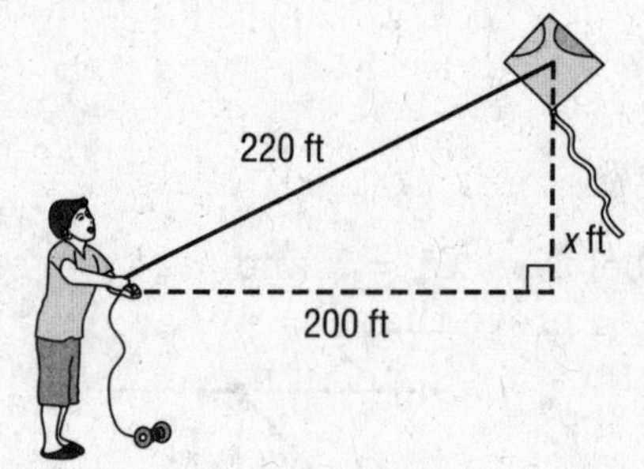

9. ____________

10. The Pentagon building in Washington, D.C., is named because it is in the shape of a regular pentagon. What is the sum of the measures of its interior angles?

10. ____________

11. Davey's log cabin is in the shape of a rectangle. Its dimensions are 40 feet by 24 feet. Find the length of the diagonal of the floor of the cabin. Round your answer to the nearest tenth if necessary.

11. ____________

12. Find the distance between points $P(3, -8)$ and $Q(7, 4)$. Round to the nearest tenth if necessary.

12. ____________

NAME _______________ DATE _______________ PERIOD _______

Test, Form 3A

SCORE _______

For Exercises 1 and 2, write and solve an equation to find each missing length. Round to the nearest tenth if necessary.

1. An observer is standing 30 feet from a flagpole. She is looking at the top of the flagpole. How tall is the flagpole? Round to the nearest tenth if necessary.

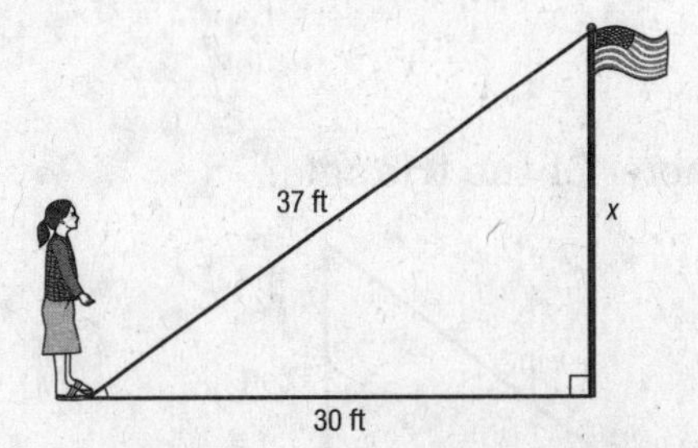

1. _______

2. Willa wants to mail a frame in the box shown at the right. Find the length of the diagonal of the box. Round to the nearest tenth if necessary.

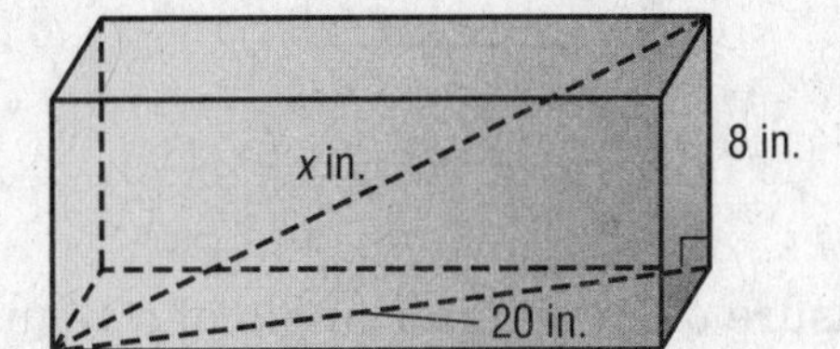

2. _______

For Exercises 3 and 4, find the distance between each pair of points with the given coordinates. Round to the nearest tenth if necessary.

3. (5.5, 0), (2.75, 4.5)

3. _______

4. (−5, 2), (5, −2)

4. _______

Find the value of x in each figure.

5.

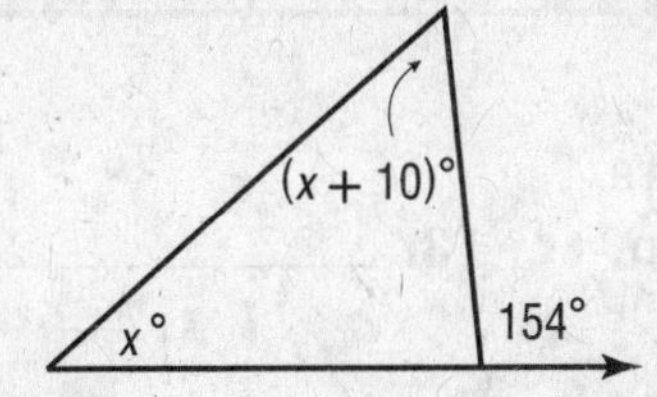

5. _______

6.

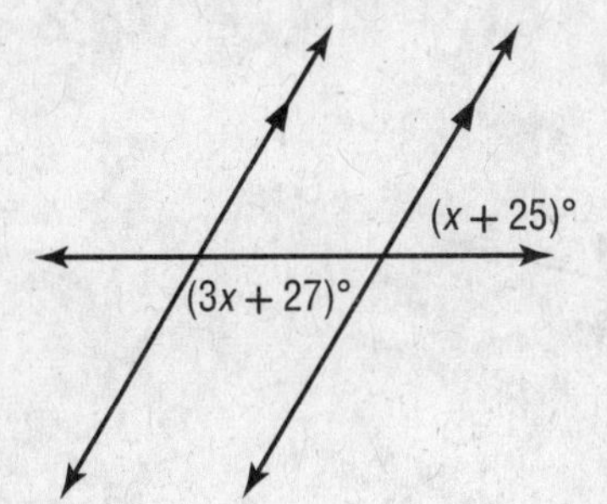

6. _______

7. The brackets for a shelf are in the shape of a triangle. Find the angles of the triangle if the measures of the angles are in the ratio $x : 2x : 3x$.

7. _______

Test, Form 3A *(continued)*

SCORE 

8. What is true about the sum of the measures of any two non-right angles in a right triangle? **8.** ________

9. What is the perimeter of the triangle?

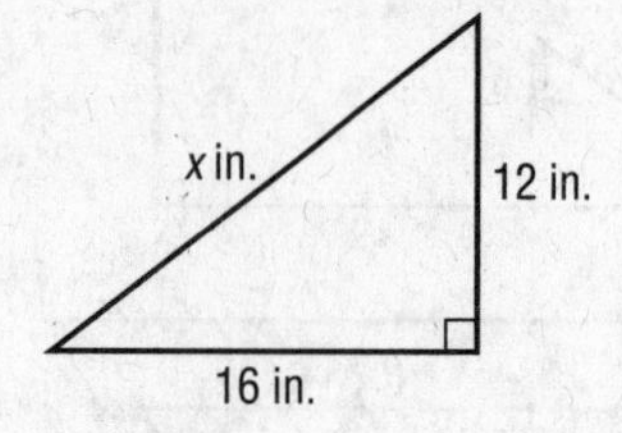

9. ________

10. Find the measure of an exterior angle of a regular nonagon. **10.** ________

11. Find the sum of the measures of the interior angles of a 30-gon. **11.** ________

12. The front view of a house (the "front elevation") is often in the shape of a pentagon. If two of the angles each measure 90°, what is the sum of the measures of the other 3 angles? **12.** ________

13. A triangle has side lengths of 10 centimeters, 24 centimeters and 28 centimeters. Is the triangle a right triangle? Explain. **13.** ________

14. Find the missing reason in the proof.

Given: two parallel lines cut by a transversal, $m\angle 6 = 3x$, and $m\angle 8 = 99°$

Prove: $x = 33$

Proof:

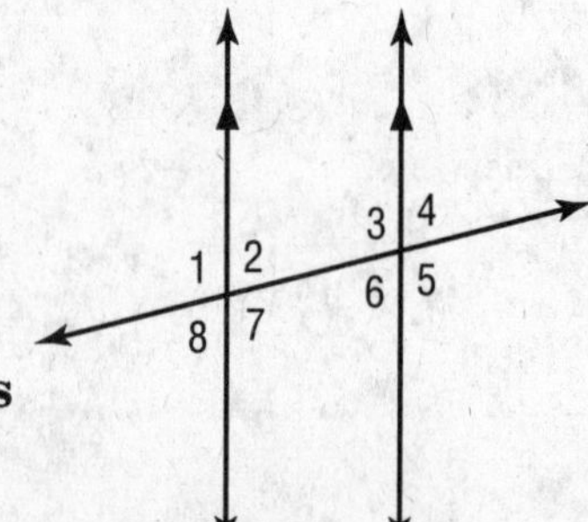

Statements	**Reasons**
1. $m\angle 6 = 3x$, $m\angle 8 = 99°$	Given
2. $m\angle 6 = m\angle 8$	?
3. $3x = 99$	Substitution
4. $x = 33$	Division Property of Equality

14. ________

NAME ______________________ DATE ______________ PERIOD ________

Test, Form 3B

SCORE ________

For Exercises 1 and 2, write and solve an equation to find each missing length. Round to the nearest tenth if necessary.

1. An observer is standing 20 feet from a tree with a hanging birdhouse. She is looking at the birdhouse. How far from the ground is the birdhouse? Round to the nearest tenth if necessary.

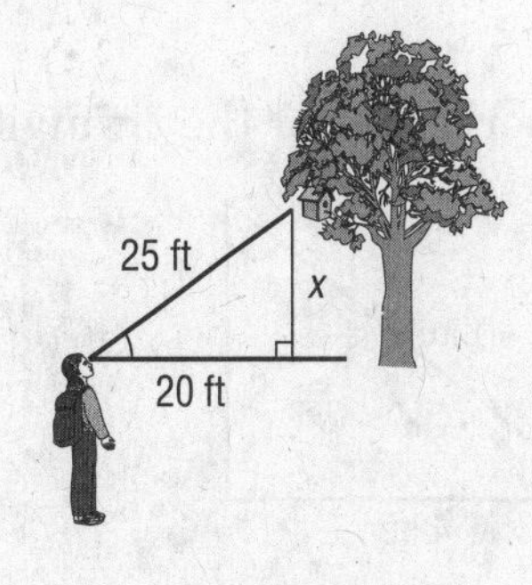

1. ________

2. Kenneth wants to wrap a collapsible fishing rod in the box shown at the right. Find the length of the diagonal of the box. Round to the nearest tenth if necessary.

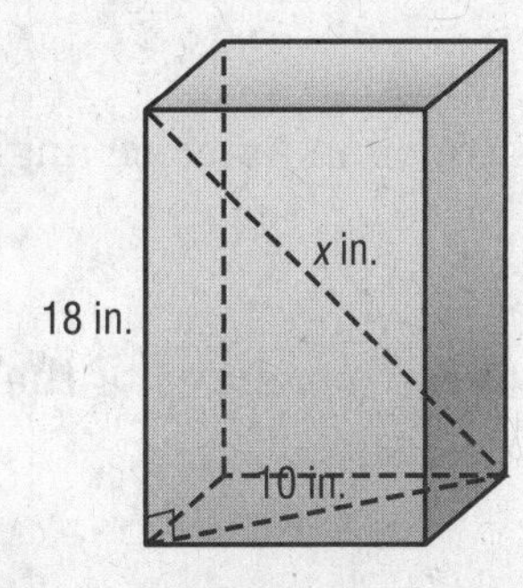

2. ________

For Exercises 3 and 4, find the distance between each pair of points with the given coordinates. Round to the nearest tenth if necessary.

3. (4.25, 5.5), (3.5, 0)   3. ________

4. (−2, 4), (−5, −2)   4. ________

Find the value of x in each figure.

5.

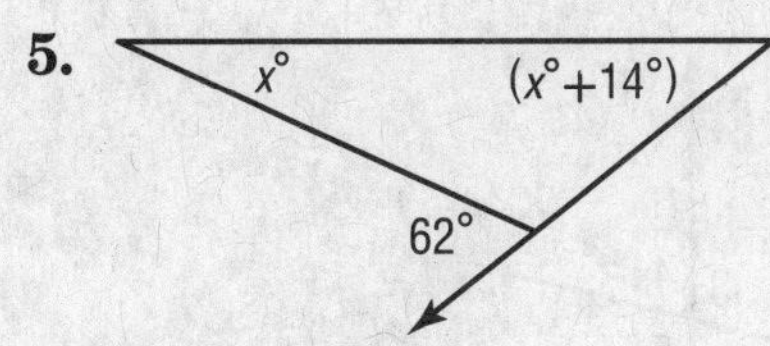

5. ________

6.

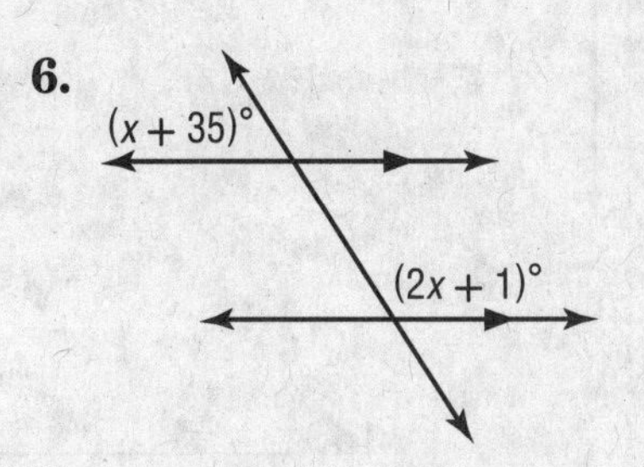

6. ________

7. A shelf is in the shape of a triangle. Find the angles of the triangle if the measures of the angles are in the ratio $x : x : 4x$.   7. ________

NAME ______________________ DATE ____________ PERIOD ________

Test, Form 3B *(continued)*

SCORE ________

8. What is true about the angles in an equilateral triangle? **8.** ____________

9. What is the perimeter of the triangle?

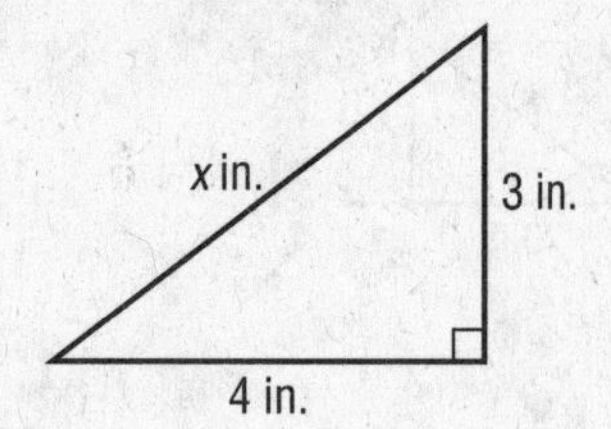

9. ____________

10. Find the measure of an exterior angle of a regular decagon. **10.** ____________

11. Find the sum of the measures of the interior angles of a 50-gon. **11.** ____________

12. The side view of a house (the "side elevation") is often in the shape of a pentagon. If the angle at the roof is 150° and there are two other angles of 90°, what is the sum of the measures of the other two angles? **12.** ____________

13. A triangle has side lengths of 10 inches, 24 inches, and 26 inches. Is the triangle a right triangle? Explain. **13.** ____________

14. Use the proof to find the missing reason.

Given: two parallel lines cut by a transversal, $m\angle 5 = 4x$, and $m\angle 7 = 96°$

Prove: $x = 24$

Proof:

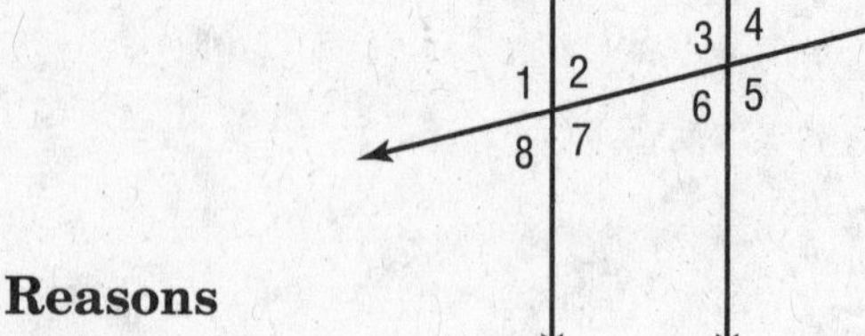

Statements	**Reasons**
1. $m\angle 5 = 4x$, $m\angle 7 = 96°$	Given
2. $m\angle 5 = m\angle 7$	?
3. $4x = 96$	Substitution
4. $x = 24$	Division Property

14. ____________

NAME ______________________ DATE ____________ PERIOD ________

Are You Ready?

Review

- To add integers with the same sign, add the absolute value of each number. The sum has the sign of the integers.
- To add integers with different signs, find the difference of their absolute values. The sum has the sign of the integer with the greater absolute value.

Example 1

Find $-22 + (-33)$.

$-22 + (-33) = -55$ $\quad |-22| + |-33| = 55$

The sum is negative because both integers are negative.

Example 2

Find $-41 + 20$.

$-41 + 20 = -21$ $\quad |-41| - |20| = 21$

The sum is negative because $|-41| > |20|$.

Exercises

Add the integers with the same sign.

1. $147 + 13$ — 1. ____________

2. $-55 + (-31)$ — 2. ____________

3. $18 + 71$ — 3. ____________

Add the integers with different signs.

4. $-14 + 21$ — 4. ____________

5. $12 + (-56)$ — 5. ____________

6. $-4 + 18$ — 6. ____________

Add the integers.

7. $-31 + (-17)$ — 7. ____________

8. $72 + (-22)$ — 8. ____________

9. $47 + 23$ — 9. ____________

NAME ______________________ DATE ____________ PERIOD ________

Are You Ready?

Practice

Add.

1. $2 + (-8)$ **1.** ________

2. $-4 + 3$ **2.** ________

3. $-3 + (-10)$ **3.** ________

4. $-5 + 10$ **4.** ________

5. $-2 + (-9)$ **5.** ________

6. $0 + (-1)$ **6.** ________

7. $-8 + 4$ **7.** ________

8. $6 + (-1)$ **8.** ________

9. $7 + (-11)$ **9.** ________

10. $1 + (-9)$ **10.** ________

11. Graph triangle EFG with vertices $E(2, 4)$ and $F(-2, 4)$ and a height of 4 units. Write the coordinates of the missing vertices. **11.** ________

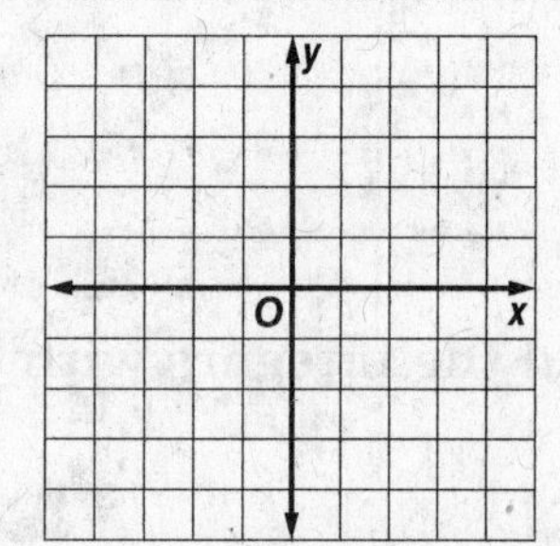

12. Graph square $JKLM$ with vertices $J(0, 6)$ and $K(6, 6)$ and side lengths of 6 units. Write the coordinates of the missing vertices. **12.** ________

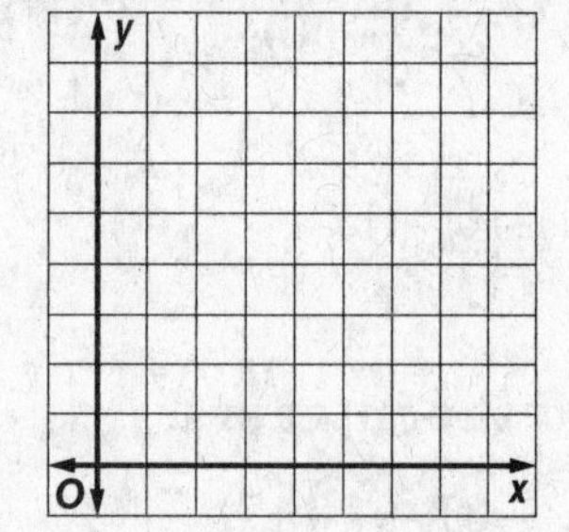

13. Graph rectangle $ABCD$ with vertices $A(-2, 0)$ and $B(-2, 4)$ and side lengths of 4 units and 6 units. Write the coordinates of the missing vertices. **13.** ________

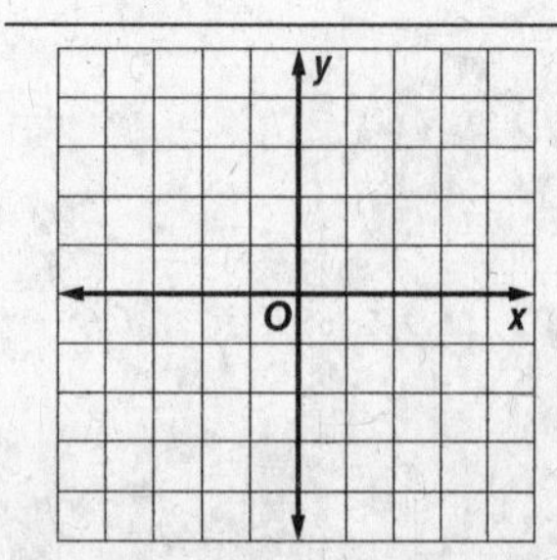

NAME ________________ DATE ________ PERIOD ________

Are You Ready?

Apply

INTERIOR DESIGN Rudell is designing his kitchen. He wants to place the appliances in his kitchen on a coordinate grid to see how they will look.

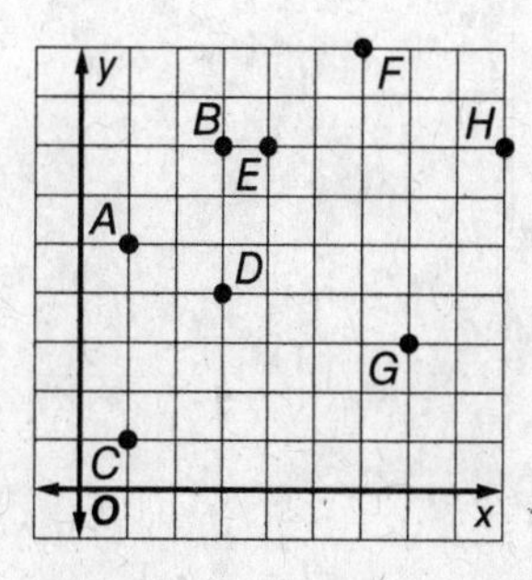

1. Two corners of a square stove are located at (1, 5) and (3, 7). Graph the figure and find the missing vertices if the side length is 2 units.	**2.** Two corners of a rectangular refrigerator are located at (1, 1) and (3, 4). Graph the figure and find the missing vertices if the side length is 3 units.
3. Two corners of a square dishwasher are located at (4, 7) and (6, 9). Graph the figure and find the missing vertices if the side length is 2 units.	**4.** Two corners of a rectangular cabinet are located at (7, 3) and (9, 7). Graph the figure and find the missing vertices if the side length is 4 units.

NAME ______________________ DATE ____________ PERIOD ________

Diagnostic Test

Add.

1. $4 + (-7)$ **1.** ________

2. $-9 + 2$ **2.** ________

3. $-5 + 3$ **3.** ________

4. $8 + (-4)$ **4.** ________

5. $-1 + (-9)$ **5.** ________

6. $0 + (-6)$ **6.** ________

7. $-7 + 4$ **7.** ________

8. $2 + (-1)$ **8.** ________

9. $3 + (-3)$ **9.** ________

10. $-1 + (-1)$ **10.** ________

11. Graph triangle *XYZ* with vertices $X(-3, 0)$ and $Y(3, 0)$ and a height of 6 units. Write the coordinates of the missing vertices. **11.** ________

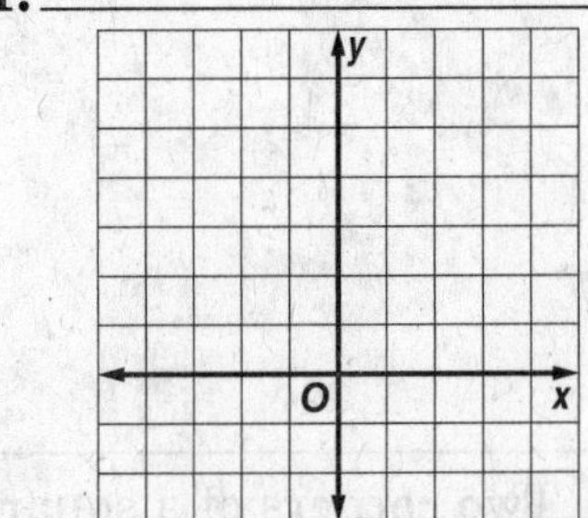

12. Graph square *ABCD* with vertices $A(0, 0)$ and $B(3, 0)$ and side lengths of 3 units. Write the coordinates of the missing vertices. **12.** ________

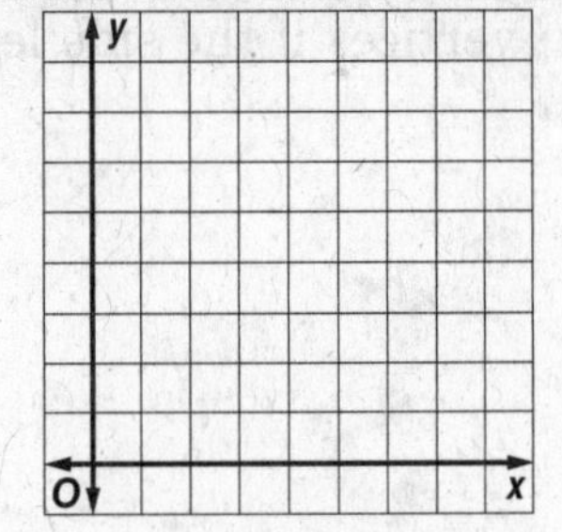

13. Graph square *EFGH* with vertices $E(0, 2)$ and $F(2, 2)$ and side length 2 units. Find the missing vertices if one of the vertices is at the origin. **13.** ________

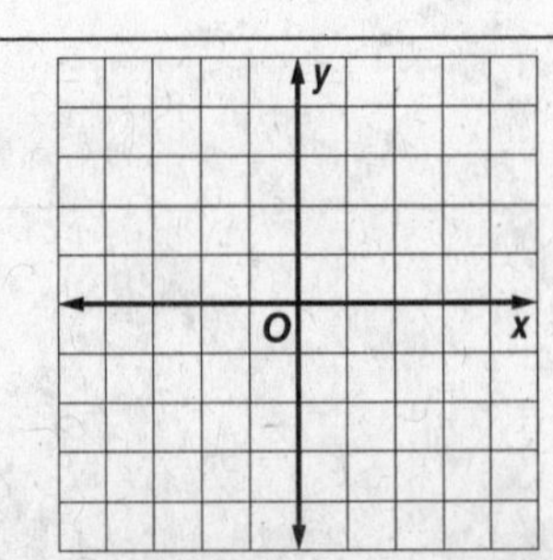

NAME ______________________ DATE ____________ PERIOD ________

Pretest

For Exercises 1 and 2, rectangle *PQRS* has coordinates *P*(2, 4), *Q*(6, 4), *R*(6, 6), and *S*(2, 6). Find the following vertices after a translation of 2 units right and 3 units up.

1. P' **1.** ____________

2. Q' **2.** ____________

For Exercises 3–5, quadrilateral *ABCD* has coordinates *A*(0, 1), *B*(3, 2), *C*(3, 4), and *D*(1, 5).

3. Graph the image of *ABCD* after a reflection over the *y*-axis on a coordinate grid. Label it $A'B'C'D'$. **3.**

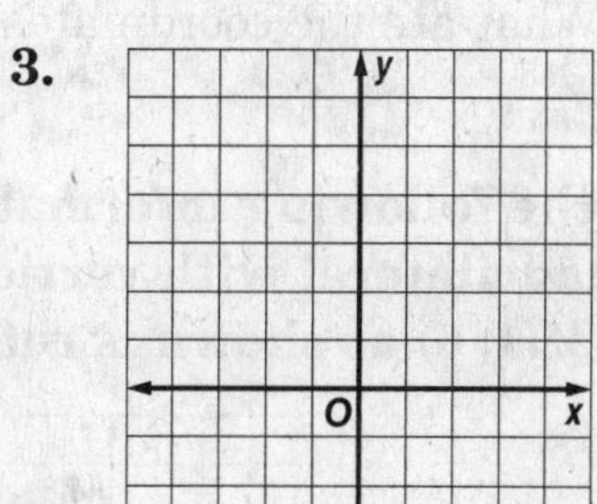

Find the following vertices after a reflection of *ABCD* over the *y*-axis.

4. A' **4.** ____________

5. B' **5.** ____________

6. **GEOMETRY** Mikhail drew parallelogram *HIJK* with coordinates *H*(1, 0), *I*(4, 0), *J*(2, 1), and *K*(−1, 1) on a coordinate plane. Find the coordinates of the image after Mikhail rotated it clockwise 180° about the origin. **6.** ____________

7. Quadrilateral *ABCD* has vertices *A*(−3, 1), *B*(2, 2), *C*(2, 4), and *D*(−3, 3). Graph the image after the given transformation. Then give the coordinates of the vertices for quadrilateral $A'B'C'D'$ after a dilation with a scale factor of $\frac{1}{2}$. **7.**

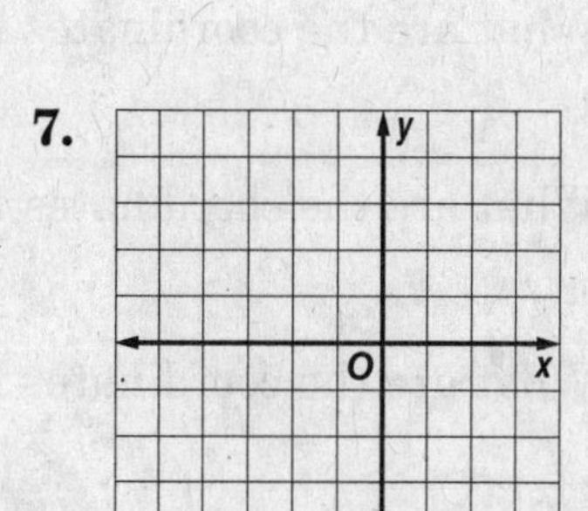

NAME ______________________ DATE ____________ PERIOD ________

Chapter Quiz

Use the following information for Exercises 1–4. A triangular wall decoration on Kelsey's wall has vertices $M(4, -1)$, $N(3, -4)$, and $R(1, -2)$. The triangle is translated 5 units left and 4 units up.

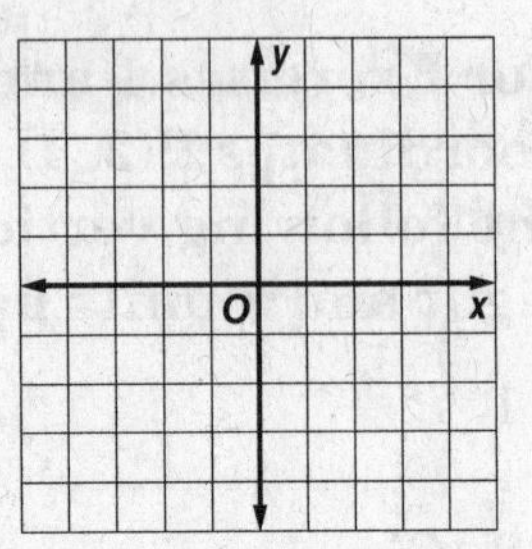

1. Graph the decoration after this translation. 1. ________

2. What are the coordinates for M'? 2. ________

3. What are the coordinates for N'? 3. ________

4. What are the coordinates for R'? 4. ________

Use the following information for Exercises 5–9. A quadrilateral with vertices $J(1, 1)$, $K(4, 3)$, $L(4, 6)$ and $M(1, 6)$ as shown is reflected over the y-axis.

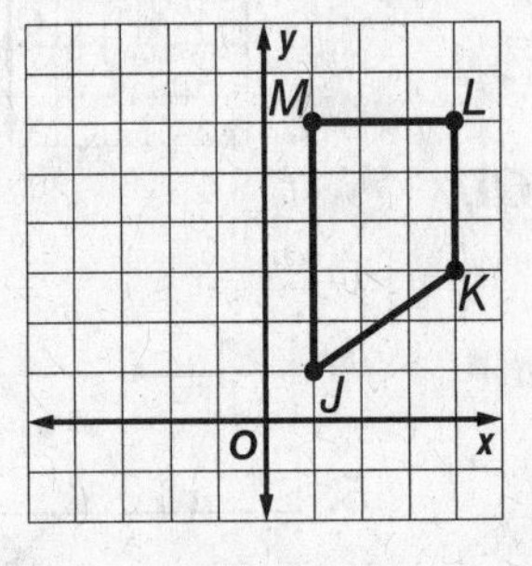

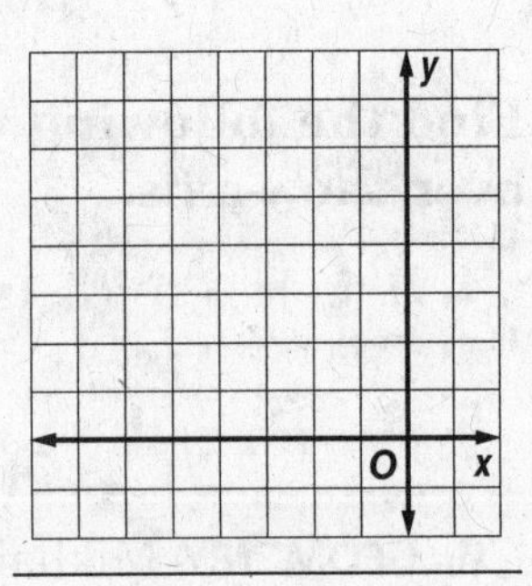

5. Graph the quadrilateral after the reflection. 5. ________

6. What are the coordinates for J'? 6. ________

7. What are the coordinates for K'? 7. ________

8. What are the coordinates for L'? 8. ________

9. What are the coordinates for M'? 9. ________

NAME ______________________ DATE ______________ PERIOD ________

Vocabulary Test

SCORE ________

center of dilation	dilation	preimage	rotational symmetry
center of rotation	image	reflection	translation
congruent	line of reflection	rotation	transformation

Choose the correct term or phrase to complete each sentence.

1. A (transformation, translation) maps one figure onto another. **1.** ____________

2. The original figure before a transformation is called a(n) (image, preimage). **2.** ____________

3. The fixed point around which shapes more in a circular motion is the center of (dilation, rotation). **3.** ____________

4. The figure after a transformation is called an (image, preimage). **4.** ____________

5. When a mirror image is produced by flipping a figure over a line, it is a (reflection, rotation). **5.** ____________

6. A (dilation, reflection) is the image produced when a figure is enlarged or reduced. **6.** ____________

7. Figures that have the same size and shape are (congruent, similar) figures. **7.** ____________

8. When a figure is moved without turning it, it is called a (rotation, translation). **8.** ____________

...ndardized Test Practice

Read each question. Then fill in the correct answer on the answer sheet provided by your teacher or on a sheet of paper.

1. Triangle *ABC* is translated 2 units right and 2 units down. What are the coordinates of *A*′?

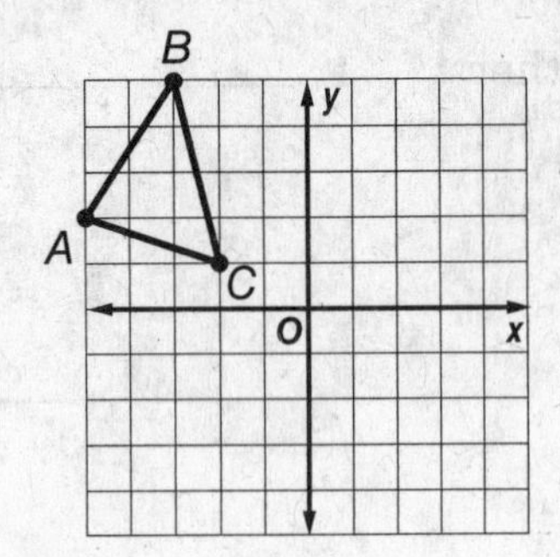

A. (0, –1)
B. (–3, 0)
C. (–1, 3)
D. (0, –3)

2. Seth has $858.60 in his savings account. He plans to spend 15% of his savings on a bicycle. Which of the following represents the amount Seth plans to spend on the bicycle?

F. $182.79
G. $171.72
H. $128.79
I. $122.79

3. Rectangle *M* is similar to rectangle *N*.

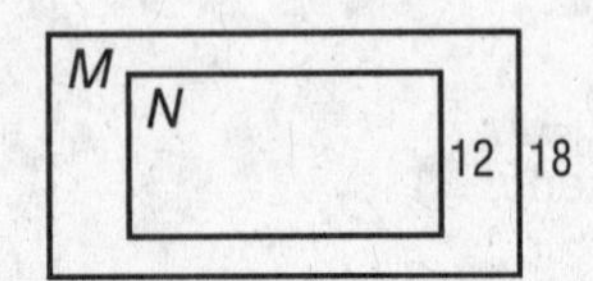

Which scale factor was used to dilate rectangle *M* to rectangle *N*?

A. $\frac{1}{4}$ **C.** $\frac{2}{3}$
B. $\frac{1}{3}$ **D.** $1\frac{1}{4}$

4. THINK SOLVE EXPLAIN **SHORT RESPONSE** The figure shown was transformed from Quadrant I to Quadrant IV. What type of transformation was applied?

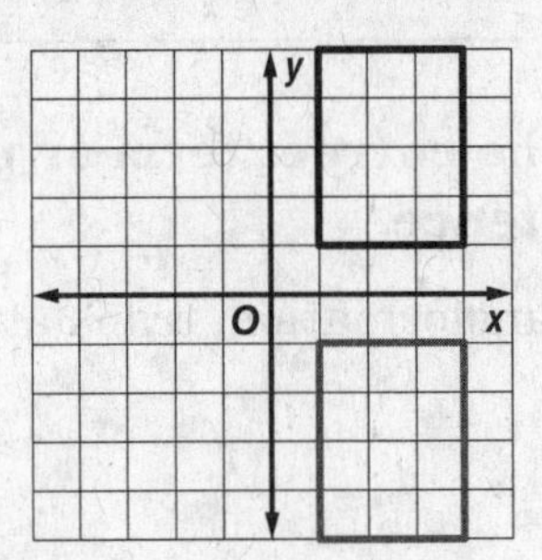

5. **GRIDDED RESPONSE** How many square feet of wrapping paper will Ashton need to cover the box shown?

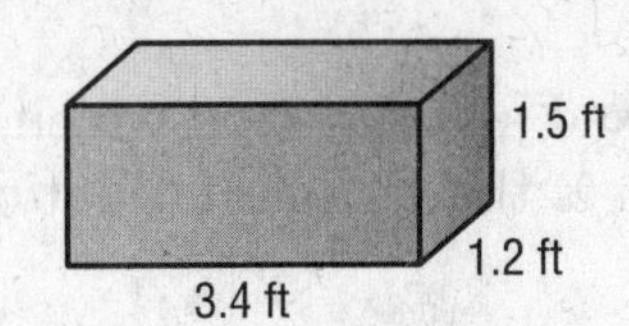

6. Carrie rotated a puzzle piece 180° clockwise to see if she could use it.

Which image represents the position of the puzzle piece after a 180° clockwise rotation?

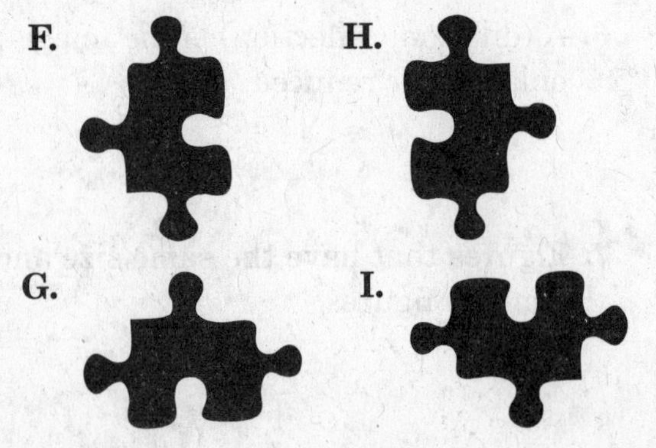

7. What is the value of x in the triangle shown below?

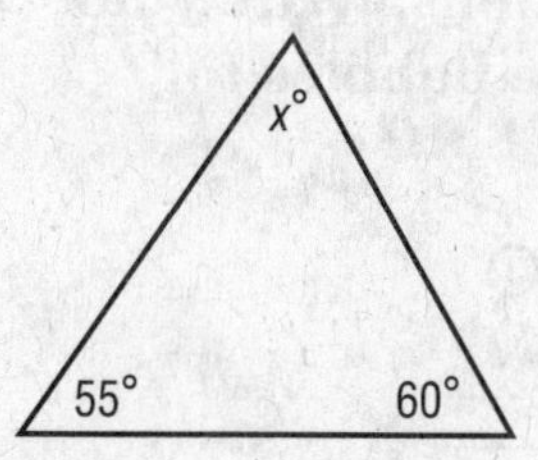

A. 180
B. 90
C. 65
D. 45

8. **GRIDDED RESPONSE** A manager took an employee to lunch. If the lunch was \$48 and she left a 20% tip, what was the total cost in dollars of the lunch?

9. Which irrational number is closest to the number 10?

F. $\sqrt{80}$

G. $\sqrt{99}$

H. $\sqrt{122}$

I. $\sqrt{200}$

10. **SHORT RESPONSE** Alfonzo drew half of a star on a coordinate plane.

If the drawing was reflected across the y-axis, what would be the reflected location of point B?

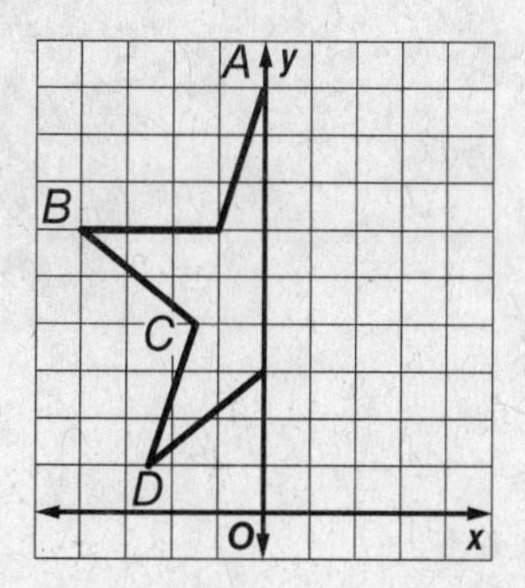

11. Which drawing best represents a reflection over the vertical line segment in the center of the rectangle?

A.
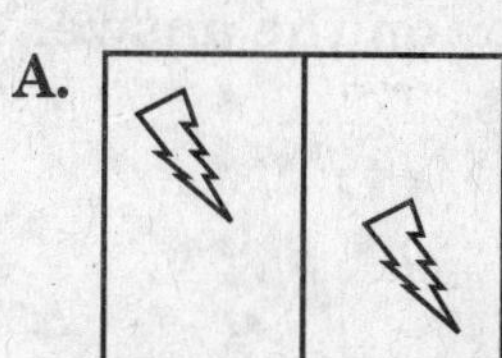

B.
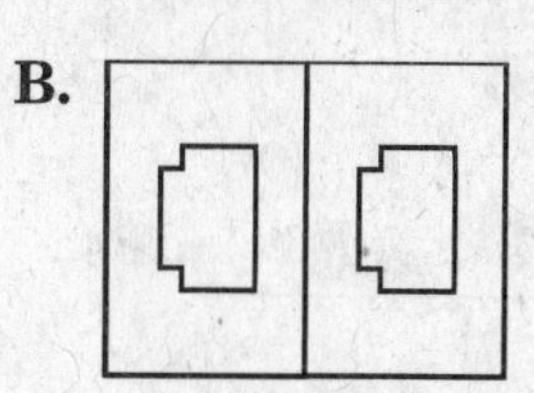

C.
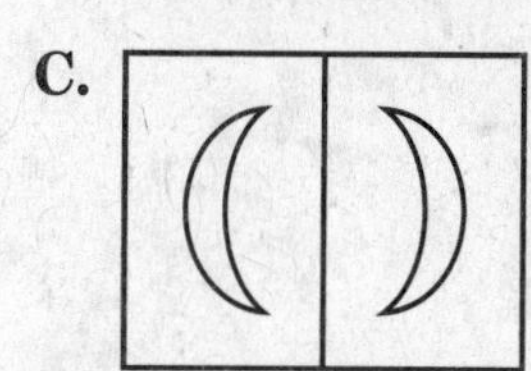

D.
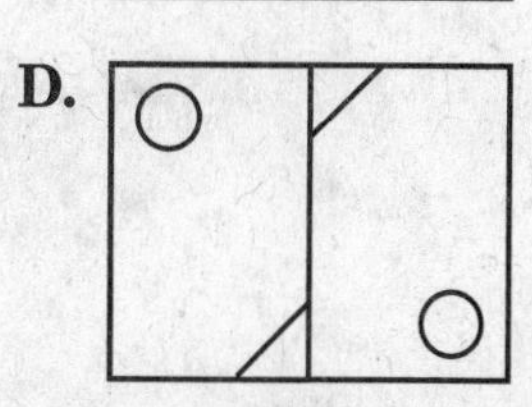

12. What is the solution of the system of equations below?

$$y = 2x - 1$$
$$y = 3x$$

F. $x = -1, y = 1$
G. $x = 1, y = 3$
H. $x = -1, y = -3$
I. $x = -1, y = 1$

13. **EXTENDED RESPONSE** Graph $\triangle XYZ$ with vertices $X(2, 1)$, $Y(7, 3)$, and $Z(3, 6)$.

Part A Translate $\triangle XYZ$ 3 units left and 4 units down. Identify the coordinates of each new vertex.

Part B Find the vertices of $\triangle X'Y'Z'$ after a dilation with a scale factor of $\frac{1}{2}$. Then graph the dilation.

Part C Rotate the dilated figure 270° clockwise around the origin. Draw the rotation.

NAME ______________________ DATE ____________ PERIOD ________

Student Recording Sheet

SCORE ________

Use this recording sheet with the Standardized Test Practice pages.

Fill in the correct answer. For gridded-response questions, write your answers in the boxes on the answer grid and fill in the bubbles to match your answers.

1. Ⓐ Ⓑ Ⓒ Ⓓ

2. Ⓕ Ⓖ Ⓗ Ⓘ

3. Ⓐ Ⓑ Ⓒ Ⓓ

4. ______________________

5.

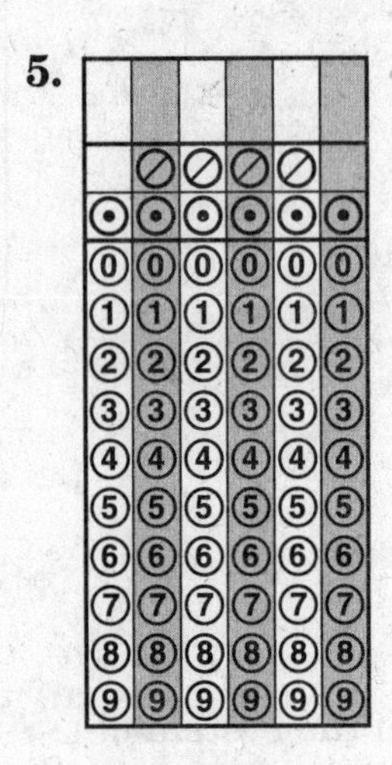

6. Ⓕ Ⓖ Ⓗ Ⓘ

7. Ⓐ Ⓑ Ⓒ Ⓓ

8.

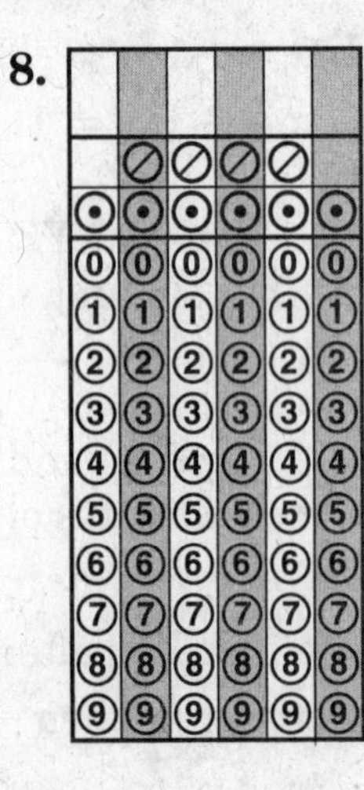

9. Ⓕ Ⓖ Ⓗ Ⓘ

10. ____________

11. Ⓐ Ⓑ Ⓒ Ⓓ

12. Ⓕ Ⓖ Ⓗ Ⓘ

Extended Response

Record your answers for Exercise 13 on the back of this paper.

NAME ______________________ DATE ____________ PERIOD ________

Extended-Response Test

SCORE ________

Demonstrate your knowledge by giving a clear, concise solution to each problem. Be sure to include all relevant drawings and justify your answers. You may show your solution in more than one way or investigate beyond the requirements of the problem. If necessary, record your answer on another piece of paper.

1. Triangle *ABC* has vertices *A*(−4, 1), *B*(−1, 1), and *C*(−1, 3) as shown. Reflect triangle *ABC* twice, first across the *y*-axis, then across the *x*-axis. Graph both reflections on one coordinate plane. Then rotate triangle *ABC* 180° about the origin. Graph the rotation. What do you notice about the two transformations? Make a conjecture about the two reflections and the 180° rotation.

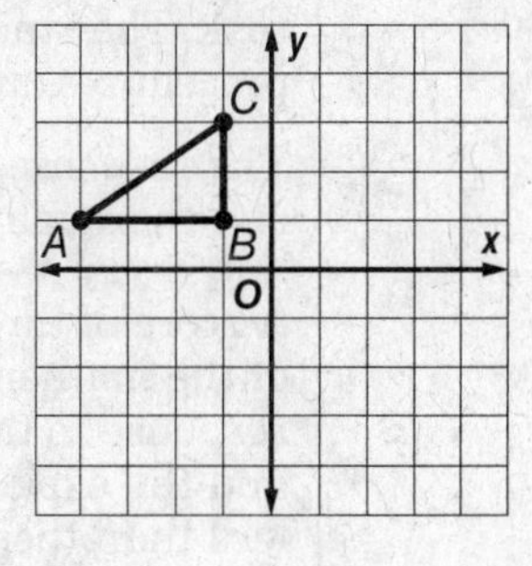

2. Use the coordinate plane on the right. Write the ordered pair that names each point. Then name the ordered pairs for *F′*, *G′*, *H′*, and *I′* after a translation 2 units right and 1 unit down.

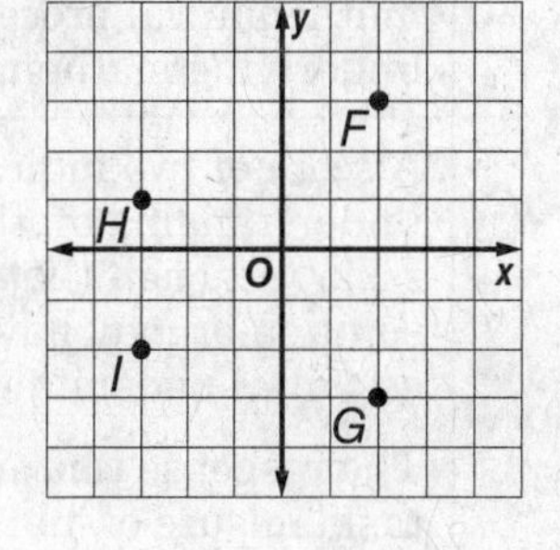

a. *F*

b. *G*

c. *H*

d. *I*

3. Graph each point on a coordinate plane. Then graph a reflection over the *x*-axis of each point, labeled *J′*, *K′*, *L′*, *M′*.

a. *J*(4, 0)

b. *K*(0, 4)

c. *L*(−4, 3)

d. *M*(4, −3)

4. a. Explain in your own words what is meant by a translation, a reflection, a rotation, and a dilation.

b. Describe and draw a translation of the triangle at the right.

c. Describe how you would draw a dilation of the original triangle in part **b** by a scale factor of 2.

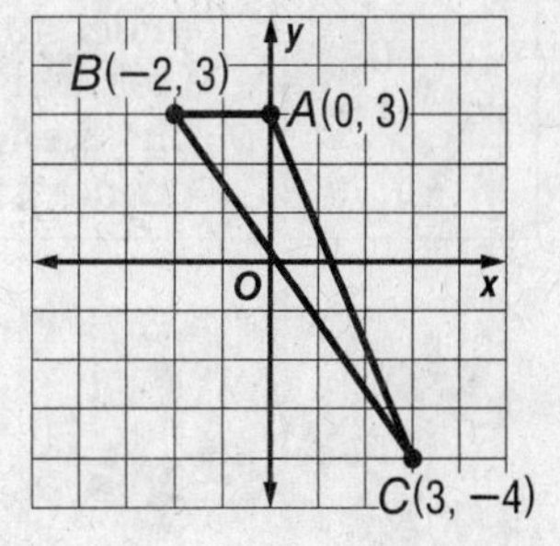

Extended-Response Rubric

SCORE ________

Score	Description
4	A score of four is a response in which the student demonstrates a thorough understanding of the mathematics concepts and/or procedures embodied in the task. The student has responded correctly to the task, used mathematically sound procedures, and provided clear and complete explanations and interpretations. The response may contain minor flaws that do not detract from the demonstration of a thorough understanding.
3	A score of three is a response in which the student demonstrates an understanding of the mathematics concepts and/or procedures embodied in the task. The student's response to the task is essentially correct with the mathematical procedures used and the explanations and interpretations provided demonstrating an essential but less than thorough understanding. The response may contain minor flaws that reflect inattentive execution of mathematical procedures or indications of some misunderstanding of the underlying mathematics concepts and/or procedures.
2	A score of two indicates that the student has demonstrated only a partial understanding of the mathematics concepts and/or procedures embodied in the task. Although the student may have used the correct approach to obtaining a solution or may have provided a correct solution, the student's work lacks an essential understanding of the underlying mathematical concepts. The response contains errors related to misunderstanding important aspects of the task, misuse of mathematical procedures, or faulty interpretations of results.
1	A score of one indicates that the student has demonstrated a very limited understanding of the mathematics concepts and/or procedures embodied in the task. The student's response is incomplete and exhibits many flaws. Although the student's response has addressed some of the conditions of the task, the student reached an inadequate conclusion and/or provided reasoning that was faulty or incomplete. The response exhibits many flaws or may be incomplete.
0	A score of zero indicates that the student has provided no response at all, or a completely incorrect or uninterpretable response, or demonstrated insufficient understanding of the mathematics concepts and/or procedures embodied in the task. For example, a student may provide some work that is mathematically correct, but the work does not demonstrate even a rudimentary understanding of the primary focus of the task.

NAME ______________________ DATE ____________ PERIOD ________

Test, Form 1A

SCORE ________

Write the letter for the correct answer in the blank at the right of each question.

For Exercises 1–5, parallelogram *JKLM* has vertices as shown.

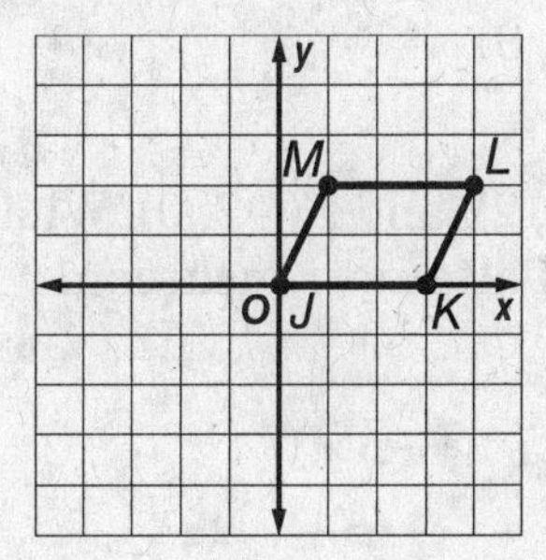

1. If the figure is translated 4 units left, what are the coordinates of J'?

A. (0, −4) **C.** (−4, 0)
B. (0, 4) **D.** (4, 0) **1.** ________

2. If the figure is translated 2 units left and 4 units down, what are the coordinates of L'?

F. (−2, 2) **H.** (2, 2)
G. (−2, −2) **I.** (2, −2) **2.** ________

3. If the figure is rotated 90° clockwise about the origin, what are the coordinates of M'?

A. (−1, 2) **C.** (−2, 1)
B. (2, −1) **D.** (−2, −1) **3.** ________

4. If the figure is reflected over the x-axis, what are the coordinates of K'?

F. (0, 3) **H.** (3, 0)
G. (0, −3) **I.** (−3, 0) **4.** ________

5. If the figure is dilated using a scale factor of $\frac{1}{2}$, what are the coordinates of L'?

A. (2, 1) **C.** (8, 4)
B. (2, −1) **D.** (−8, 4) **5.** ________

6. Five runners are entered in a race. Assuming there are no ties, in how many different ways can first and second places be awarded?

F. 9 **H.** 20
G. 15 **I.** 25 **6.** ________

NAME ______________________ DATE ____________ PERIOD ________

Test, Form 1A *(continued)*

SCORE ________

7. Triangle JKL has vertices $J(0, 2)$, $K(-1, 2)$, and $L(0, -3)$. What are the coordinates of the image of point K after a dilation with a scale factor of 4?

A. $K'(8, 4)$ **C.** $K'(4, 8)$
B. $K'(8, -4)$ **D.** $K'(-4, 8)$

7. ________

8. Triangle XYZ has vertices $X(-4, 3)$, $Y(-1, 2)$, and $Z(-2, 0)$. What are the coordinates of the image of $\triangle XYZ$ after a translation 3 units to the right and 1 unit up?

F. $X'(-4, 4)$, $Y'(-1, 3)$ and $Z'(-2, 1)$
G. $X'(-1, 4)$, $Y'(2, 3)$ and $Z'(1, 1)$
H. $X'(-1, 3)$, $Y'(2, 2)$ and $Z'(1, 0)$
I. $X'(-7, 4)$, $Y'(-4, 3)$ and $Z'(-5, 1)$

8. ________

9. Quadrilateral $ABCD$ has vertices $A(-1, 3)$, $B(-1, 0)$, $C(4, 0)$, and $D(4, 3)$. What are the coordinates of the image of point A after a reflection across the y-axis?

A. $A'(3, -1)$ **C.** $A'(1, 3)$
B. $A'(3, 1)$ **D.** $A'(-1, -3)$

9. ________

10. Triangle MNP has vertices $M(5, 4)$, $N(5, 9)$, and $P(-1, 4)$. What are the coordinates of the image of point P after the triangle is rotated 180° clockwise about the origin?

F. $P'(-4, 1)$ **H.** $P'(-1, -4)$
G. $P'(1, -4)$ **I.** $P'(1, 4)$

10. ________

11. The graph shows segment $M'N'$ is a dilation of segment MN. What is the scale factor of the dilation?

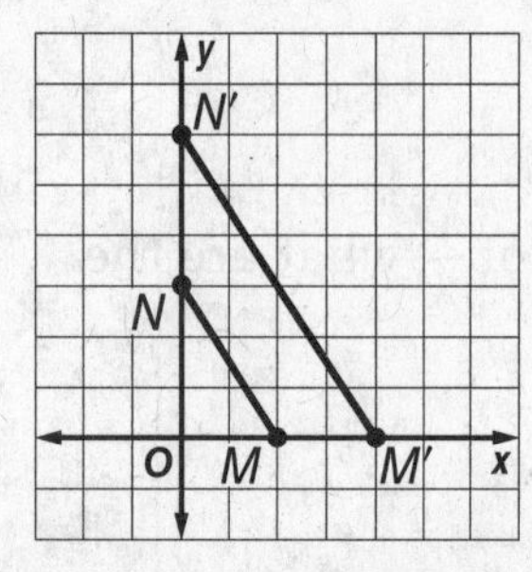

A. 4 **C.** $\frac{1}{2}$
B. 2 **D.** $\frac{1}{4}$

11. ________

NAME ________________________ DATE ______________ PERIOD ________

Test, Form 1B

SCORE ________

Write the letter for the correct answer in the blank at the right of each question.

For Exercises 1–3, triangle *PQR* has vertices as shown.

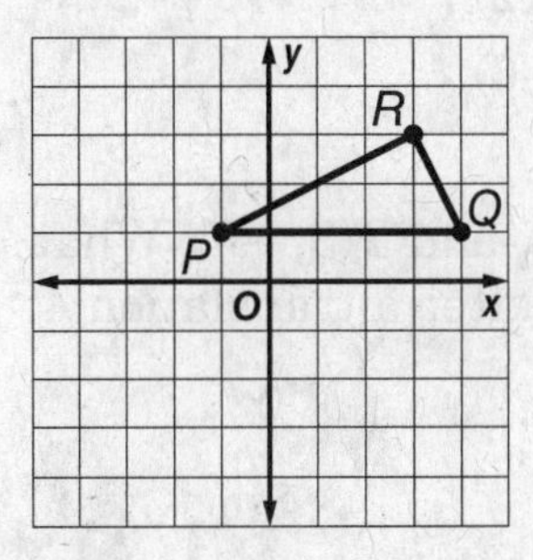

1. If the figure is translated 6 units down, what are the coordinates of R'?

A. (−3, 3) **C.** (−3, −3)
B. (3, −3) **D.** (3, 3)

1. ________

2. If the figure is translated 1 unit left and 2 units down, what are the coordinates of Q?

F. (3, 1) **H.** (−3, 1)
G. (3, −1) **I.** (−3, −1)

2. ________

3. If the figure is rotated 270° clockwise about the origin, what are the coordinates of Q?

A. (−4, 1) **C.** (−1, −4)
B. (4, −1) **D.** (−1, 4)

3. ________

4. If the figure is reflected over the *y*-axis, what are the coordinates of P'?

F. (−1, −1) **H.** (1, −1)
G. (−1, 1) **I.** (1, 1)

4. ________

5. If the figure is dilated using a scale factor of 3, what are the coordinates of R'?

A. (−9, 9) **C.** (−9, −9)
B. (9, −9) **D.** (9, 9)

5. ________

6. Six people are running for class president. Assuming there are no ties, in how many different ways can first and second places be awarded?

F. 6 **H.** 30
G. 11 **I.** 35

6. ________

NAME ______________________ DATE ____________ PERIOD ________

Test, Form 1B *(continued)*

SCORE ________

7. Triangle *LMN* has vertices $L(-1, 4)$, $M(-2, 4)$, and $N(-1, -1)$. What are the coordinates of the image of point *M* after a dilation with a scale factor of 3?

A. $M'(-6, 12)$ **C.** $M'(-2, 12)$

B. $M'(-6, 4)$ **D.** $M'(12, -6)$

7. ________

8. Triangle *RST* has vertices $R(-1, 1)$, $S(3, 1)$, and $T(3, -4)$. What are the coordinates of the image of $\triangle RST$ after a translation 4 units to the left and 2 units down?

F. $R'(-5, 1)$, $S'(-1, 1)$, and $T'(-1, -4)$

G. $R'(-5, -1)$, $S'(3, -1)$, and $T'(-1, -4)$

H. $R'(-1, -1)$, $S'(3, -1)$, and $T'(3, -6)$

I. $R'(-5, -1)$, $S'(-1, -1)$, and $T'(-1, -6)$

8. ________

9. Quadrilateral *WXYZ* has vertices $W(-3, 4)$, $X(-3, 1)$, $Y(2, 1)$, and $Z(2, 4)$. What are the coordinates of the image of point *W* after a reflection across the *y*-axis?

A. $W'(3, 4)$ **C.** $W'(-3, -4)$

B. $W'(3, -4)$ **D.** $W'(4, -3)$

9. ________

10. Triangle *PQR* has vertices $P(6, 5)$, $Q(6, 10)$, and $R(0, 5)$. What are the coordinates of the image of point *P* after the triangle is rotated 180° clockwise about the origin?

F. $P'(-6, 5)$ **H.** $P'(5, 6)$

G. $P'(-6, -5)$ **I.** $P'(5, -6)$

10. ________

11. The graph shows segment $M'N'$ is a dilation of segment *MN*. What is the scale factor of the dilation?

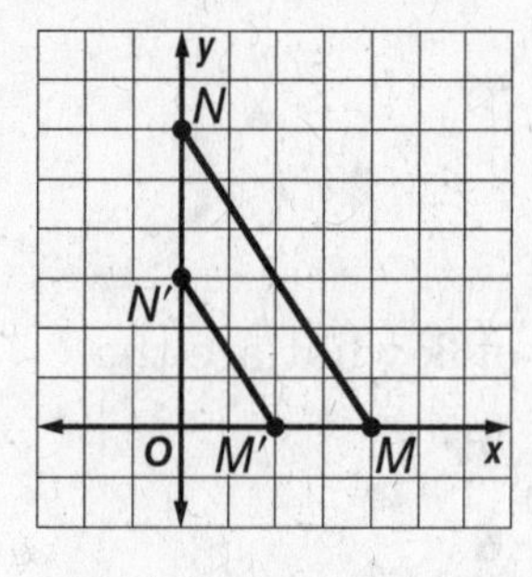

A. 4 **C.** $\frac{1}{2}$

B. 2 **D.** $\frac{1}{4}$

11. ________

NAME ______________________ DATE ______________ PERIOD ________

Test, Form 2A

SCORE ________

Write the letter for the correct answer in the blank at the right of each question.

For Exercises 1–4, trapezoid *GHIJ* has vertices as shown.

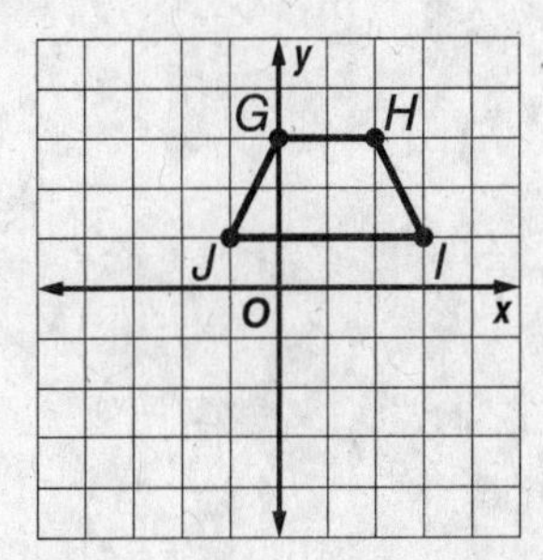

1. If the figure is translated 3 units left and 4 units down, what are the coordinates of G'?

A. (−1, −3) **C.** (−3, 1)
B. (−3, −1) **D.** (1, −3)

1. ________

2. If the figure is rotated 90° clockwise about the origin, what are the coordinates of H'?

F. (−3, 2) **H.** (2, −3)
G. (3, −2) **I.** (−3, −2)

2. ________

3. If the figure is reflected over the *y*-axis, what are the coordinates of I'?

A. (3, 1) **C.** (−3, −1)
B. (3, −1) **D.** (−3, 1)

3. ________

4. After a transformation, the coordinates of J' are (0, −1). Which of the following best represents the transformation?

F. a reflection over the *y*-axis
G. a translation of 1 unit right and 2 units down
G. a 90° clockwise rotation about the origin
I. a dilation with a scale factor of 2

4. ________

5. The ordered pair *R*(−2, 7) is translated 5 units to the right and down 2 units. Which of the following describes the translation using translation notation?

A. $(x - 2, y + 5)$ **C.** $(x - 5, y + 2)$
B. $(x + 2, y - 5)$ **D.** $(x + 5, y - 2)$

5. ________

6. Which type of transformation enlarges or reduces a figure?

F. dilation **H.** rotation
G. reflection **I.** translation

6. ________

NAME ______________________ DATE ____________ PERIOD ________

Test, Form 2A *(continued)*

SCORE ________

For Exercises 7 and 8, segment $A'B'$ is a dilation of segment AB.

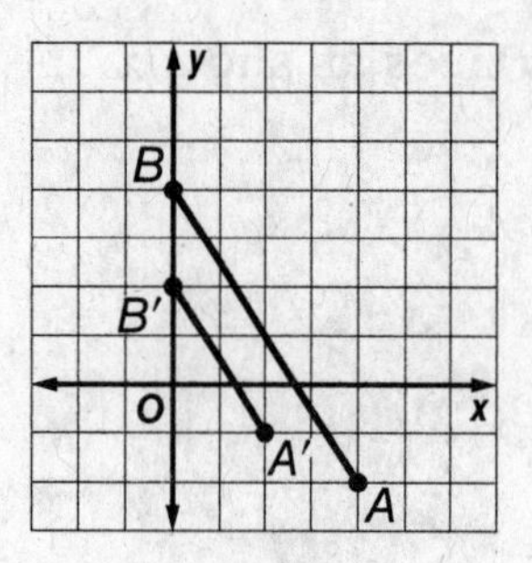

7. What is the scale factor of the dilation? **7.** ________

8. Classify the dilation as an enlargement or a reduction. **8.** ________

9. Triangle FGH has vertices $F(3, -1)$, $G(5, -1)$, and $H(5, 2)$. What are the coordinates of the image of point H after a translation 1 unit to the right and 3 units down? **9.** ________

10. Quadrilateral $JKLM$ has vertices $J(-4, 4)$, $K(-4, 1)$, $L(1, 1)$, and $M(1, 4)$. What are the coordinates of the image of point K after a reflection across the x-axis? **10.** ________

11. What are the coordinates of the image of point P after $\triangle PQR$ is rotated 90° counterclockwise about point Q?

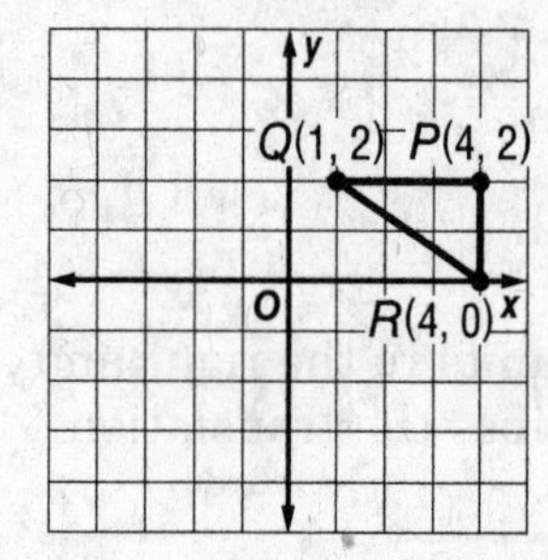

11. ________

12. Triangle ABC has vertices $A(0, 4)$, $B(-1, 4)$, and $C(0, -3)$. What are the coordinates of the image of point A after a dilation with a scale factor of 2? **12.** ________

NAME ______________________ DATE ______________ PERIOD ________

Test, Form 2B

SCORE ________

Write the letter for the correct answer in the blank at the right of each question.

For Exercises 1–4, quadrilateral *MNQP* has vertices as shown.

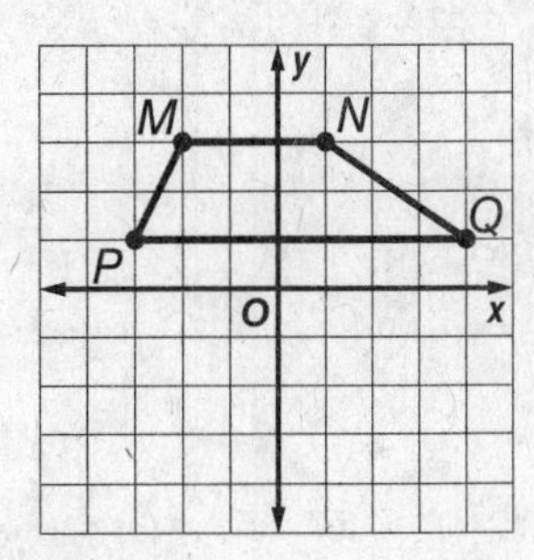

1. If the figure is translated 4 units left and 3 units down, what are the coordinates of N'?

A. (0, 3) **C.** (3, 0)

B. (0, –3) **D.** (–3, 0)

1. ________

2. If the figure is rotated 180° clockwise about the origin, what are the coordinates of M'?

F. (2, –3) **H.** (–2, 3)

G. (–2, –3) **I.** (2, 3)

2. ________

3. If the figure is reflected over the y-axis, what are the coordinates of Q'?

A. (4, –1) **C.** (–4, –1)

B. (–4, 1) **D.** (4, 1)

3. ________

4. After a transformation, the coordinates of N' are (3, –1). Which of the following best represents the transformation?

F. a reflection over the x-axis

G. a translation of 2 units right and 4 units down

H. a 90° clockwise rotation about the origin

I. a dilation with a scale factor of 3

4. ________

5. The ordered pair $D(0, -5)$ is translated 2 units left and 6 units up. Which of the following describes the translation using translation notation?

A. $(x - 2, y - 6)$ **C.** $(x + 6, y - 2)$

B. $(x - 6, y + 2)$ **D.** $(x - 2, y + 6)$

5. ________

6. Which type of transformation turns a figure around a fixed point?

F. dilation **H.** rotation

G. reflection **I.** translation

6. ________

NAME ______________________ DATE ____________ PERIOD ________

Test, Form 2B *(continued)*

SCORE ________

For Exercises 7 and 8, segment $D'R'$ is a dilation of segment DR.

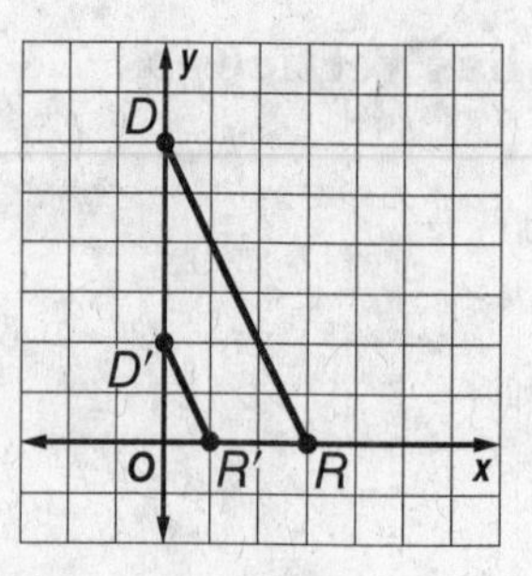

7. What is the scale factor of the dilation? **7.** ________

8. Classify the dilation as an enlargement or a reduction. **8.** ________

9. Triangle BDF has vertices $B(4, 3)$, $D(6, 3)$, and $F(6, 1)$. What are the coordinates of the image of point F after a translation 2 units to the left and 4 units down? **9.** ________

10. Quadrilateral $MNOP$ has vertices $M(-2, 4)$, $N(-2, 1)$, $O(3, 1)$, and $P(3, 4)$. What are the coordinates of the image of point P after a reflection across the x-axis? **10.** ________

11. What are the coordinates of the image of point K after $\triangle GHK$ is rotated 90° counterclockwise about point G?

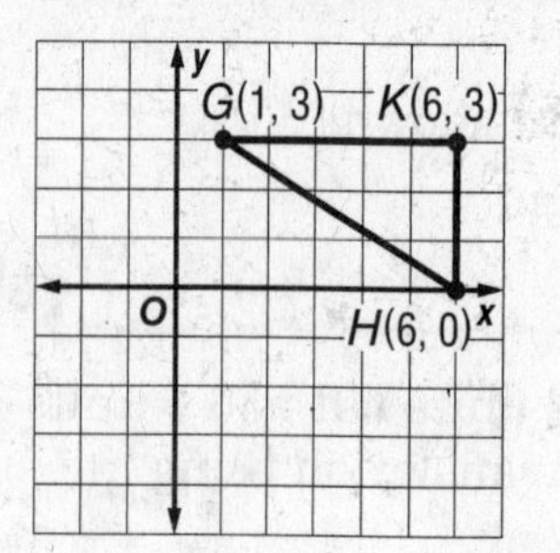

11. ________

12. Triangle QRS has vertices $Q(0, -6)$, $R(3, 0)$, and $S(3, -6)$. What are the coordinates of the image of point S after a dilation with a scale factor of $\frac{1}{3}$? **12.** ________

NAME ______________________ DATE ____________ PERIOD ________

Test, Form 3A

SCORE ________

Write the correct answer in the blank at the right of each question.

1. The triangle $N'L'M'$ shown was reflected over the x-axis. Find the original coordinates of the triangle NLM.

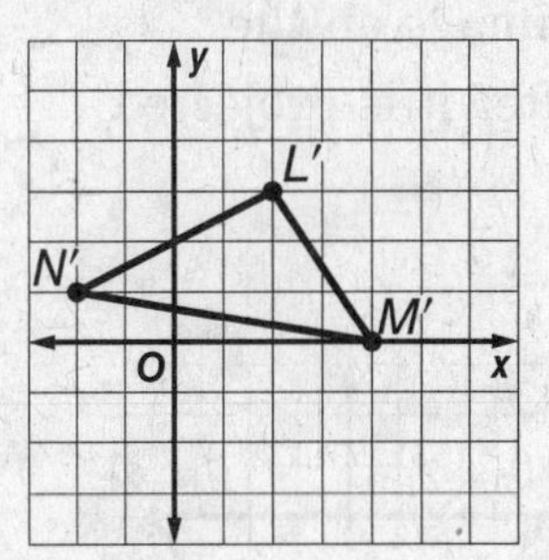

1. ________

2. In the figure at the right, $\triangle X'Y'Z'$ is a dilation of $\triangle XYZ$. Find the scale factor of the dilation, and classify it as an enlargement or a reduction.

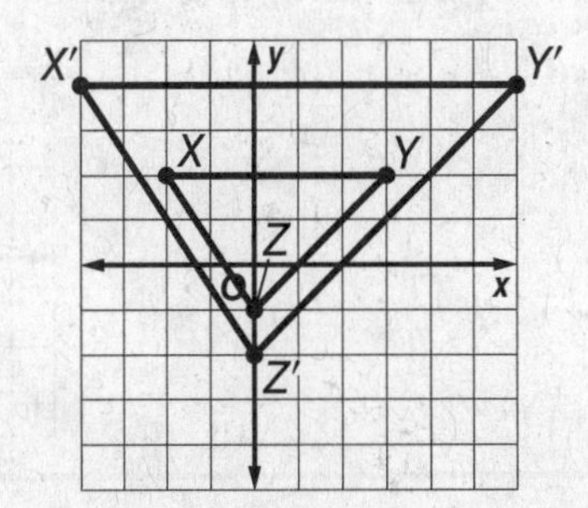

2. ________

For Exercises 3 and 4, triangle PQR is rotated 90° clockwise about the origin. The vertices of the triangle are $P(3, 1)$, $Q(1, 4)$, and $R(2, -5)$.

3. Find the coordinates of P', Q', and R'.

3. ________

4. What is true about triangles PQR and $P'Q'R'$?

4. ________

5. The point $M'(4, -5)$ is the result of a translation of 4 units left and 2 units up. Use translation notation to describe the translation.

5. ________

NAME ______________________ DATE ____________ PERIOD ________

Test, Form 3A *(continued)*

SCORE ________

6. What are the coordinates of the image of $A(2, 5)$ after it is rotated 180° clockwise about the origin?

6. ____________

7. A projector transforms the image on a computer screen so that it is dilated by a scale factor of $\frac{7}{2}$. If the original image on the screen is 10 inches wide, find the new width after it is projected on the wall.

7. ____________

For Exercises 8–10, refer to the graph of $\triangle YZW$ at the right.

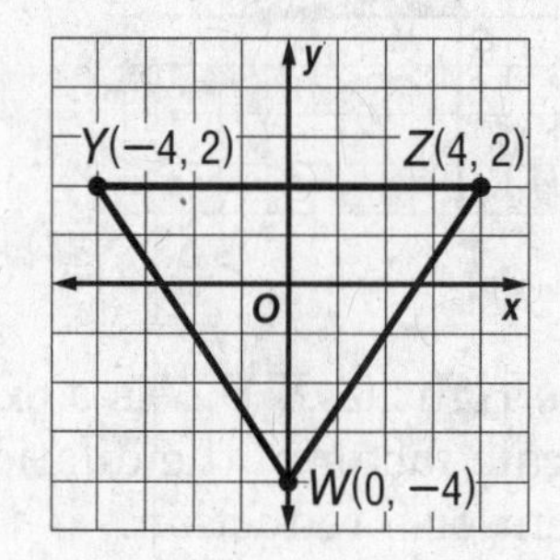

8. Graph and label the image of $\triangle YZW$ after a translation 2 units right and 1 unit down.

8. ____________

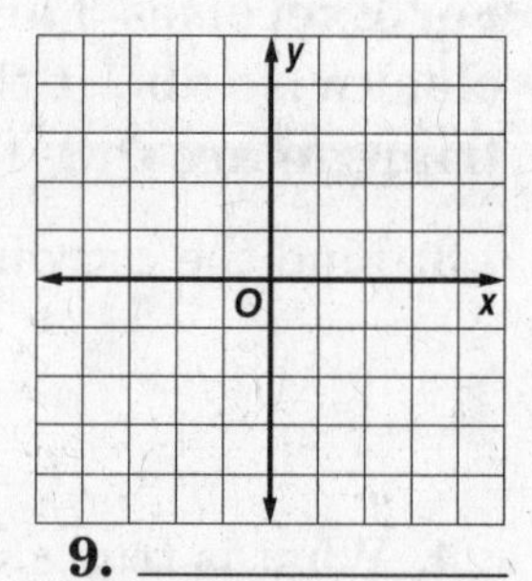

9. Graph and label the image of $\triangle YZW$ after a reflection over the y-axis.

9. ____________

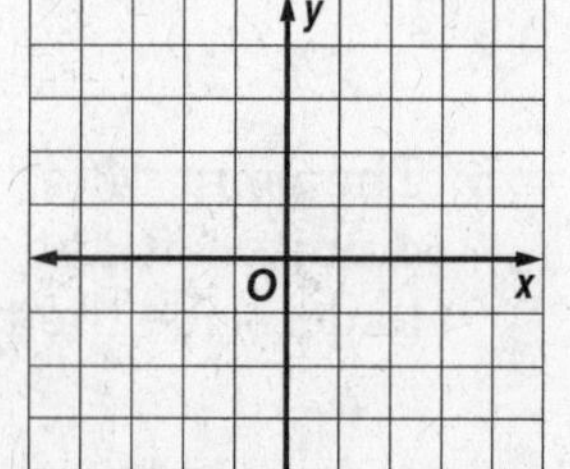

10. Graph and label the image of $\triangle YZW$ after a dilation by a scale factor of $\frac{1}{2}$.

10. ____________

NAME ______________________ DATE ______________ PERIOD __________

Test, Form 3B

SCORE __________

Write the correct answer in the blank at the right of each question.

1. The triangle $R'K'G'$ shown was reflected over the x-axis. Find the original coordinates of the triangle RKG.

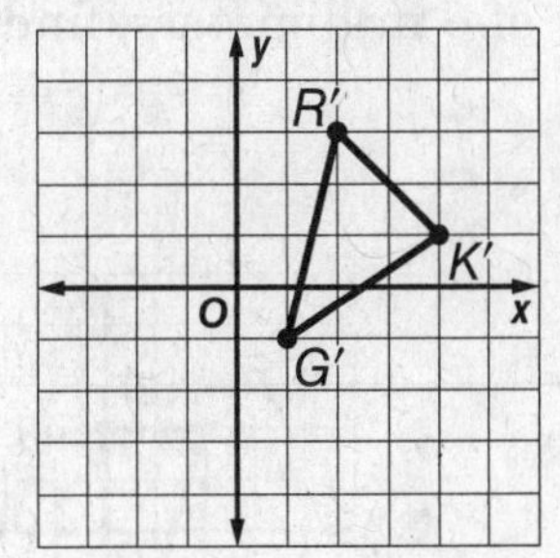

1. __________

2. In the figure at the right, $\triangle D'E'F'$ is a dilation of $\triangle DEF$. Find the scale factor of the dilation, and classify it as an enlargement or a reduction.

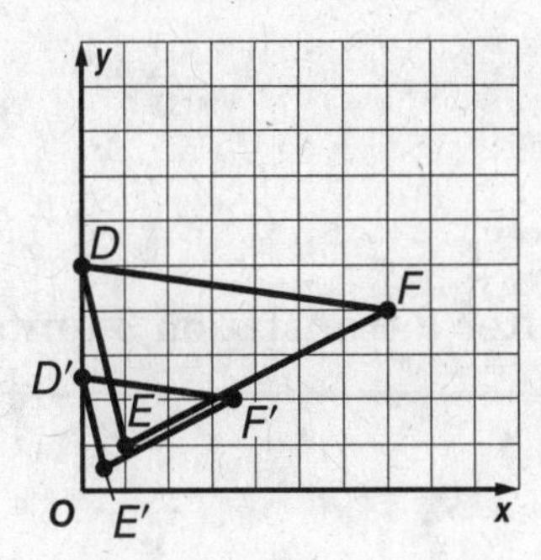

2. __________

For Exercises 3 and 4, triangle *LPT* is rotated 270° clockwise about the origin. The vertices of the triangle are *L*(−3, −1), *P*(1, 2), and *T*(3, 0).

3. Find the coordinates of L', P', and T'. **3.** __________

4. What is true about triangles LPT and $L'P'T'$? **4.** __________

5. The point $H'(-3, 4)$ is the result of a translation of 6 units left and 2 units up. Use translation notation to describe the translation. **5.** __________

Test, Form 3B *(continued)*

SCORE ________

6. What are the coordinates of the image of $Y(-2, 5)$ after it is rotated 180° clockwise about the origin?

6. ________

7. An optometrist dilates a patient's pupils by a scale factor of $\frac{5}{3}$. If the pupil before dilation has a diameter of 6 millimeters, find the new diameter after the pupil is dilated.

7. ________

For Exercises 8–10, refer to the graph of $\triangle TRS$ at the right.

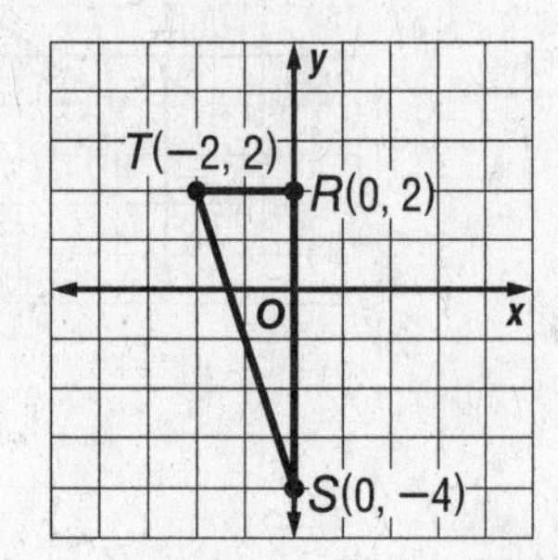

8. Graph and label the image of $\triangle TRS$ after a translation 3 units left and 2 units down.

8. ________

9. Graph and label the image of $\triangle TRS$ after a reflection over the y-axis.

9. ________

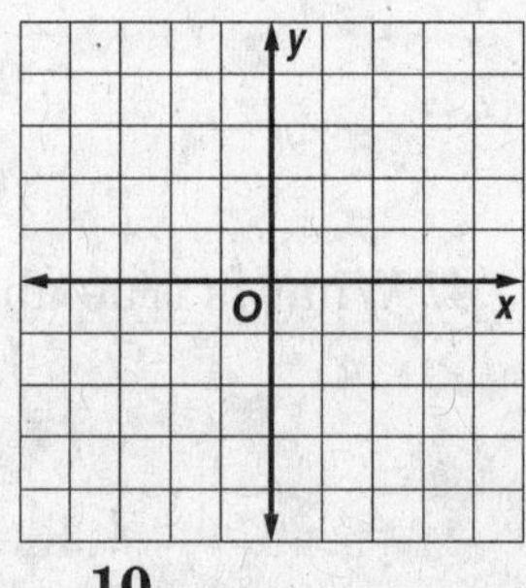

10. Graph and label the image of $\triangle TRS$ after a dilation by a scale factor of $\frac{1}{2}$.

10. ________

NAME ______________________ DATE ______________ PERIOD __________

Are You Ready?

Review

Example

Solve $\frac{w}{16} = \frac{3}{4}$.

$\frac{w}{16} = \frac{3}{4}$	Write the proportion.
$w \cdot 4 = 16 \cdot 3$	Find cross products.
$4w = 48$	Simplify.
$w = 12$	Division Property of Equality

Solve each proportion.

1. $\frac{75}{w} = \frac{5}{6}$ — 1. ______________

2. $\frac{1}{5} = \frac{11}{p}$ — 2. ______________

3. $\frac{9}{z} = \frac{3}{13}$ — 3. ______________

4. $\frac{7}{15} = \frac{m}{30}$ — 4. ______________

5. $\frac{n}{11} = \frac{19}{22}$ — 5. ______________

6. $\frac{9}{10} = \frac{18}{p}$ — 6. ______________

7. $\frac{4}{5} = \frac{x}{25}$ — 7. ______________

8. $\frac{z}{24} = \frac{3}{4}$ — 8. ______________

9. $\frac{10}{17} = \frac{4}{y}$ — 9. ______________

10. $\frac{x}{8} = \frac{15}{16}$ — 10. ______________

NAME ______________________ DATE ______________ PERIOD ________

Are You Ready?

Practice

Solve each proportion.

1. $\frac{3}{w} = \frac{18}{42}$ 1. ______

2. $\frac{x}{12} = \frac{1}{2}$ 2. ______

3. $\frac{5}{z} = \frac{10}{22}$ 3. ______

4. $\frac{7}{8} = \frac{m}{48}$ 4. ______

5. $\frac{n}{5} = \frac{12}{20}$ 5. ______

6. $\frac{3}{8} = \frac{18}{p}$ 6. ______

7. $\frac{x}{12} = \frac{59}{60}$ 7. ______

8. $\frac{z}{12} = \frac{12}{4}$ 8. ______

9. **PAINTING** Kira is mixing a purple paint that has a ratio of 2 parts blue to 3 parts red. If she had 18 parts of red paint, how much blue paint will she need? 9. ______

10. **RUNNING** Raul is training for a marathon. He can run 16 laps in 9 minutes. At this rate, how long will it take to run 64 laps? 10. ______

Find the slope of the line that passes through each pair of points.

11. $(-2, -8), (4, 4)$ 11. ______

12. $(2, 2), (10, 6)$ 12. ______

13. $(3, 6), (-12, 9)$ 13. ______

NAME ______________________ DATE ______________ PERIOD ________

Are You Ready?

Apply

1. BAKERY In the window at a local bakery the ratio of muffins to cookies is 3 to 5. If there are 75 cookies in the window, how many muffins are in the window?	**2. FIELD TRIPS** For the field trip to the museum, the ratio to adults to students is 2 to 9. If there are 108 students going on the field trip, how many adults will there be?
3. BLOOD TYPES For every 10 people that donated blood, 3 had blood type A positive. If 51 people who had type A positive donated blood, how many people donated blood?	**4. RECIPES** A recipe for punch calls for 2.5 parts cranberry juice to 3 parts orange juice. If you use 12 parts orange juice, how many parts of cranberry juice do you need?
5. DRIVING A family drove 585 miles in 9 hours. At this rate, how many miles will the family drive in 7 hours?	**6. TEXTING** Jin can type 90 words in 2.5 minutes on his cell phone. At this rate, how many words can he type in 7.5 minutes?

NAME ______________________ DATE ______________ PERIOD ________

Diagnostic Test

Solve each proportion.

1. $\frac{16}{w} = \frac{2}{3}$ **1.** ________________

2. $\frac{x}{72} = \frac{1}{4}$ **2.** ________________

3. $\frac{36}{z} = \frac{16}{40}$ **3.** ________________

4. $\frac{5}{14} = \frac{m}{28}$ **4.** ________________

5. $\frac{n}{18} = \frac{5}{3}$ **5.** ________________

6. $\frac{81}{90} = \frac{p}{20}$ **6.** ________________

7. $\frac{x}{3} = \frac{4}{5}$ **7.** ________________

8. $\frac{45}{75} = \frac{z}{3}$ **8.** ________________

9. **ART** An art teacher is mixing orange paint that has a ratio of 3 parts red to 4 parts yellow. If she had 24 parts of yellow paint, how much red paint will she need? **9.** ________________

10. **TRAINING** A runner is training for a marathon. He can run 8 miles in 64 minutes. At this rate, how many miles can he run in 88 minutes? **10.** ________________

Find the slope of the line that passes through each pair of points.

11. (3, 5), (1, 1) **11.** ________________

12. (−1, 0), (3, 2) **12.** ________________

13. (−6, −4), (10, 8) **13.** ________________

NAME ______________________ DATE ____________ PERIOD ________

Pretest

1. Determine if the two figures below are congruent by using transformations.

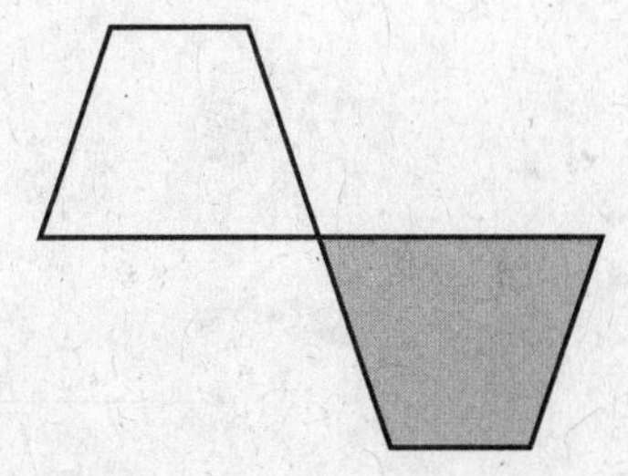

1. ______________

2. The pair of polygons is similar. Find the missing side measure.

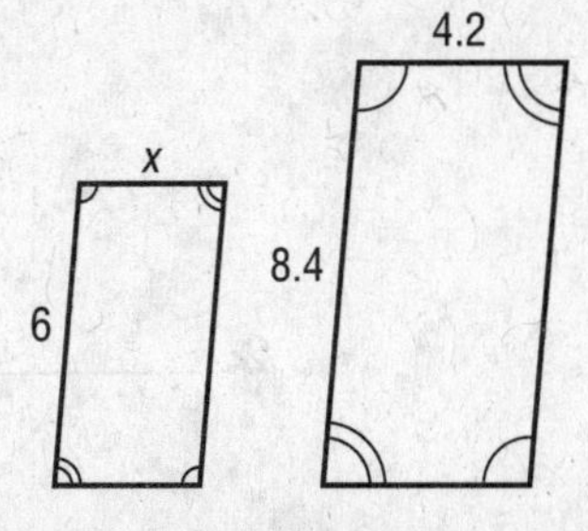

2. ______________

3. Find the value of x in the pair of similar figures.

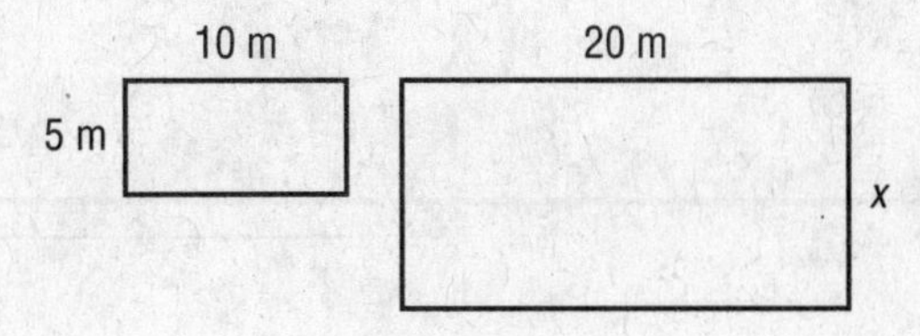

3. ______________

4. Triangle ABC is similar to triangle DEF. If the perimeter of $\triangle ABC$ is 24 inches, what is the perimeter of $\triangle DEF$?

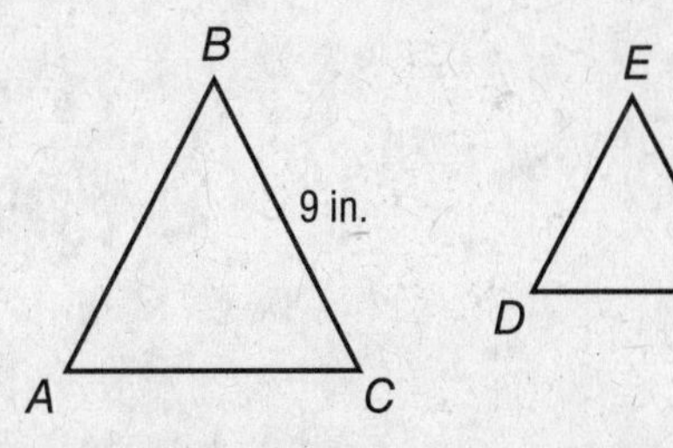

4. ______________

5. Neil is painting two squares on his bedroom wall. The length of one side of the first square is 12 inches. The second square will be 3 times the size of the first square. What is the area of the second square?

5. ______________

Chapter Quiz

Determine if the two figures are congruent by using transformations. Explain your reasoning.

1.

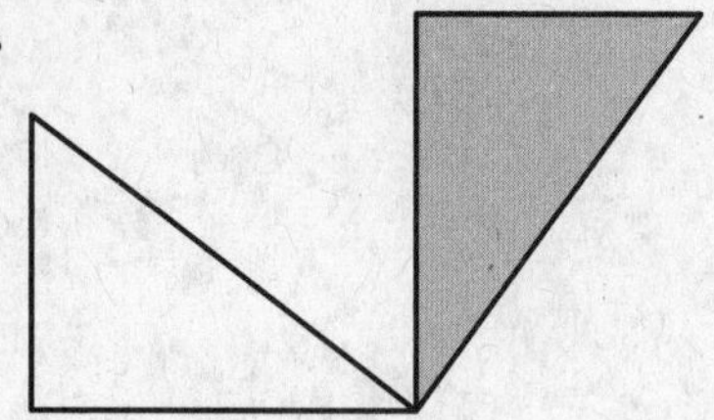

1. ____________________

2.

2. ____________________

3.

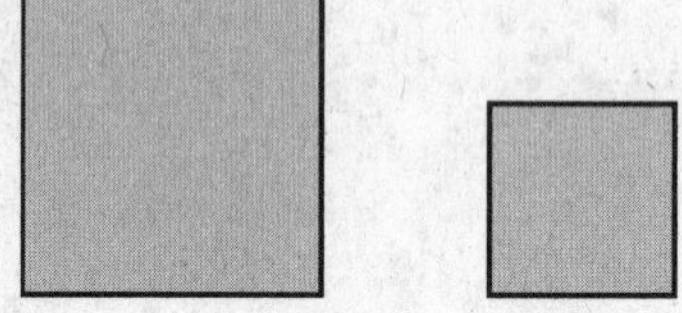

3. ____________________

4. Write congruence statements comparing the corresponding parts in the congruent triangles shown below.

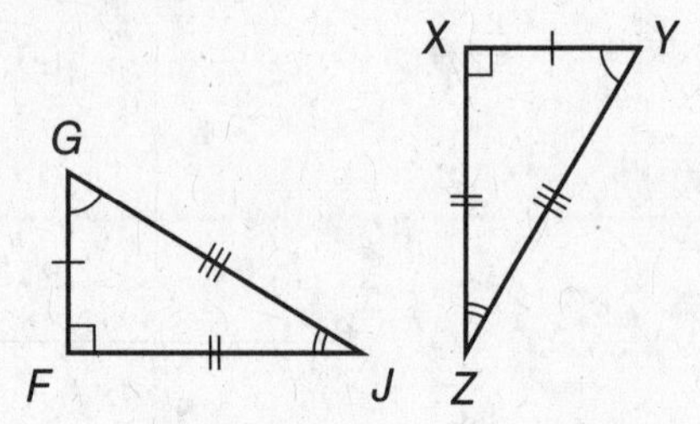

4. ____________________

NAME ______________________ DATE ____________ PERIOD ________

Vocabulary Test

composition of transformations	scale factor
corresponding parts	similar
indirect measurement	similar polygons

Choose from the terms above to complete each sentence.

1. Polygons that have the same shape are called ____________. **1.** ____________

2. Two figures are ____________ if the second can be obtained from the first by a sequence of transformations or dilations. **2.** ____________

3. The parts of congruent figures that match or correspond are called ____________. **3.** ____________

4. The ratio of the lengths of two corresponding sides of two similar polygons is called the ____________. **4.** ____________

Define each term in your own words

5. composition of transformations **5.** ____________

6. indirect measurement **6.** ____________

NAME ______________________________ DATE ______________ PERIOD ________

Standardized Test Practice

Read each question. Then fill in the correct answer on the answer document provided by your teacher or on a sheet of paper.

1. **GRIDDED RESPONSE** The triangles in the figure below are similar.

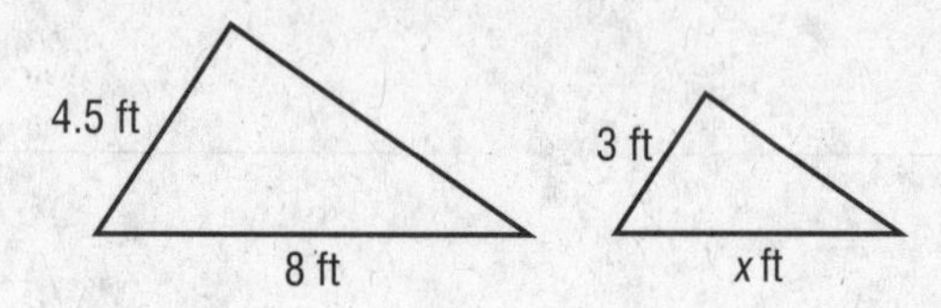

What is the value of x to the nearest tenth?

2. How many ways are there to arrange six family members for a photo if mom and dad must be seated in front with the other 4 members behind them?

A. 24

B. 36

C. 48

D. 56

3. **GRIDDED RESPONSE** Niles casts a shadow that is 8 feet long at the same time his dog casts a shadow that is 30 inches long. If Niles is $5\frac{1}{2}$ feet tall, how many inches tall is his dog? Round to the nearest tenth.

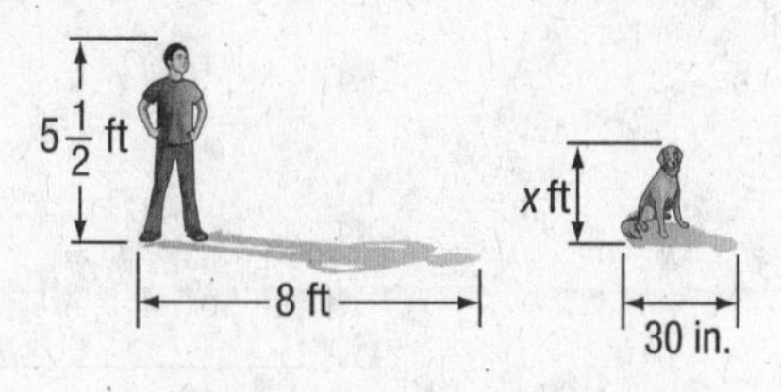

4. A purse regularly priced at $70 is on sale for 20% off. What is the sale price of the purse?

F. $63

G. $56

H. $14

I. $7

5. THINK SOLVE EXPLAIN **SHORT RESPONSE** A telephone pole is supported by a guy wire. How tall is the telephone pole in feet? Round to the nearest tenth if necessary.

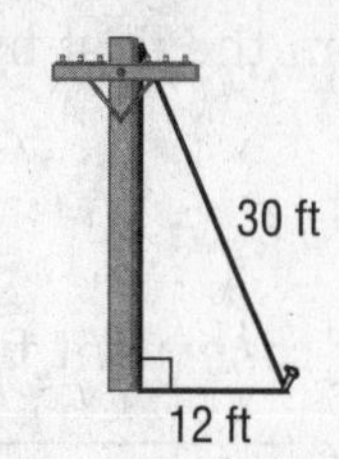

6. What is the solution of the equation $\frac{x}{2} + 10 = -8$?

A. −36

B. −4

C. 4

D. 16

7. Which of the following is equivalent to the expression shown below?

$$(3x^2y)^3$$

F. $3x^8y$

G. $9x^6y^3$

H. $27x^6y^3$

I. $9x^8y^3$

8. What is the approximate value of the missing measure on the triangle below?

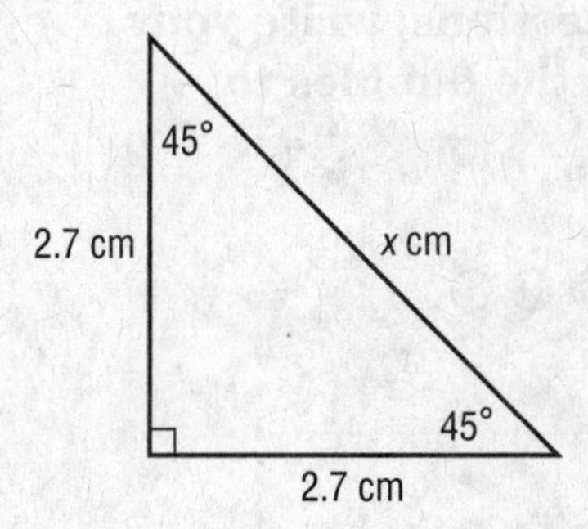

A. 5.4 cm
B. 3.8 cm
C. 4.7 cm
D. 2.7 cm

9. THINK SOLVE EXPLAIN **SHORT ANSWER** Fifteen students were surveyed to see whether they liked the color red or blue. There were three more students who preferred blue over red. Write and solve a system of equations to find how many students preferred blue and how many preferred red.

10. Rory covers her boat with a square plastic tarp. The tarp has an area of 900 square feet. What is the length of a side of the tarp in feet?

F. 15
G. 27
H. 30
I. 35

11. Adar, Chi, and Melissa each left the party by themselves. In how many different orders can they leave?

A. 5
B. 6
C. 7
D. 9

12. The scale factor between two figures is $\frac{5}{6}$. What is the ratio of areas of the two figures?

F. $\frac{1}{6}$
G. $\frac{5}{6}$
H. $\frac{5}{11}$
I. $\frac{25}{36}$

13. The scale factor between two figures is $\frac{8}{9}$. What is the ratio of their perimeters?

A. $\frac{1}{9}$
B. $\frac{5}{9}$
C. $\frac{8}{9}$
D. $\frac{64}{81}$

14. THINK SOLVE EXPLAIN **EXTENDED RESPONSE** Refer to the figures below.

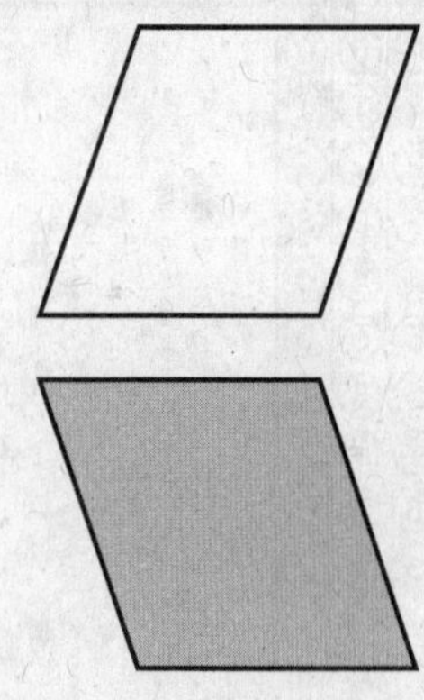

Part A Determine if the two figures are congruent by using transformations.

Part B Explain your reasoning to Part A.

NAME ______________________ DATE ____________ PERIOD ________

Student Recording Sheet

Use this recording sheet with the Standardized Test Practice pages.

Fill in the correct answer. For gridded-response questions, write your answers in the boxes on the answer grid and fill in the bubbles to match your answers.

1.

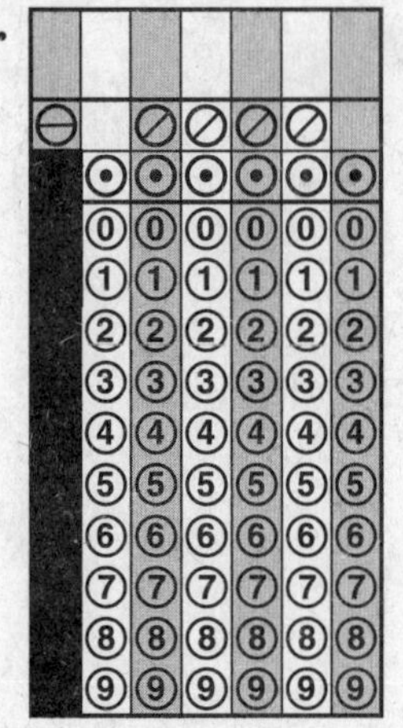

2. Ⓐ Ⓑ Ⓒ Ⓓ

3.

4. Ⓕ Ⓖ Ⓗ Ⓘ

5. ____________

6. Ⓐ Ⓑ Ⓒ Ⓓ

7. Ⓕ Ⓖ Ⓗ Ⓘ

8. Ⓐ Ⓑ Ⓒ Ⓓ

9. ____________

10. Ⓕ Ⓖ Ⓗ Ⓘ

11. Ⓐ Ⓑ Ⓒ Ⓓ

12. Ⓕ Ⓖ Ⓗ Ⓘ

13. Ⓐ Ⓑ Ⓒ Ⓓ

Extended Response

Record your answers for Exercise 14 on the back of this paper.

NAME ______________________ DATE ______________ PERIOD ________

Extended-Response Test

SCORE ________

Demonstrate your knowledge by giving a clear, concise solution to each problem. Be sure to include all relevant drawings and justify your answers. You may show your solution in more than one way or investigate beyond the requirements of the problem. If necessary, record your answer on another piece of paper.

1. a. Polygons *ABCD* and *EFGH* are similar. Explain what this means.

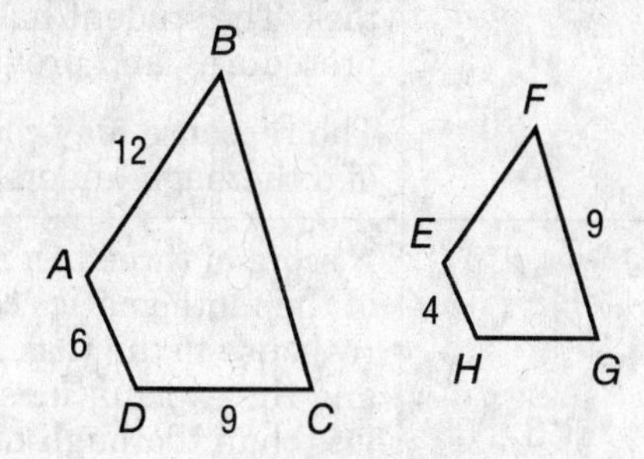

b. In the polygons, identify the corresponding sides. How can you use this information to find the lengths of the sides $\overline{BC}$, $\overline{EF}$, and $\overline{GH}$?

c. Use your method from part **b** to find the lengths of $\overline{BC}$, $\overline{EF}$, and $\overline{GH}$. Show your work.

d. Draw two similar rectangles. Label the lengths of each side. Name the corresponding parts.

2. a. Explain in your own words what is meant by a composition of transformation.

b. Draw a figure on the coordinate plane in the first quadrant.

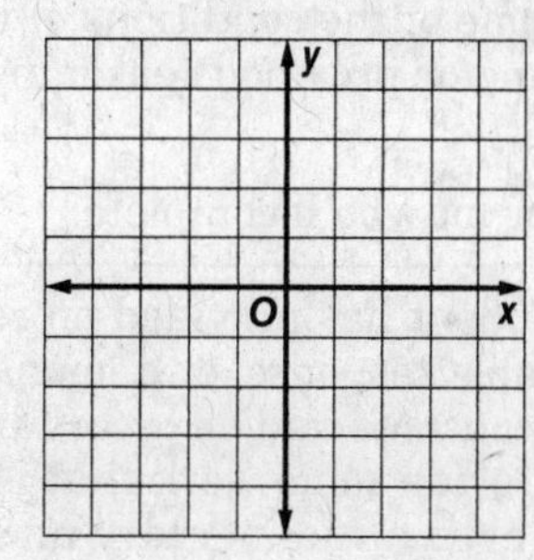

c. Reflect the figure over the y-axis. Then translate the figure.

d. Is the transformed figure congruent to the original figure? Explain.

NAME ______________________________ DATE ______________ PERIOD ________

Extended-Response Rubric

SCORE ________

Score	Description
4	A score of four is a response in which the student demonstrates a thorough understanding of the mathematics concepts and/or procedures embodied in the task. The student has responded correctly to the task, used mathematically sound procedures, and provided clear and complete explanations and interpretations. The response may contain minor flaws that do not detract from the demonstration of a thorough understanding.
3	A score of three is a response in which the student demonstrates an understanding of the mathematics concepts and/or procedures embodied in the task. The student's response to the task is essentially correct with the mathematical procedures used and the explanations and interpretations provided demonstrating an essential but less than thorough understanding. The response may contain minor flaws that reflect inattentive execution of mathematical procedures or indications of some misunderstanding of the underlying mathematics concepts and/or procedures.
2	A score of two indicates that the student has demonstrated only a partial understanding of the mathematics concepts and/or procedures embodied in the task. Although the student may have used the correct approach to obtaining a solution or may have provided a correct solution, the student's work lacks an essential understanding of the underlying mathematical concepts. The response contains errors related to misunderstanding important aspects of the task, misuse of mathematical procedures, or faulty interpretations of results.
1	A score of one indicates that the student has demonstrated a very limited understanding of the mathematics concepts and/or procedures embodied in the task. The student's response is incomplete and exhibits many flaws. Although the student's response has addressed some of the conditions of the task, the student reached an inadequate conclusion and/or provided reasoning that was faulty or incomplete. The response exhibits many flaws or may be incomplete.
0	A score of zero indicates that the student has provided no response at all, or a completely incorrect or uninterpretable response, or demonstrated insufficient understanding of the mathematics concepts and/or procedures embodied in the task. For example, a student may provide some work that is mathematically correct, but the work does not demonstrate even a rudimentary understanding of the primary focus of the task.

NAME ______________ DATE ______________ PERIOD ______

Test, Form 1A

SCORE ______

Write the letter for the correct answer in the blank at the right of each question.

1. A 25,000 gallon swimming pool is being filled. Two hundred and fifty gallons are in it after 30 minutes. How many hours will it take to fill the pool? Use the *draw a diagram* strategy.

A. 200 h **C.** 50 h

B. 100 h **D.** 25 h **1.** ______

2. The triangles are similar. Which series of transformations maps $\triangle ABC$ onto $\triangle DEF$?

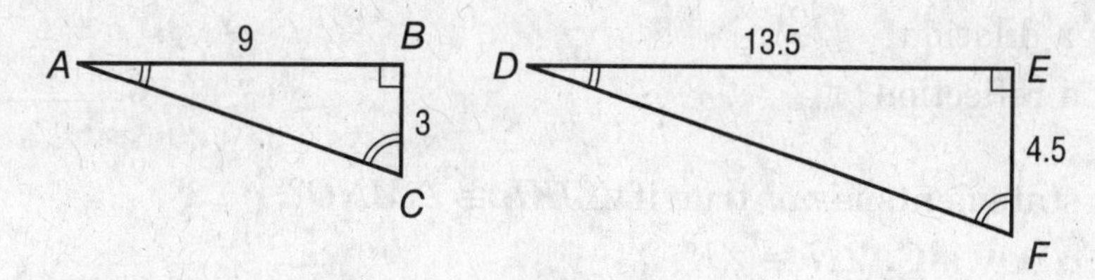

F. translation followed by a rotation

G. translation followed by a dilation

H. rotation followed by a dilation

I. reflection followed by a dilation **2.** ______

3. The length and width of a rectangle are 5 feet and 2 feet, respectively. A similar rectangle has a width of 8 feet. What is the length of the second rectangle?

A. 8 ft **C.** 16 ft

B. 14 ft **D.** 20 ft **3.** ______

4. Which statement about the triangles at the right is true?

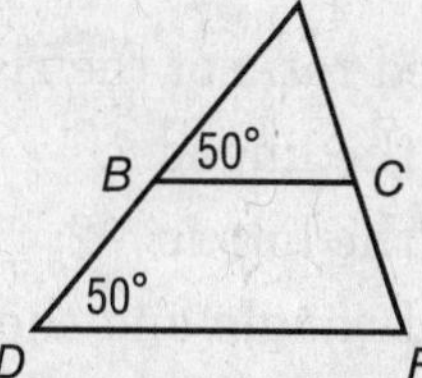

F. $\triangle ABC$ is not similar to $\triangle ADF$

G. $\triangle ABC$ is similar to $\triangle ADF$

H. $\angle BAC$ is not congruent to $\angle DAF$

I. $\triangle ABC$ is congruent to $\triangle ADF$ **4.** ______

5. Rectangle *DEFG* is similar to rectangle *JKLM*. Rectangle *DEFG* has a length of 5 units and a perimeter of 16 units. Rectangle *JKLM* has a length of 10 units. What is the perimeter of rectangle *JKLM*?

A. 8 units **C.** 32 units

B. 20 units **D.** 64 units **5.** ______

NAME ______________________ DATE ____________ PERIOD ________

Test, Form 1A *(continued)*

SCORE ________

6. Triangle ABC is congruent to triangle DEF. Which series of transformations maps $\triangle DEF$ onto $\triangle ABC$?

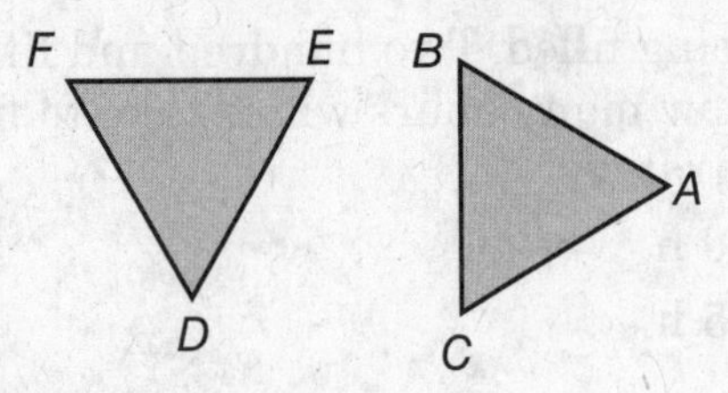

F. rotation followed by a translation
G. translation followed by a dilation
H. rotation followed by a dilation
I. dilation followed by a reflection

6. ________

7. Which of the following statements is *not* true if $\triangle JKL \cong \triangle MNO$?

A. $\angle J \cong \angle M$ **C.** $\angle N \cong \angle K$
B. $\angle L \cong \angle O$ **D.** $\angle L \cong \angle N$

7. ________

8. Which of the following statements is *not* true about the graph shown?

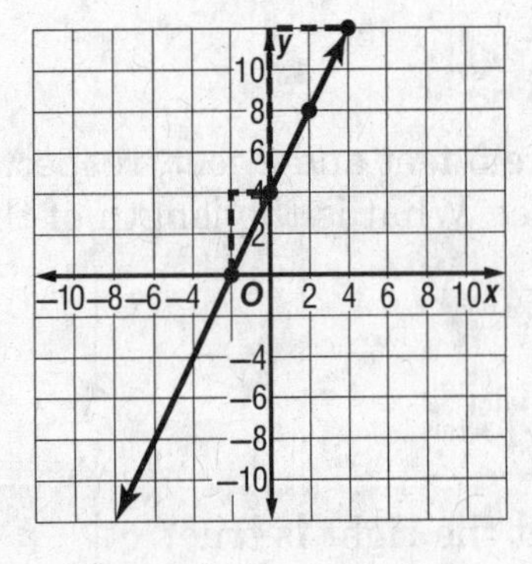

F. The simplified ratio of the rise to the run of each triangle is 2.
G. The slope of the line is 2.
H. The slope of the line is −2.
I. The smaller triangle and the larger triangle shown are similar.

8. ________

9. Which statement is *not* true concerning any non-vertical line on the coordinate plane?

A. All of the slope triangles on the line are similar.
B. The slope is the same between any two distinct points on the line.
C. In the slope triangles, the ratios of the rise to the run are equal to the slope.
D. The slope varies between any two distinct points on the line.

9. ________

NAME _______________ DATE __________ PERIOD ________

Test, Form 1B

SCORE ________

Write the letter for the correct answer in the blank at the right of each question.

1. A 72,000 gallon water tower is being drained. Two thousand gallons are drained in the first hour. How many hours will it take to drain the water tower? Use the *draw a diagram* strategy.

A. 72 h **B.** 36 h **C.** 18 h **D.** 9 h

1. ________

2. The triangles are similar. Which series of transformations maps $\triangle ABC$ onto $\triangle DEF$?

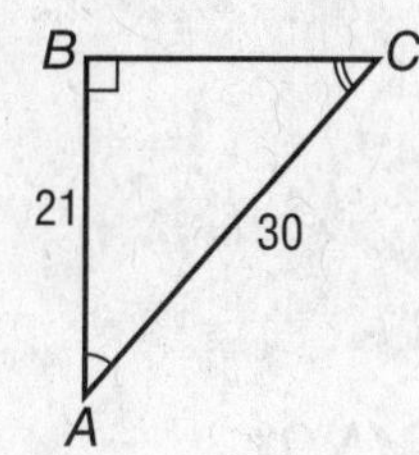

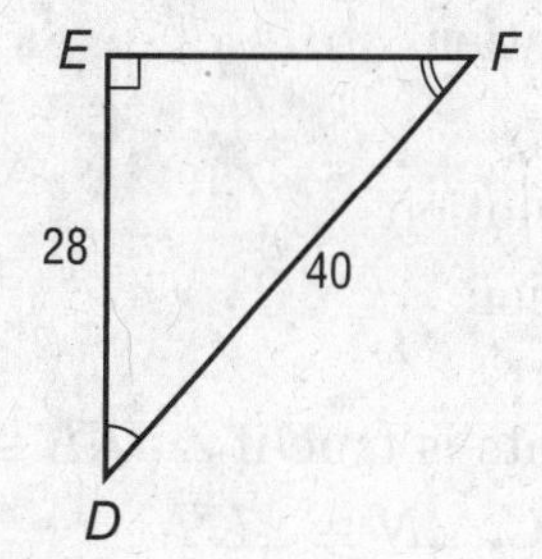

F. translation followed by a rotation
G. translation followed by a dilation
H. rotation followed by a dilation
I. reflection followed by a dilation

2. ________

3. The length and width of a rectangle are 4 feet and 3 feet, respectively. A similar rectangle has a width of 9 feet. What is the length of the second rectangle?

A. 9 ft **C.** 14 ft
B. 12 ft **D.** 16 ft

3. ________

4. Which statement about the triangles at the right is true?

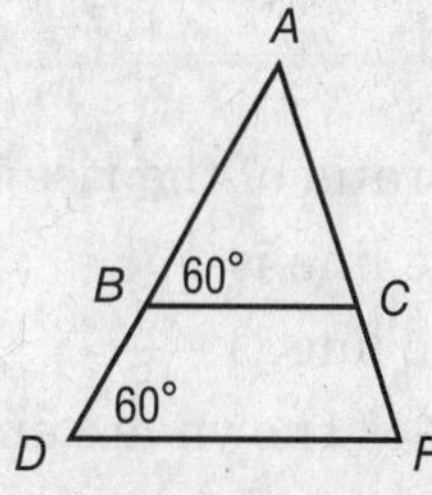

F. $\triangle ABC$ is similar to $\triangle ADF$
G. $\triangle ABC$ is not similar to $\triangle ADF$
H. $\angle BAC$ is not congruent to $\angle DAF$
I. $\triangle ABC$ is congruent to $\triangle ADF$

4. ________

5. Rectangle *RSTU* is similar to rectangle *WXYZ*. Rectangle *RSTU* has a length of 6 units and a perimeter of 18 units. Rectangle *WXYZ* has a length of 12 units. What is the perimeter of rectangle *WXYZ*?

A. 18 units **C.** 36 units
B. 24 units **D.** 72 units

5. ________

NAME ______________________ DATE ______________ PERIOD ________

Test, Form 1B *(continued)*

SCORE ________

6. The figures below are congruent. Which series of transformations maps figure *ABCD* onto *EFGH*?

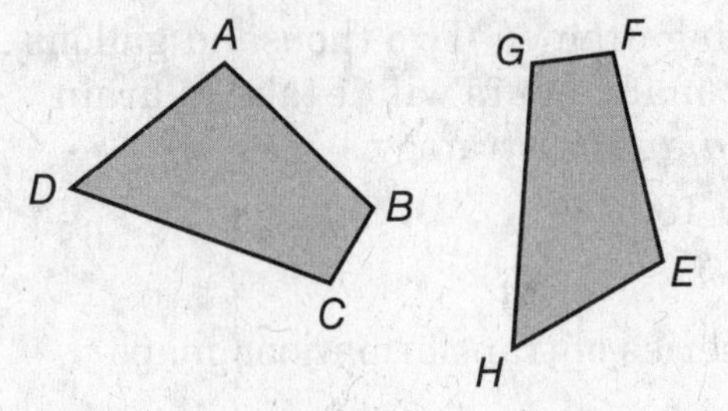

F. rotation followed by a translation
G. rotation followed by a dilation
H. reflection followed by a translation
I. reflection followed by a rotation

6. ________

7. Which of the following statements is true if $\triangle JKL \cong \triangle MNO$?

A. $\angle J \cong \angle N$ **C.** $\angle N \cong \angle K$
B. $\angle L \cong \angle M$ **D.** $\angle L \cong \angle N$

7. ________

8. Which of the following statements is *not* true about the graph shown?

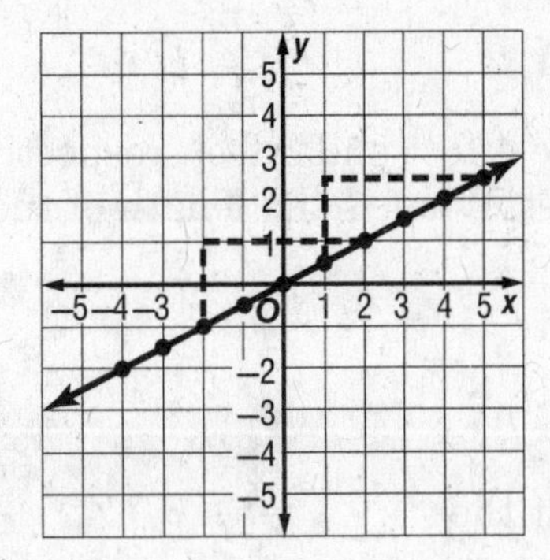

F. The simplified ratio of the rise to the run of each triangle is $\frac{1}{2}$.
G. The slope of the line is $\frac{1}{2}$.
H. The slope of the line is $-\frac{1}{2}$.
I. The two triangles shown are similar.

8. ________

9. Which statement is true concerning any non-vertical line on the coordinate plane?

A. All of the slope triangles on the line are congruent.
B. The slope is the same between any two distinct points on the line.
C. In the slope triangles, the ratios of the rise to the run are equal to the absolute value of the *y*-coordinate.
D. The slope varies between any two distinct points on the line.

9. ________

NAME ______________________ DATE ______________ PERIOD ________

Test, Form 2A

SCORE ________

Write the letter for the correct answer in the blank at the right of each question.

1. A survey of 12 students showed that 7 liked football, 10 liked basketball, and 5 liked both. How many students just liked basketball? Use the *draw a diagram* strategy.

A. 12 **C.** 5

B. 10 **D.** 2

1. ________

2. Debbie is painting an image on a piece of art canvas. The image she is reproducing is 3 inches by 5 inches. She enlarges the dimensions 4 times. Which of the following statements is *not* true?

F. The perimeter of the original image and the perimeter of the new image are related by a scale factor of 4.

G. The area of the new image is 4 times the area of the original image.

H. The area of the original image and the area of the new image are related by a scale factor of 16.

I. The perimeter of the original image is $\frac{1}{4}$ the perimeter of the new image.

2. ________

3. Which pair of polygons is similar?

A. 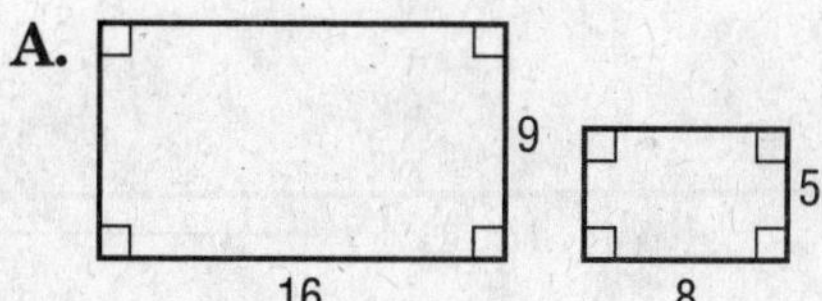

C. 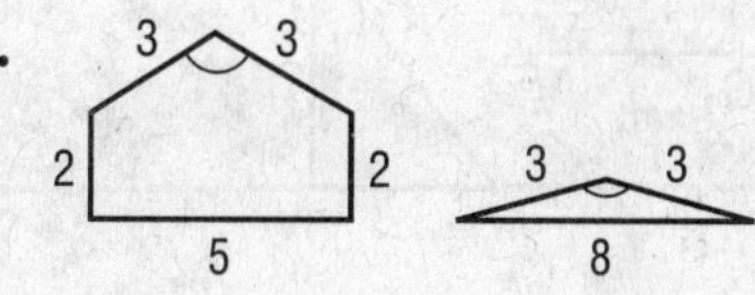

B. 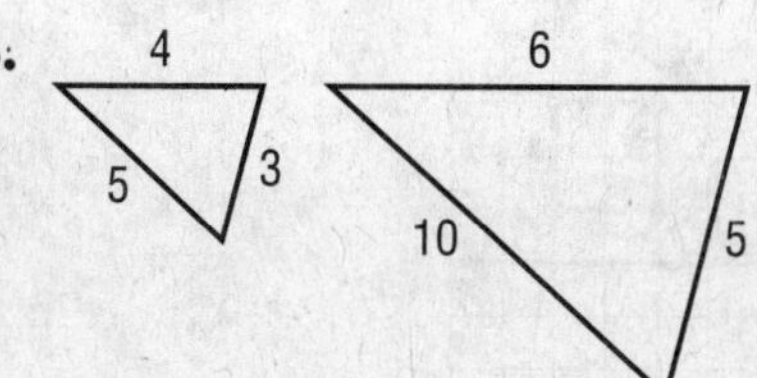

D. 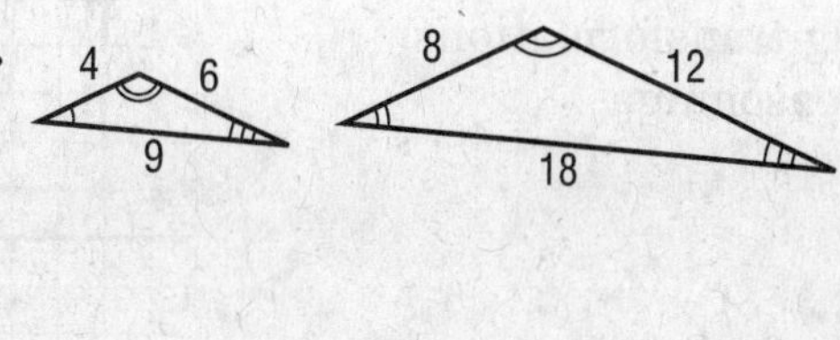

3. ________

4. Mitzi is 64 inches tall and casts a 48 inch shadow. Her daughter, who is standing next to her, casts a 30 inch shadow. How tall is her daughter?

F. 47.5 in. **H.** 35 in.

G. 40 in. **I.** 22.5 in.

4. ________

5. Which of the following statements is *not* true if quadrilateral *ABCD* is congruent to quadrilateral *RSTU*?

A. $\overline{AB} \cong \overline{RS}$ **C.** $\angle T \cong \angle C$

B. $\overline{CD} \cong \overline{TU}$ **D.** $\angle A \cong \angle U$

5. ________

Test, Form 2A *(continued)*

SCORE ________

6. The length of a rectangle is 18 centimeters and the width is 6 centimeters. A similar rectangle has a width of 2 centimeters. What is the length of the second rectangle?

6. 

7. Determine whether the triangles are similar. If so, write a similarity statement.

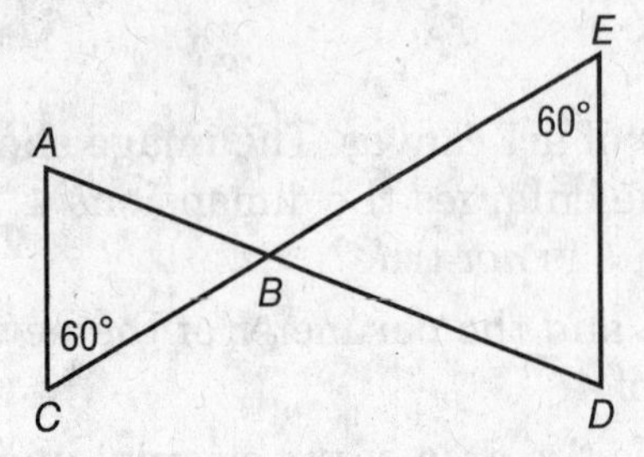

7. ____________

8. Determine if the two figures are congruent by using transformations. Explain your reasoning.

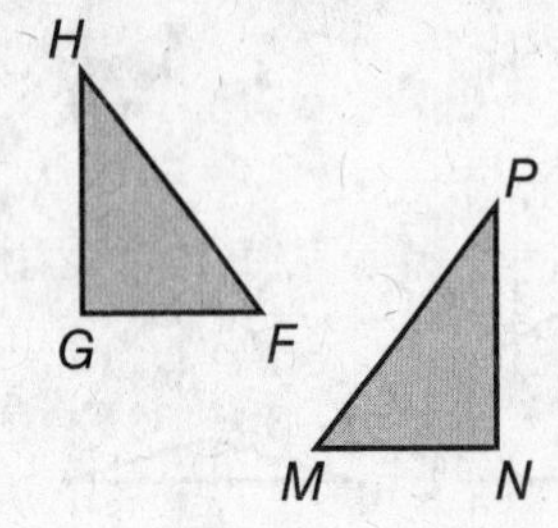

8. ____________

9. Determine if the two figures are similar by using transformations. Explain your reasoning.

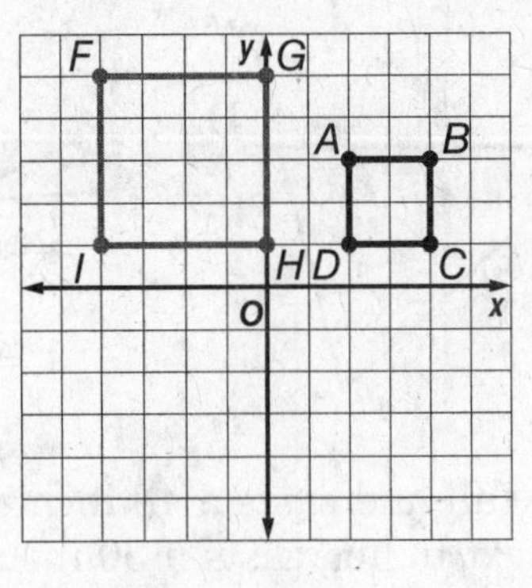

9. ____________

10. Write a proportion comparing the rise to the run for each of the similar slope triangles shown at the right. Then find the numeric value.

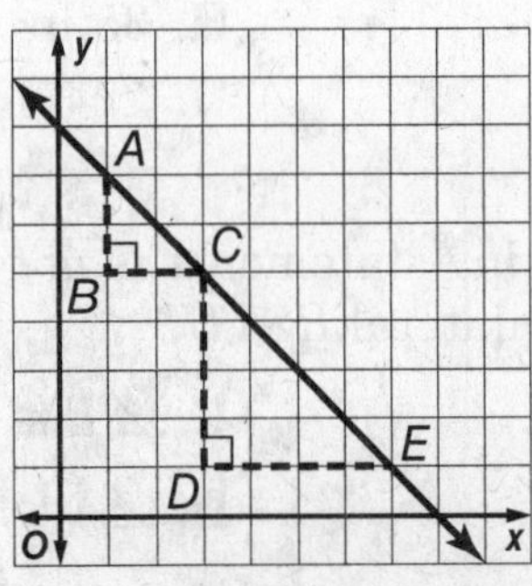

10. ____________

NAME ______________________ DATE ______________ PERIOD __________

Test, Form 2B

SCORE __________

Write the letter for the correct answer in the blank at the right of each question.

1. A survey of 11 students showed that 8 liked science, 7 liked mathematics, and 4 liked both. How many students just liked science? Use the *draw a diagram* strategy.

A. 8
B. 7
C. 5
D. 4

1. __________

2. Selena is painting an image on a piece of art canvas. The image she is reproducing is 4 inches by 6 inches. She enlarges the dimensions 3 times. Which of the following statements is *not* true?

F. The perimeter of the original image and the perimeter of the new image are related by a scale factor of 3.
G. The perimeter of the original image is $\frac{1}{3}$ the perimeter of the new image.
H. The area of the new image is 3 times the area of the original image.
I. The area of the original image and the area of the new image are related by a scale factor of 9.

2. __________

3. Which pair of polygons is similar?

A.

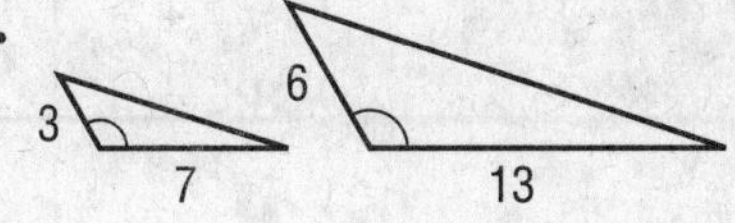

C.

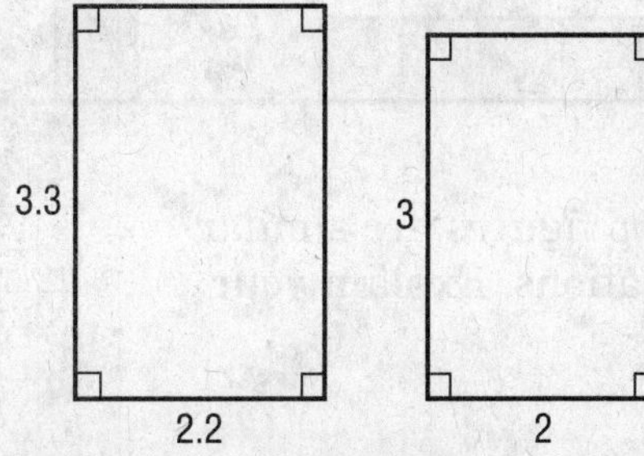

B.

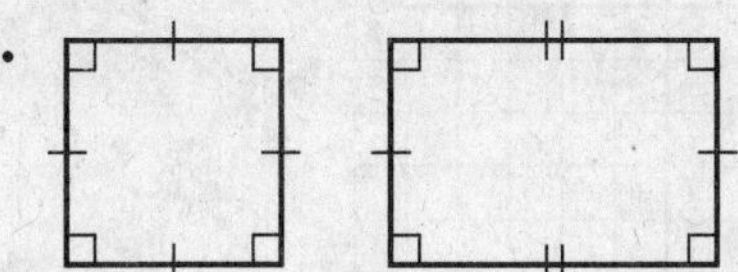

D.

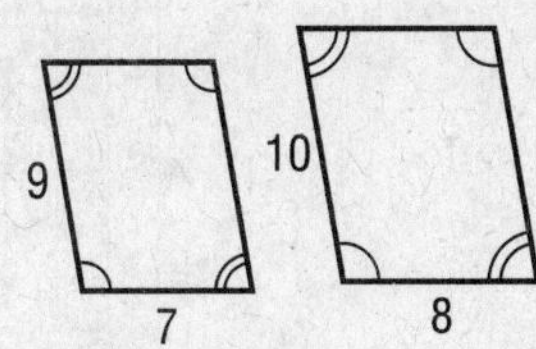

3. __________

4. Dominic is 72 inches tall and casts a 60 inch shadow. His son, who is standing next to him, casts a 50 inch shadow. How tall is his son?

F. 41.7 in.
G. 60 in.
H. 68 in.
I. 86.4 in.

4. __________

5. Which of the following statements is *not* true if $\triangle JKL$ is congruent to $\triangle RST$?

A. $\angle J \cong \angle R$
B. $\angle K \cong \angle T$
C. $\overline{JK} \cong \overline{RS}$
D. $\overline{KL} \cong \overline{ST}$

5. __________

NAME ______________________ DATE ____________ PERIOD ________

Test, Form 2B *(continued)*

SCORE 

6. The length of a rectangle is 14 centimeters and the width is 5 centimeters. A similar rectangle has a width of 2.5 centimeters. What is the length of the second rectangle?

6. ____________

7. Determine whether the triangles are similar. If so, write a similarity statement.

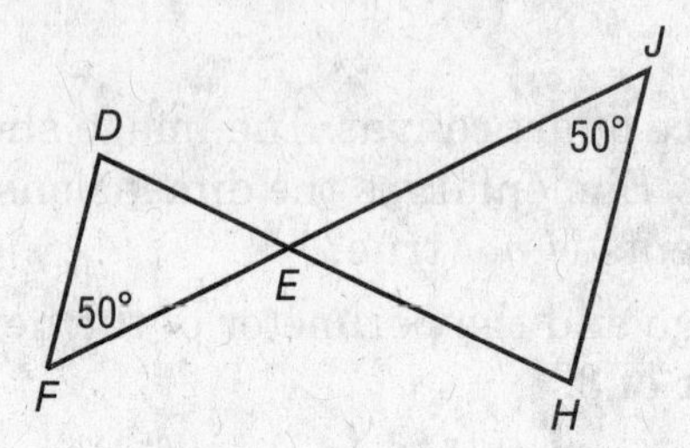

7. ____________

8. Determine if the two figures are congruent by using transformations. Explain your reasoning.

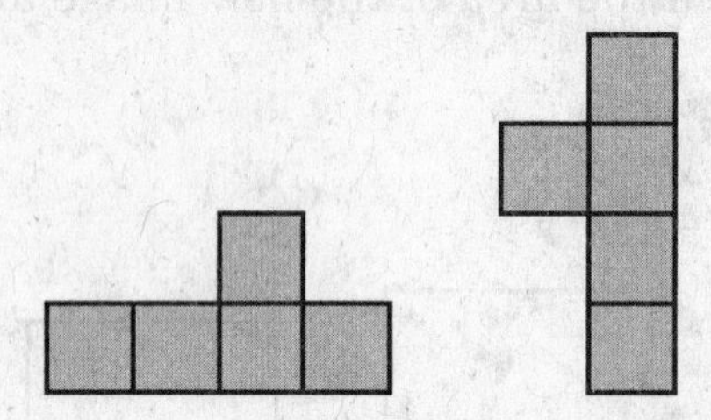

8. ____________

9. Determine if the two figures are similar by using transformations. Explain your reasoning.

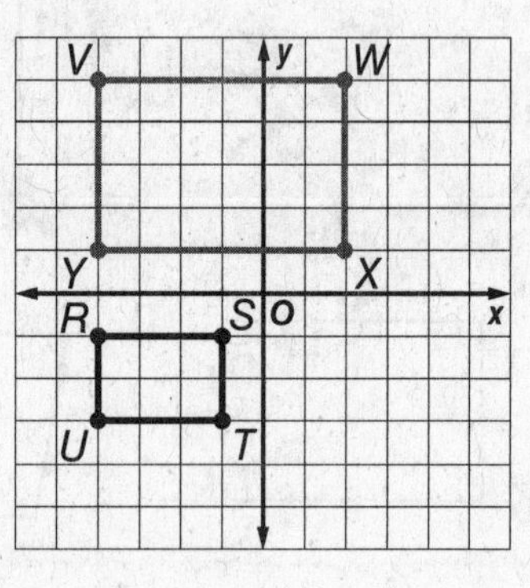

9. ____________

10. Write a proportion comparing the rise to the run for each of the similar slope triangles shown at the right. Then find the numeric value.

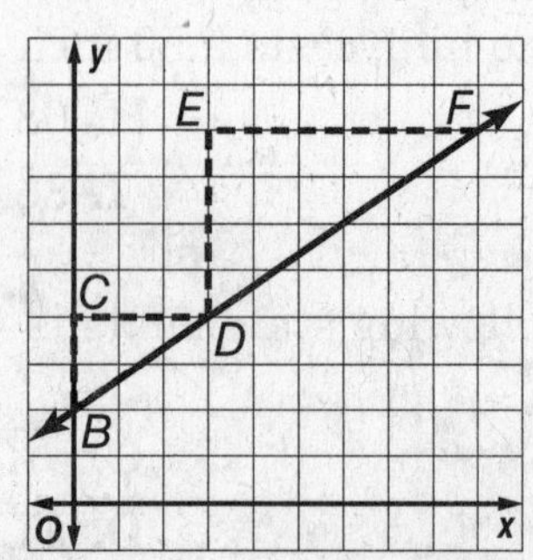

10. ____________

NAME ______________________ DATE ____________ PERIOD ________

Test, Form 3A

SCORE ________

1. The floors of houses in Japan are traditionally covered by tatami. Tatami are rectangular-shaped straw mats that measure about 6 feet by 3 feet. If a room is 48 feet by 24 feet, how many tatami are needed to cover the floor? Use the *draw a diagram* strategy.

1. ________

2. Determine whether the pair of polygons is similar using properties of similar polygons. Explain your reasoning.

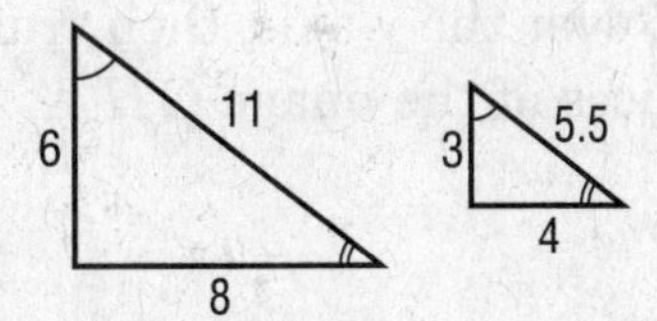

2. ________

3. A road sign casts a shadow that is 4 feet long. At the same time, a 6-foot man standing next to the sign casts a shadow that is 2.4 feet long. How tall is the sign?

3. ________

4. The length of a rectangle is 22 centimeters and the width is 4 centimeters. A similar rectangle has a width of 6 centimeters. What is the perimeter of the second rectangle?

4. ________

5. Determine whether the triangles are similar. If so, write a similarity statement.

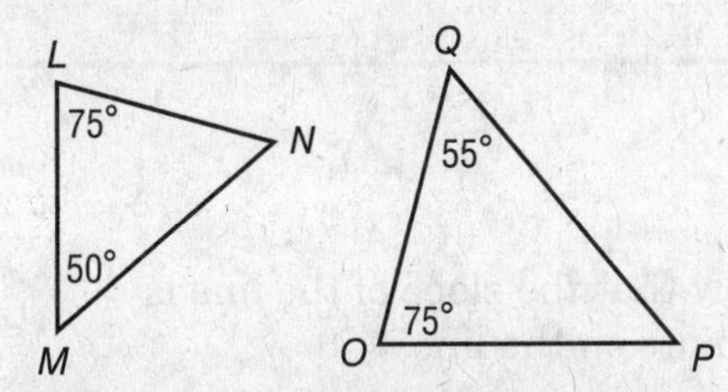

5. ________

6. The triangles below are congruent.

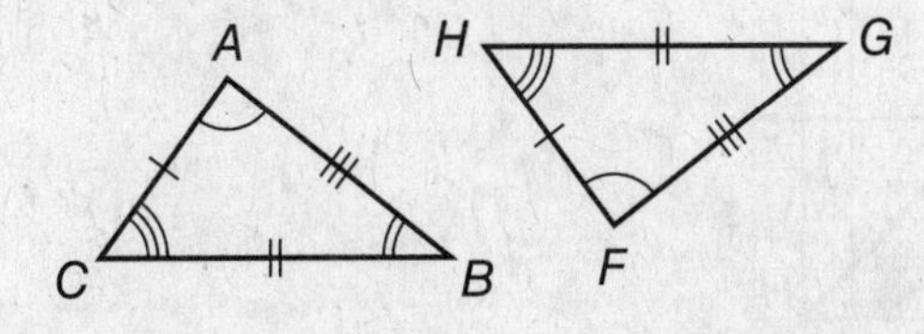

a. Write congruency statements comparing the corresponding parts.

6a. ________

b. Describe a series of transformations that maps $\triangle ABC$ onto $\triangle FGH$.

6b. ________

NAME ______________________ DATE ____________ PERIOD ________

Test, Form 3A *(continued)*

SCORE ________

7. Triangle GHJ has vertices at (0, 1), (4, 0), and (4, 1).

a. Graph $\triangle GHJ$.

7a, b.

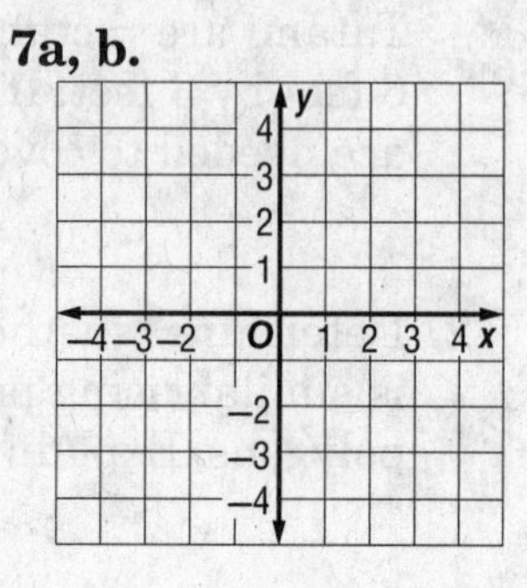

b. Reflect $\triangle GHJ$ over the y-axis, then translate it 3 units up. Label the vertices of the image $G'H'J'$.

c. Are the two triangles congruent? Justify your response.

7c. ______________

8. Determine if the two figures are similar by using transformations. Explain your reasoning.

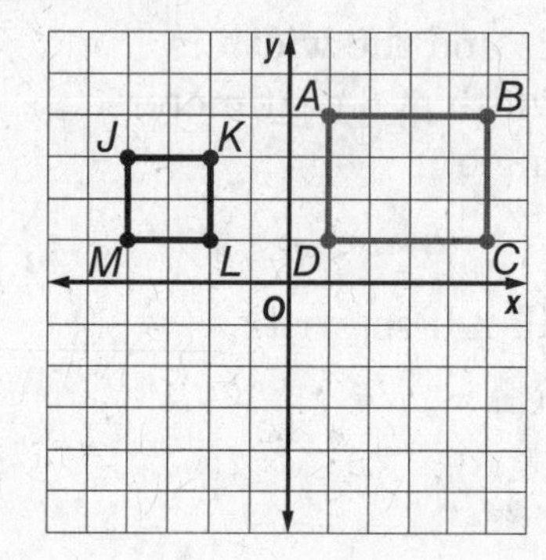

8. ______________

9. Use the similar slope triangles to show that the slope of the line is the same between any two distinct points on the line.

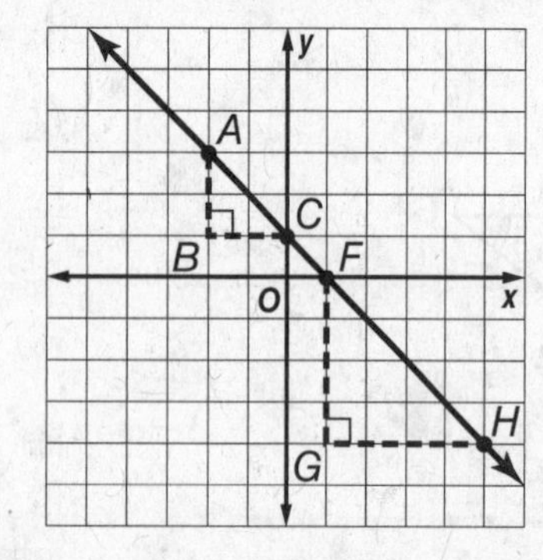

9. ______________

10. Two rectangles are similar. The length and width of the first rectangle is 8 meters by 6 meters. The second rectangle is similar by a scale factor 5. What is the area of the second rectangle?

10. ______________

NAME ______________________ DATE __________ PERIOD ______

Test, Form 3B

SCORE ______

1. The floors of houses in Japan are traditionally covered by tatami. Tatami are rectangular-shaped straw mats that measure about 6 feet by 3 feet. If the width of the room is 9 feet and the area of the room is 216 square feet, how many tatami are needed to cover the floor? Use the *draw a diagram* strategy.

1. 

2. Determine whether the pair of polygons is similar using properties of similar polygons. Explain your reasoning.

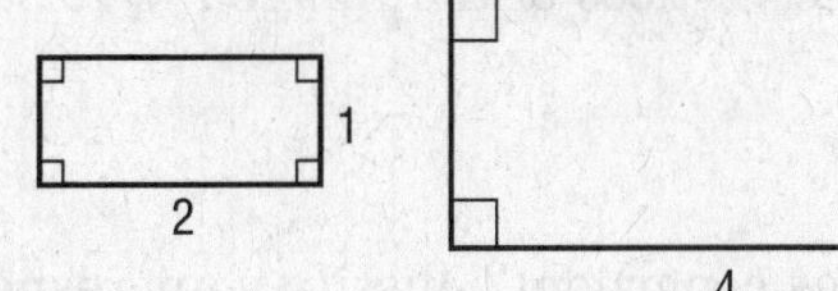

2. 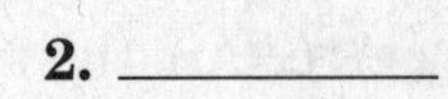

3. A statue casts a shadow 25 feet long. A boy standing next to the statue is 4.5 feet tall and casts a shadow that is 3.6 feet long. How tall is the statue?

3. 

4. The length of a rectangle is 45 inches and the width is 8 inches. A similar rectangle has a width of 24 inches. What is the perimeter of the second rectangle?

4. 

5. Determine whether the triangles are similar. If so, write a similarity statement.

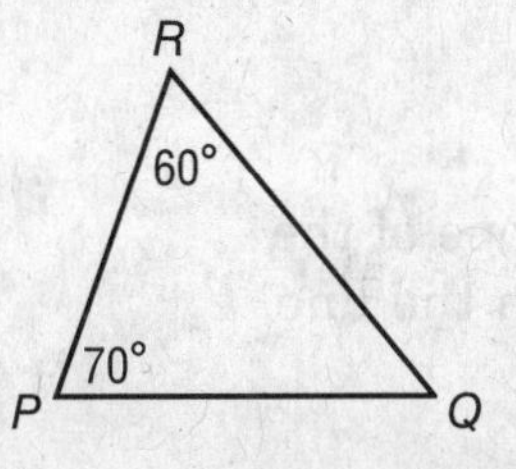

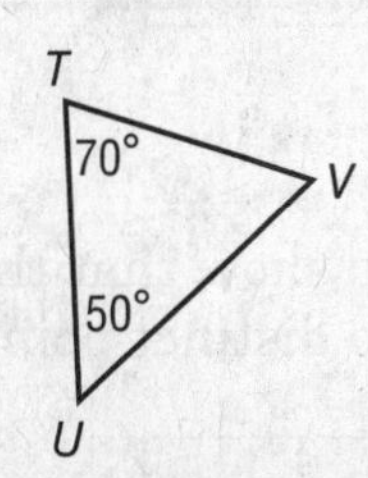

5. ______

6. The triangles below are congruent.

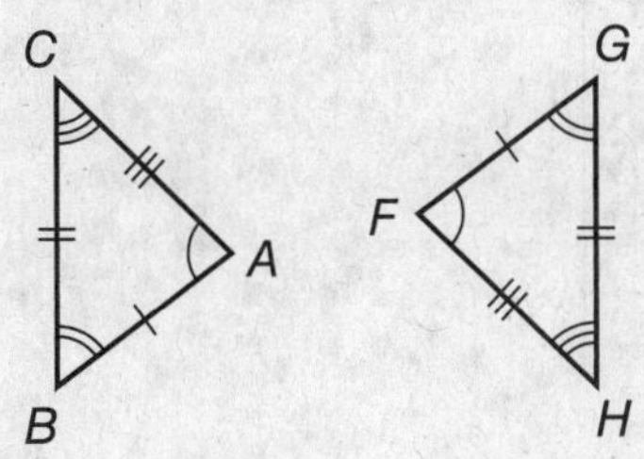

a. Write congruency statements comparing the corresponding parts.

6a. 

b. Describe a series of transformations that maps $\triangle ABC$ onto $\triangle FGH$.

6b. 

NAME ______________________ DATE ______________ PERIOD ________

Test, Form 3B *(continued)*

SCORE ________

7. Triangle PQR has vertices at (0, 0), (2, 0), and (0, 2).

a. Graph $\triangle PQR$.

7a, b.

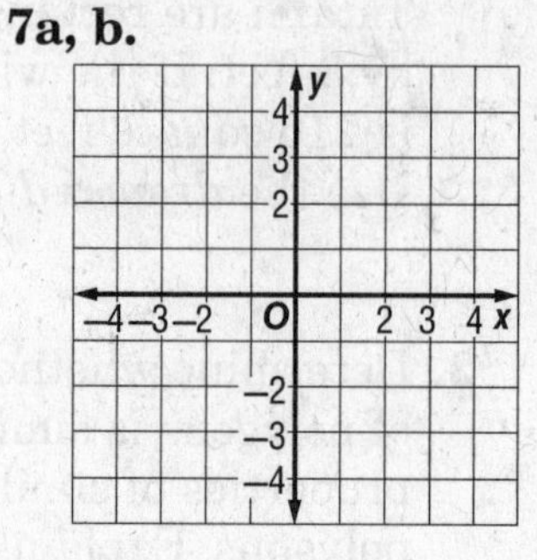

b. Reflect $\triangle PQR$ over the x-axis, then dilate it by a scale factor of 2. Label the vertices of the image $P'Q'R'$.

c. Are the two triangles congruent? Justify your response.

7c. ________________

8. Determine if the two figures are similar by using transformations. Explain your reasoning.

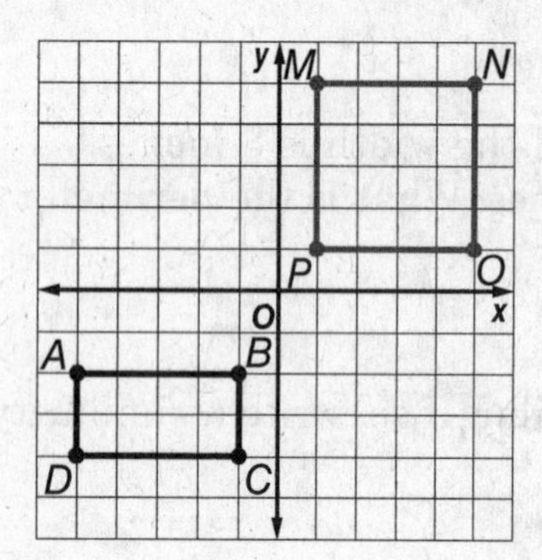

8. ________________

9. Use the similar slope triangles to show that the slope of the line is the same between any two distinct points on the line.

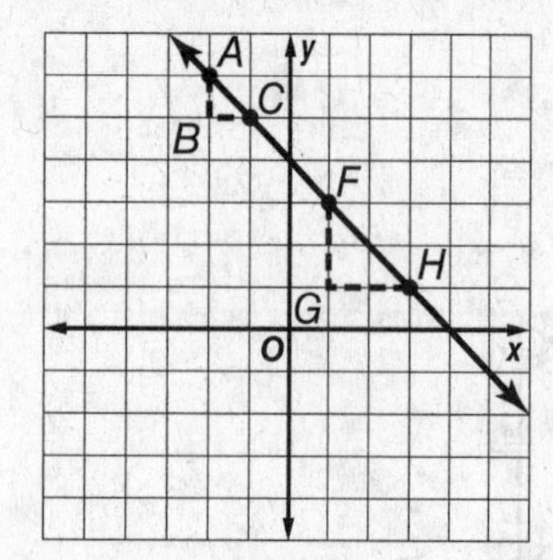

9. ________________

10. Two rectangles are similar. The length and width of the first rectangle is 4 meters by 5 meters. The second rectangle is similar by a scale factor 4. What is the area of the second rectangle?

10. ________________

NAME ______________________ DATE ____________ PERIOD ________

Are You Ready?

Review

The area of a triangle can be found using the formula $A = \frac{1}{2}bh$.

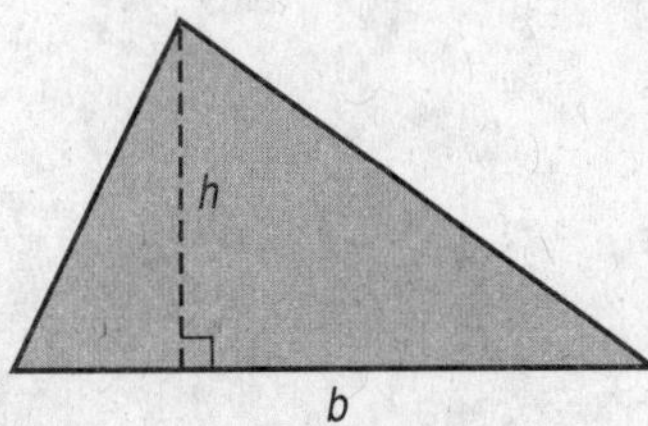

Find the area of the triangle below.

Example

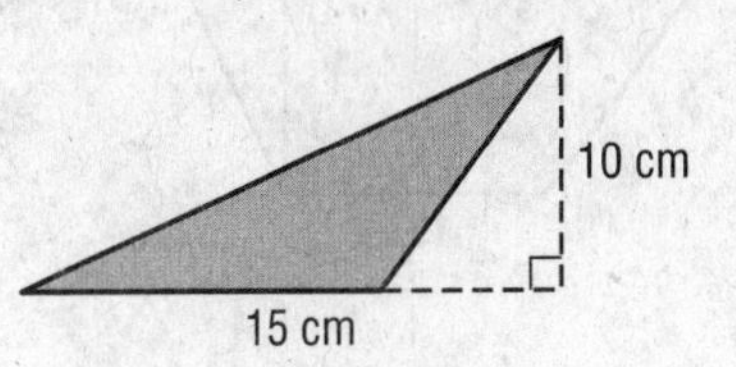

$A = \frac{1}{2}bh$	Formula for the area of a triangle
$A = \frac{1}{2} \cdot 15 \cdot 10$	Replace b with 15 and h with 10.
$A = 75$	Simplify.

The area is 75 square centimeters.

Exercises

Find the area of each triangle.

1.

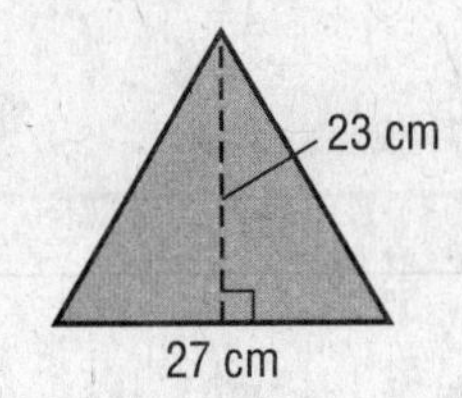

2.

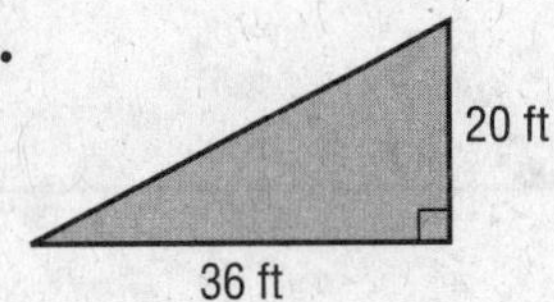

3.

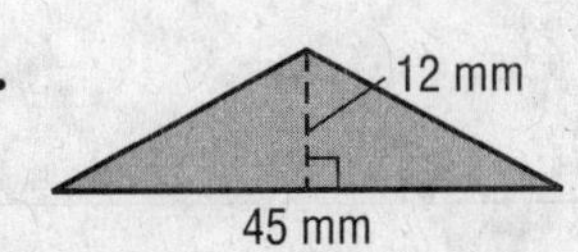

4.

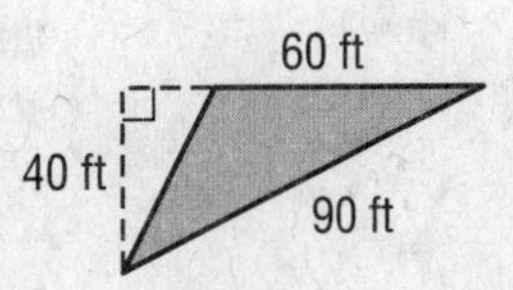

5.

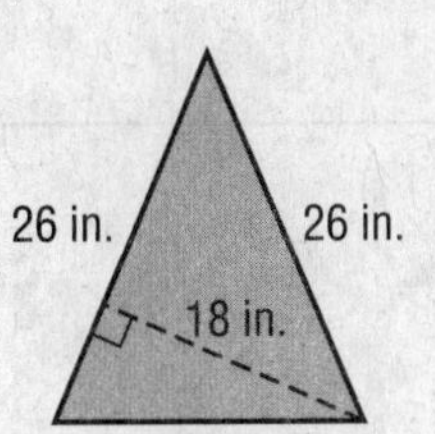

6.

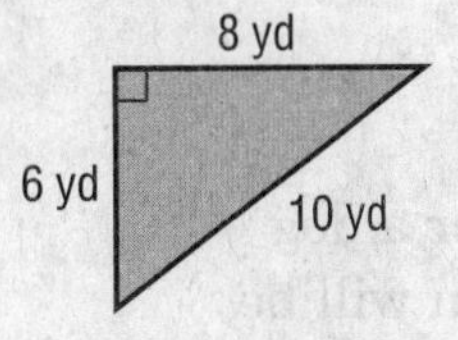

1. 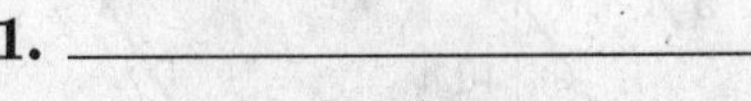

2. ____________________

3. ____________________

4. 

5. ____________________

6. ____________________

NAME ______________________ DATE ____________ PERIOD ________

Are You Ready?

Practice

Find the area of each figure.

1.

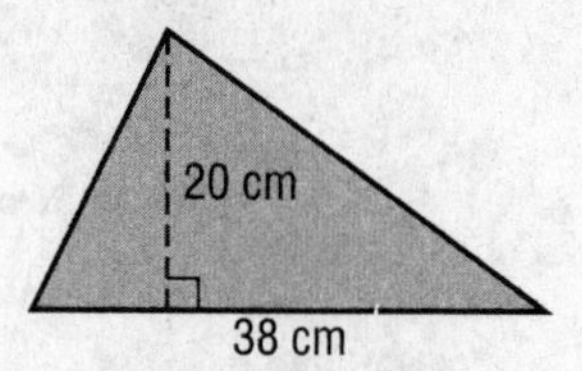

1. ____________

2.

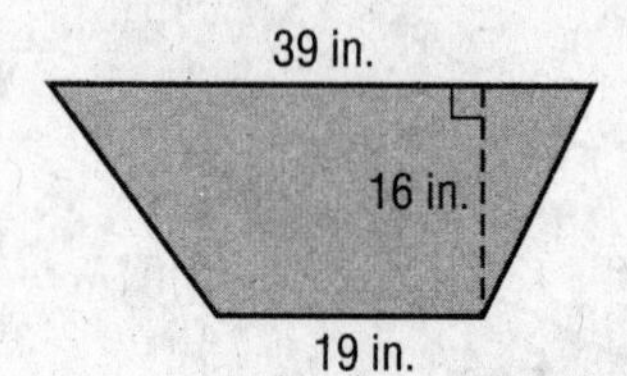

2. ____________

3.

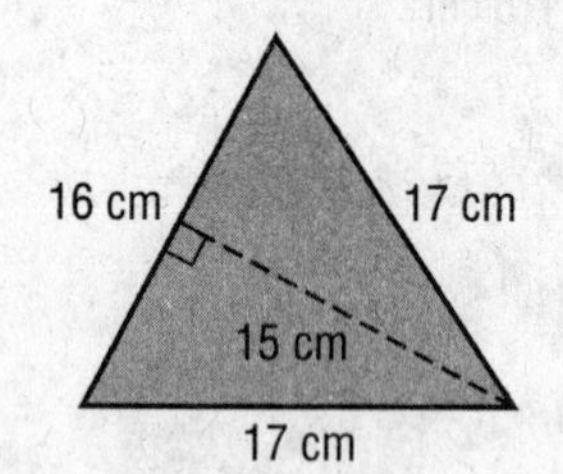

3. ____________

4.

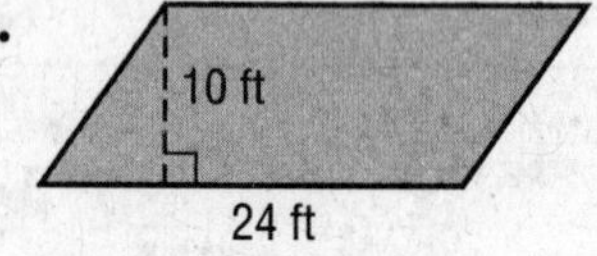

4. ____________

5.

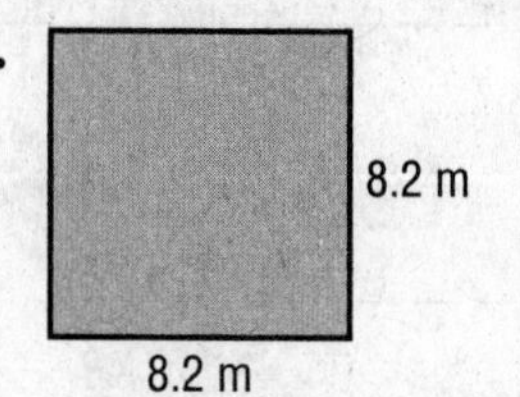

5. ____________

6.

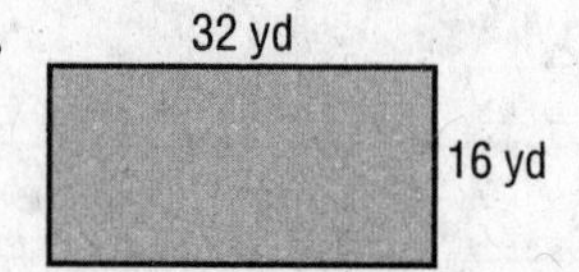

6. ____________

7. **DOG RUN** Rowena is installing a fence for a dog run on the side of her house. The dog run will be 12 feet long and cover an area of 84 square feet. How wide is the dog run?

7. ____________

NAME ______________________ DATE ____________ PERIOD ________

Are You Ready?

Apply

1. **FLAG** A flag of Florida measures 12 inches by 18 inches. What is the area of the flag?

2. **TILES** Rondell covered a table top with 36 square tiles. Each tile measures 8 inches by 8 inches. What is the area of the table top?

3. **ROAD SIGN** Cullen saw the sign below on his walk. The triangle is 12 inches on each side and the height is 10.4 inches. Find the area of the sign.

4. **HEXAGONS** The patio shown below is made up of hexagons, which are two trapezoids joined at one base. If each trapezoid has a height of 1.7 feet and base lengths of 2 feet and 4 feet, what is the area of one hexagon?

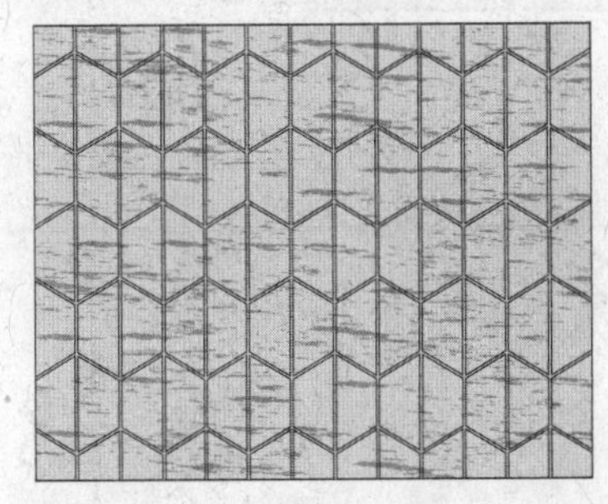

5. **LOGO** Josh used the figure below for a logo contest. If the height of the figure is 1.8 inches and the base is 3.8 inches, what is the area of the figure?

6. **DESIGN** Russell made the design below on his computer. Each triangle has a height of 12 millimeters and a base of 42 millimeters. Find the area of the design.

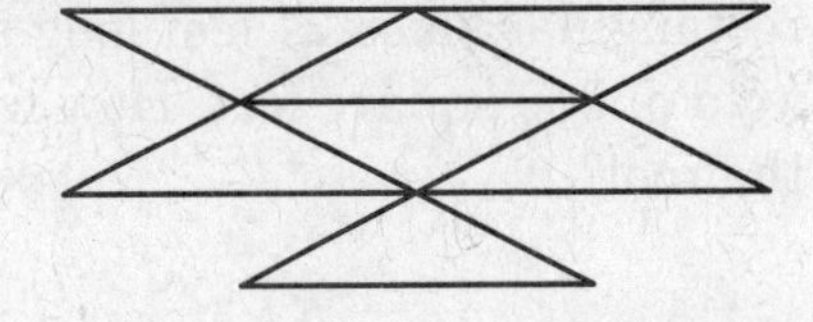

NAME ______________________ DATE ______________ PERIOD ________

Diagnostic Test

Find the area of each figure.

1.

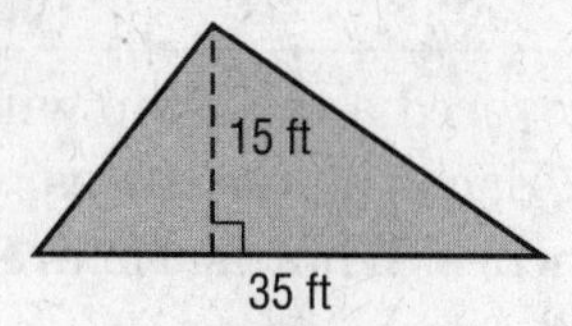

2.

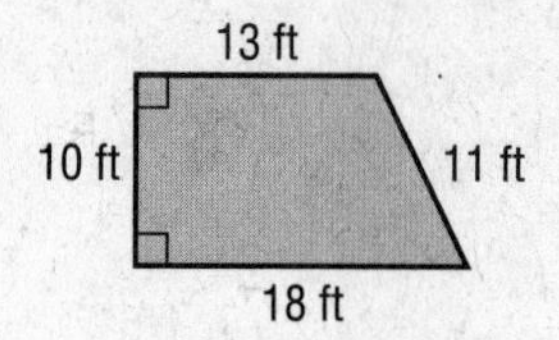

3.

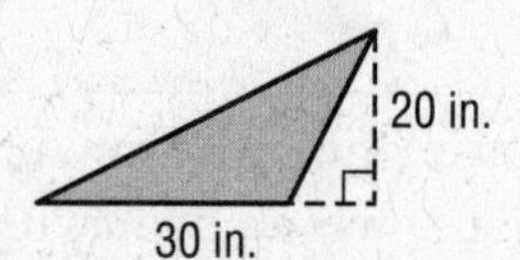

4.

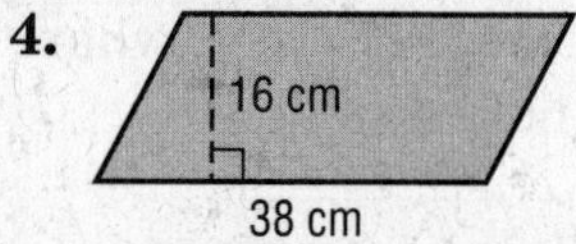

5.

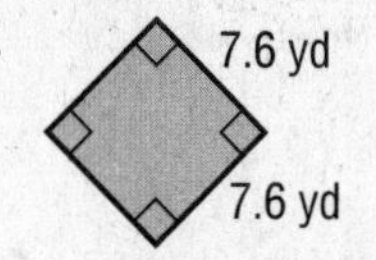

6.

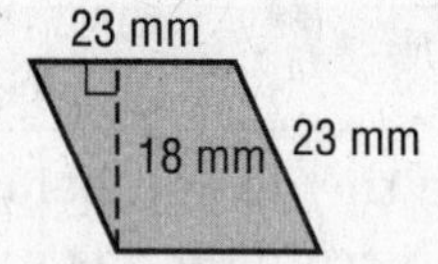

7. ROOF Brendan is adding an addition to his house. The addition will have a roof with rectangular sides 24 feet long and will cover an area of 360 square feet. How wide is the side of the roof?

1. ______________

2. ______________

3. ______________

4. ______________

5. ______________

6. ______________

6. ______________

NAME ______________________ DATE ____________ PERIOD ________

Pretest

Find the volume of each figure. Round to the nearest tenth if necessary.

1\.

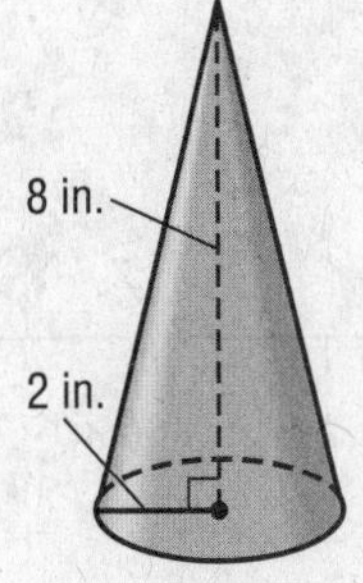

2\.

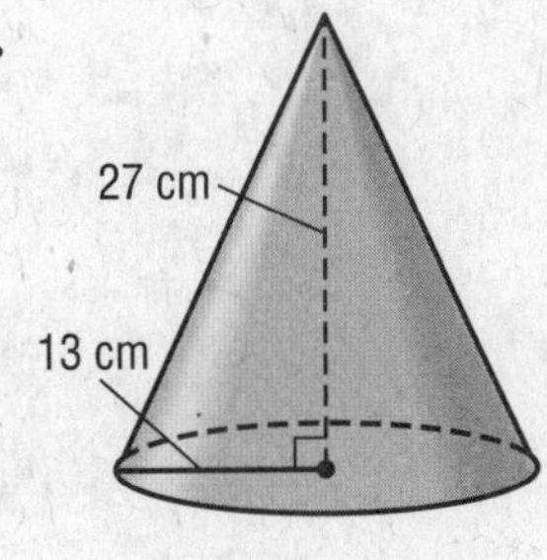

1\. ______________

2\. ______________

3\.

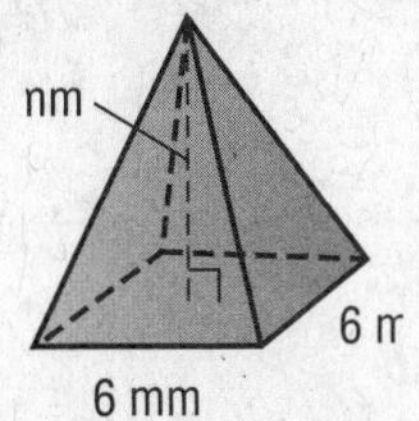

4\.

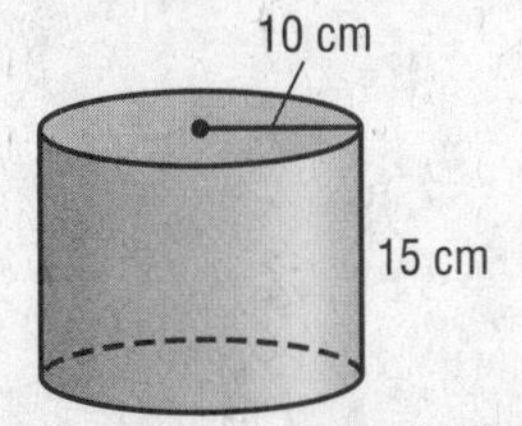

3\. ______________

4\. ______________

Find the surface area of each figure. Round to the nearest tenth if necessary.

5\.

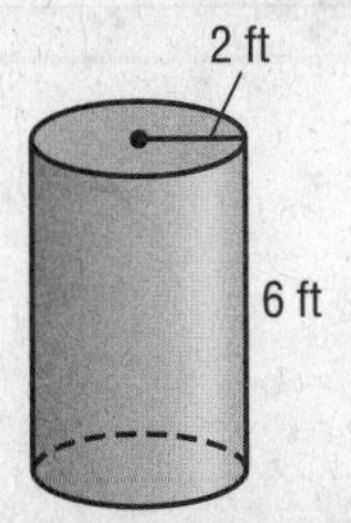

6\.

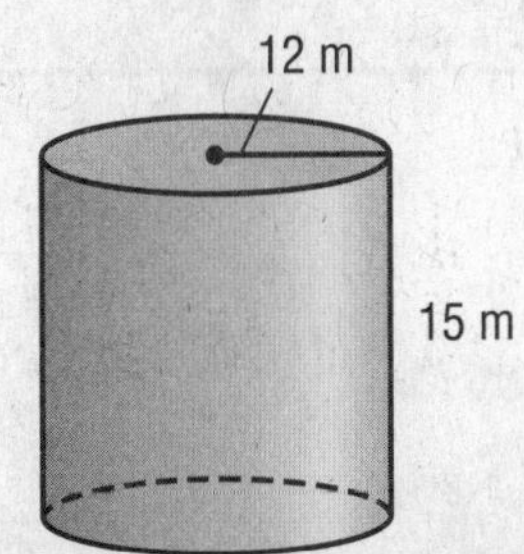

5\. ______________

6\. ______________

7\.

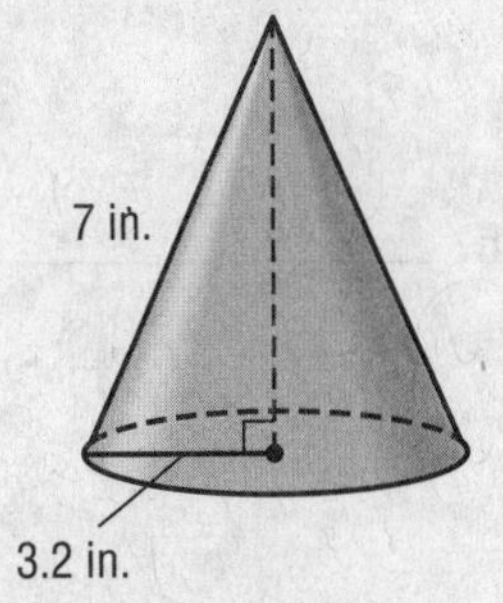

8\.

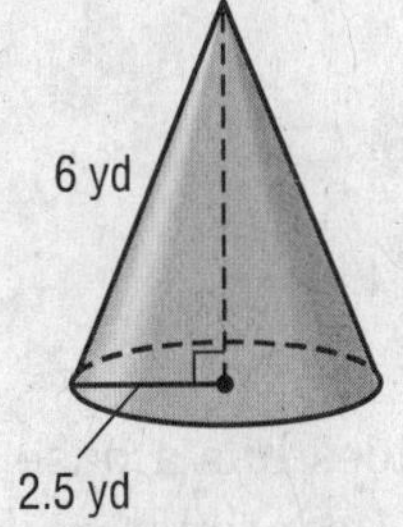

7\. ______________

8\. ______________

NAME ______________________ DATE ____________ PERIOD ________

Chapter Quiz

Find the volume of each solid. Round to the nearest tenth if necessary.

1.

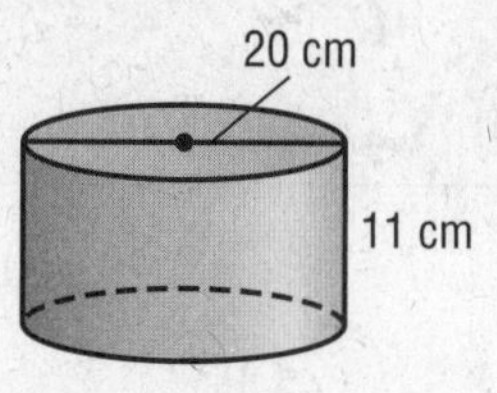

1. ________________

2.

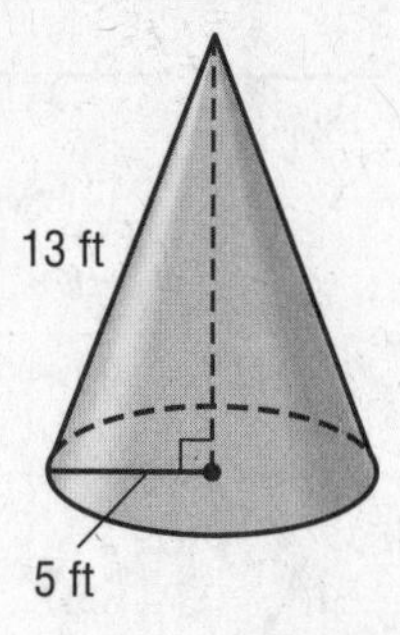

2. ________________

3.

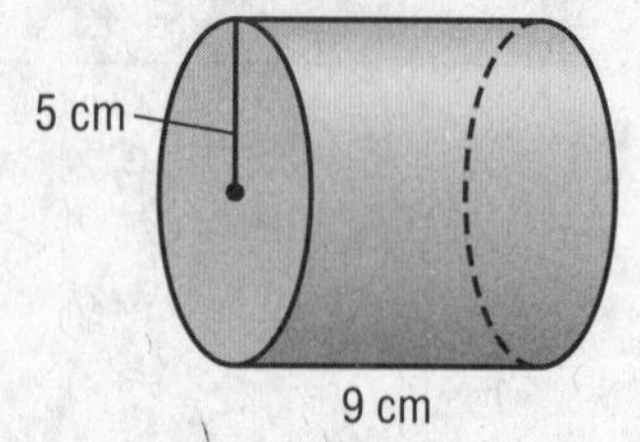

3. ________________

4.

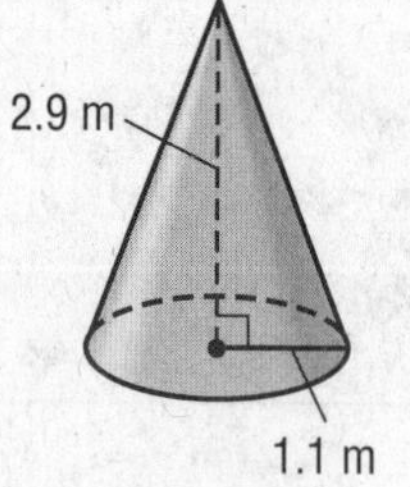

4. ________________

5.

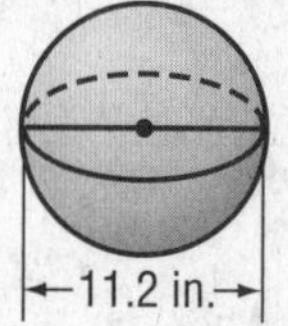

5. ________________

6. **CANNED VEGETABLES** A can of tomatoes has a base diameter of 9.7 centimeters and a height of 11 centimeters. What is the volume of the can? Round to the nearest tenth if necessary.

6. ________________

NAME ______________________ DATE ______________ PERIOD ________

Vocabulary Test

SCORE ________

composite solid	lateral area	similar solids
cone	polyhedron	sphere
cylinder	precision	total surface area
hemisphere	prism	volume

Choose from the terms above to complete each sentence.

1. A(n) __________ is a three-dimensional figure with two parallel congruent circular bases. **1.** __________

2. A __________ is a three-dimensional figure with faces that are polygons. **2.** __________

3. An object that is made up of more than one type of solid is called a(n) __________. **3.** __________

4. A circle separates a sphere into two congruent halves each called a __________. **4.** __________

5. A __________ is a set of all points in space that are a given distance from a given point. **5.** __________

6. __________ is the measure of the space occupied by a solid. **6.** __________

7. A __________ is a three-dimensional figure with one base connected by a curved surface to a single vertex. **7.** __________

Define the term in your own words.

8. lateral area **8.** __________

9. similar solids **9.** __________

Standardized Test Practice

Read each question. Then fill in the correct answer on the answer document provided by your teacher or on a sheet of paper.

1. The figure shows a circle inside a square.

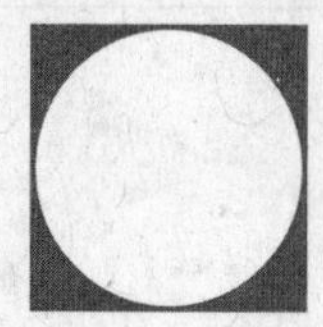

Which procedure should be used to find the area of the shaded region?

A. Find the area of the square and then subtract the area of the circle.

B. Find the area of the circle and then subtract the area of the square.

C. Find the perimeter of the square and then subtract the circumference of the circle.

D. Find the circumference of the circle and then subtract the perimeter of the square.

2. What is the value of x in the figure below?

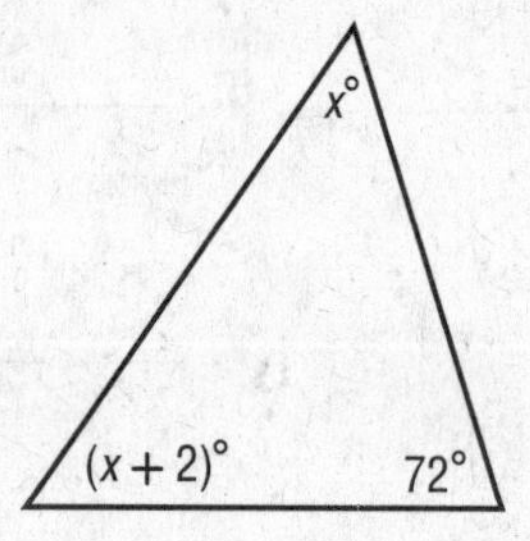

F. 53

G. 60

H. 90

I. 180

3. THINK SOLVE EXPLAIN **SHORT RESPONSE** The hypotenuse of a right triangle measures 15 feet. If one of the legs measures 9 feet, what is the length of the other leg?

4. **GRIDDED RESPONSE** A rectangular prism has a length of 7.5 inches, a width of 1.4 inches, and a volume of 86.4 cubic inches. What is the height, in inches, of the rectangular prism? Round to the nearest tenth.

5. Which of the following conclusions about the number of rebounds per game and the height of a player is **best** supported by the scatter plot below?

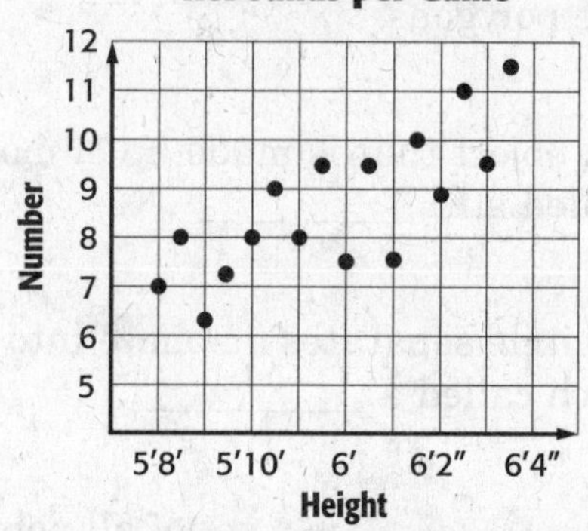

A. The number of rebounds increases as the player's height decreases.

B. The number of rebounds is unchanged as the player's height increases.

C. The number of rebounds increases as the player's height increases.

D. There is no relationship between the number of rebounds and the player's height.

6. What is the volume of a sphere with a diameter of 15 inches?

F. 2,000 in^3

G. 1,767.1 in^3

H. 1,325.4 in^3

I. 421.9 in^3

7. Which point on the number line below is closest to $\sqrt{50}$?

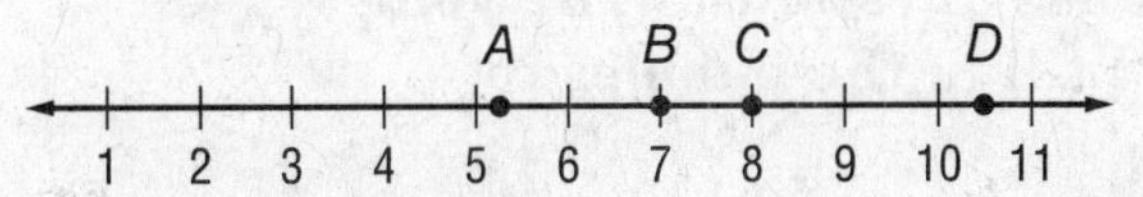

A. point A

B. point B

C. point C

D. point D

8. **GRIDDED RESPONSE** In the figure below, every angle is a right angle. What is the area of the figure in square units?

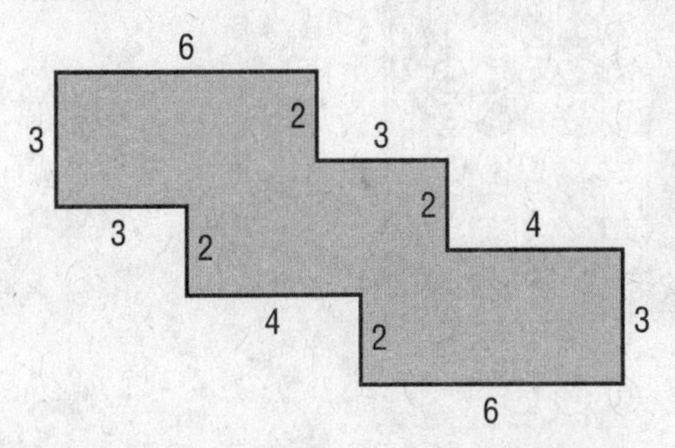

9. Mr. Brauen's farm has a square cornfield. Which of the following is a possible area for the cornfield if the sides are measured in whole numbers?

F. 164,000 ft^2

G. 170,150 ft^2

H. 170,586 ft^2

I. 174,724 ft^2

10. Allison, Carl, and Theo drove from Austin, Texas, to Los Angeles, California, a distance of 1,224 miles. Allison drove $\frac{1}{3}$ of the total distance, Carl drove 40%, and Theo drove the remainder. How many miles were driven by the person who drove the greatest distance?

A. 326.4 mi

B. 408 mi

C. 489.6 mi

D. 897.6 mi

11. Which of the following is the approximate surface area of the figure shown below?

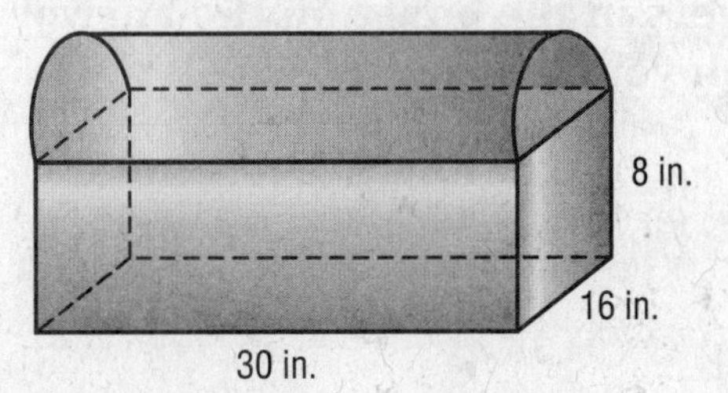

F. 1,170.6 in^2

G. 1,170.6 in^3

H. 2,171.0 in^3

I. 2,171.0 in^2

12. Two similar prisms have surface areas that are 95 square inches and 1,520 square inches, respectively. How many times larger is the second prism than the first prism?

A. 2 times as large

B. 3 times as large

C. 4 times as large

D. 16 times as large

13. THINK SOLVE EXPLAIN **EXTENDED RESPONSE** A movie theater has two different containers for popcorn.

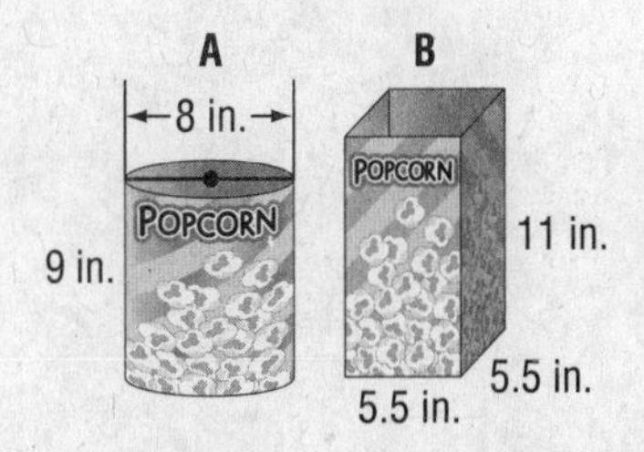

Part A Which container will hold more popcorn? Justify your selection.

Part B Which container requires less packaging to construct? Explain your reasoning.

NAME ______________________ DATE ____________ PERIOD ________

Student Recording Sheet

SCORE ________

Use this recording sheet with the Standardized Test Practice pages.

Fill in the correct answer. For gridded-response questions, write your answers in the boxes on the answer grid and fill in the bubbles to match your answers.

1. Ⓐ Ⓑ Ⓒ Ⓓ

2. Ⓕ Ⓖ Ⓗ Ⓘ

3. __________

4.

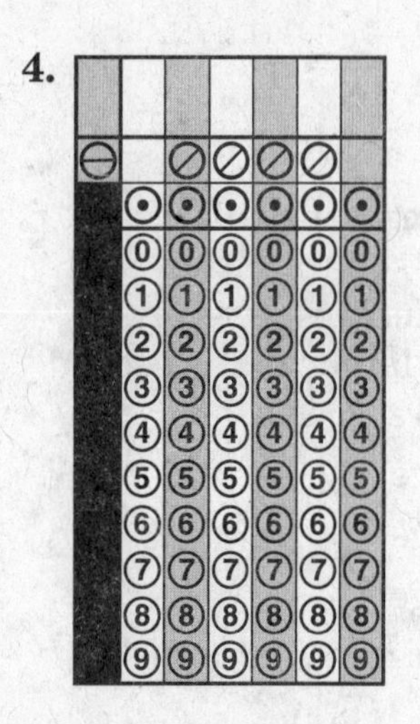

5. Ⓐ Ⓑ Ⓒ Ⓓ

6. Ⓕ Ⓖ Ⓗ Ⓘ

7. Ⓐ Ⓑ Ⓒ Ⓓ

8.

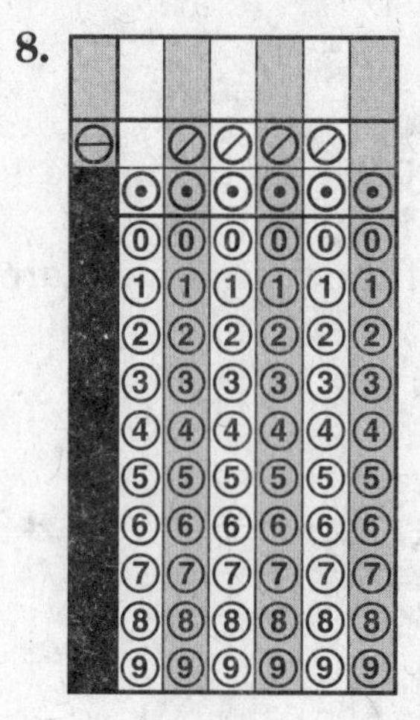

9. Ⓕ Ⓖ Ⓗ Ⓘ

10. Ⓐ Ⓑ Ⓒ Ⓓ

11. Ⓕ Ⓖ Ⓗ Ⓘ

12. Ⓐ Ⓑ Ⓒ Ⓓ

Extended Response

Record your answers for Exercise 13 on the back of this paper.

NAME ______________________ DATE __________ PERIOD ________

Extended-Response Test

SCORE ________

Demonstrate your knowledge by giving a clear, concise solution to each problem. Be sure to include all relevant drawings and justify your answers. You may show your solution in more than one way or investigate beyond the requirements of the problem. If necessary, record your answers on another piece of paper.

1. An organic foods company is planning packaging for a new snack mix consisting of dried fruits and nuts. The president of the company has narrowed the choices to the three containers shown below.

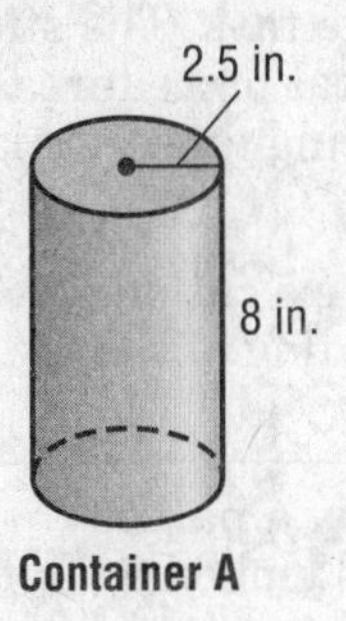

Container A

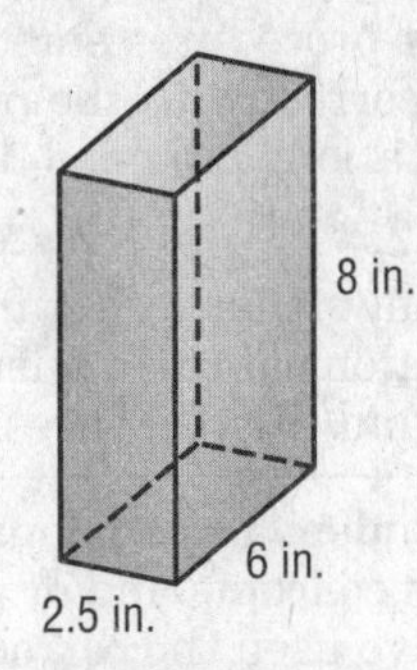

Container B

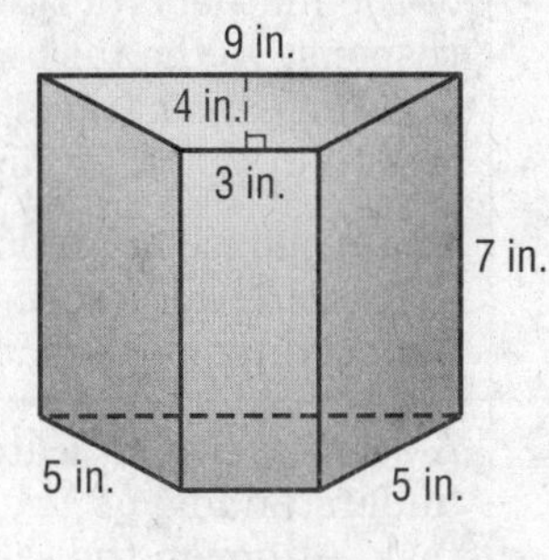

Container C

a. Identify each container as either a *prism,* a *pyramid,* a *cylinder,* or a *cone.* Explain your reasoning.

b. For Container B, name the number of faces, edges, and vertices. Explain how you counted the faces.

c. The base of Container A is a circle. Explain how to find the area of a circle. Then find the area of the base of Container A. Round to the nearest tenth.

d. If the dimensions of Container B are halved, what will the new surface area be?

e. Explain how computing the volumes of cylinders and prisms are similar. Then compute the volumes of all three containers. Round to the nearest tenth if necessary. Which container will hold the most snack mix?

f. Compute the surface areas of Containers A and B. Round to the nearest tenth if necessary. Which container has less surface area?

g. Each container will be made of a material that costs $0.0015 per square inch. Explain how to determine the cost of each container. Then find the cost of Container B. Round to the nearest cent.

NAME ______________________ DATE ____________ PERIOD ________

Extended-Response Rubric

SCORE ________

Score	Description
4	A score of four is a response in which the student demonstrates a thorough understanding of the mathematics concepts and/or procedures embodied in the task. The student has responded correctly to the task, used mathematically sound procedures, and provided clear and complete explanations and interpretations. The response may contain minor flaws that do not detract from the demonstration of a thorough understanding.
3	A score of three is a response in which the student demonstrates an understanding of the mathematics concepts and/or procedures embodied in the task. The student's response to the task is essentially correct with the mathematical procedures used and the explanations and interpretations provided demonstrating an essential but less than thorough understanding. The response may contain minor flaws that reflect inattentive execution of mathematical procedures or indications of some misunderstanding of the underlying mathematics concepts and/or procedures.
2	A score of two indicates that the student has demonstrated only a partial understanding of the mathematics concepts and/or procedures embodied in the task. Although the student may have used the correct approach to obtaining a solution or may have provided a correct solution, the student's work lacks an essential understanding of the underlying mathematical concepts. The response contains errors related to misunderstanding important aspects of the task, misuse of mathematical procedures, or faulty interpretations of results.
1	A score of one indicates that the student has demonstrated a very limited understanding of the mathematics concepts and/or procedures embodied in the task. The student's response is incomplete and exhibits many flaws. Although the student's response has addressed some of the conditions of the task, the student reached an inadequate conclusion and/or provided reasoning that was faulty or incomplete. The response exhibits many flaws or may be incomplete.
0	A score of zero indicates that the student has provided no response at all, or a completely incorrect or uninterpretable response, or demonstrated insufficient understanding of the mathematics concepts and/or procedures embodied in the task. For example, a student may provide some work that is mathematically correct, but the work does not demonstrate even a rudimentary understanding of the primary focus of the task.

NAME ______________________ DATE ____________ PERIOD ________

Test, Form 1A

SCORE ________

Write the letter for the correct answer in the blank at the right of each question.

For Exercises 1–4, what is the volume of each solid? Round to the nearest tenth if necessary.

1. **A.** 4,188.8 in^3 **C.** 418.8 in^3
   B. 3,141.6 in^3 **D.** 314.2 in^3

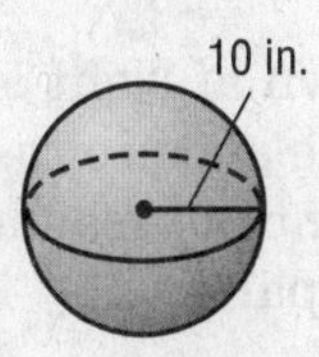

1. ________

2. **F.** 37.7 cm^3 **H.** 113 cm^3
   G. 56.5 cm^3 **I.** 169.9 cm^3

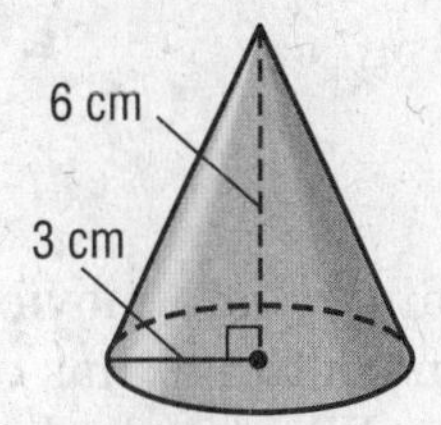

2. ________

3. **A.** 49.1 in^3 **C.** 196.3 in^3
   B. 65.4 in^3 **D.** 200 in^3

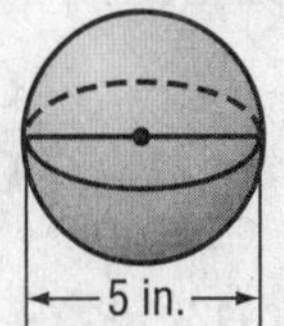

3. ________

4. **F.** 94.2 in^3 **H.** 282.7 in^3
   G. 188.4 in^3 **I.** 1,130.4 in^3

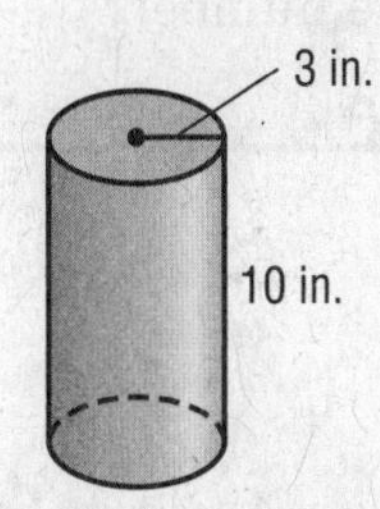

4. ________

5. What is the surface area of the cylinder? Round to the nearest tenth.
   A. 62.8 ft^2 **C.** 150.8 ft^2
   B. 113.0 ft^2 **D.** 251.2 ft^2

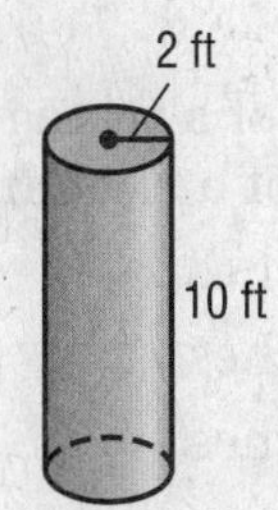

5. ________

6. What is the surface area of a cone with a radius of 3 meters and a slant height of 4 meters? Round to the nearest tenth.
   F. 113.0 m^2 **H.** 226.0 m^2
   G. 131.9 m^2 **I.** 263.8 m^2

6. ________

Test, Form 1A *(continued)*

SCORE ________

7. A can of juice is 6 inches high, and its base has a diameter of 4 inches. What is the volume of the can? Round to the nearest tenth.

A. 37.7 in³ **B.** 75.4 in³ **C.** 150.7 in³ **D.** 301.4 in³

7. ________

8. The storage tank shown at right is to be painted. What is the area of the surface to be painted to the nearest whole number? Assume that the bottom does not need painting.

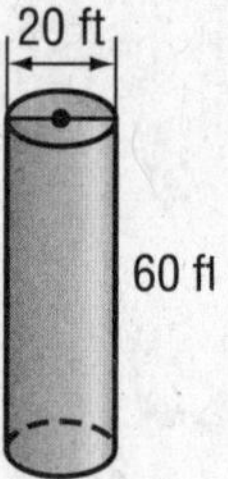

F. 7,536 ft² **H.** 4,084 ft²
G. 4,396 ft² **I.** 3,768 ft²

8. ________

9. The popcorn containers at a movie theater are in the shape of cones. Suppose a popcorn container has a radius of 6 inches and a slant height of 12 inches. What is the lateral area of the popcorn container rounded to the nearest inch?

A. 113 in² **C.** 339 in²
B. 226 in² **D.** 1,357 in²

9. ________

10. What is the volume of the composite shape to the nearest whole number?

F. 2,138 in³
G. 1,178 in³
H. 1,035 in³
I. 960 in³

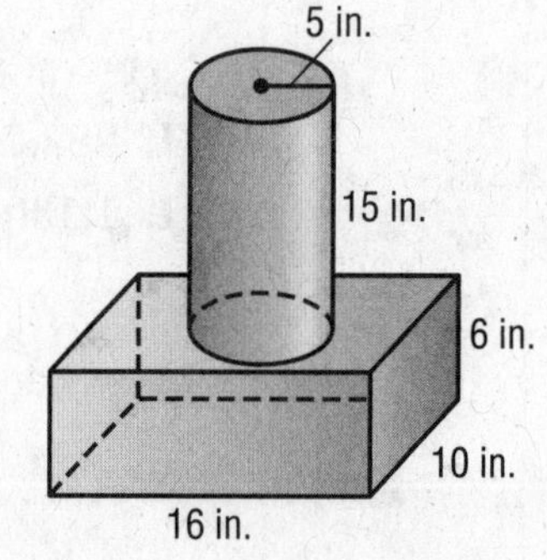

10. ________

11. The surface area of a triangular prism is 78 square inches. What is the surface area of a similar prism that is three times as large?

A. 156 square inches
B. 234 square inches
C. 312 square inches
D. 702 square inches

11. ________

12. A cone has a volume that is 350 cubic meters. What is the volume of a similar cone that is twice as large as the first cone?

F. 2,800 cubic meters
G. 1,400 cubic meters
H. 700 cubic meters
I. 175 cubic meters

12. ________

NAME ______________________ DATE ____________ PERIOD ________

Test, Form 1B

SCORE ________

Write the letter for the correct answer in the blank at the right of each question.

For Exercises 1–4, what is the volume of each solid? Round to the nearest tenth if necessary.

1. A. 329.9 yd^3 C. 813.7 yd^3
   B. 769.7 yd^3 D. 2,309.1 yd^3

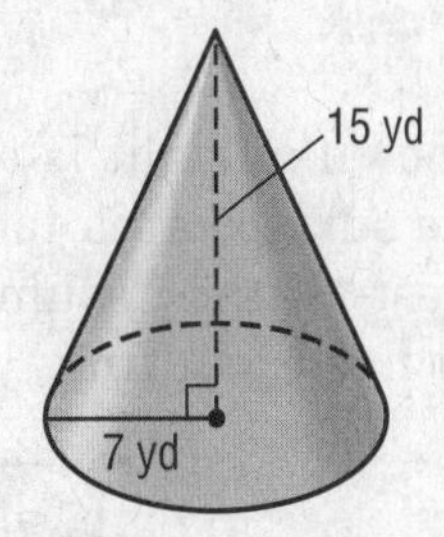

1. ________

2. F. 339.3 in^3 H. 84.9 in^3
   G. 113.1 in^3 I. 9 in^3

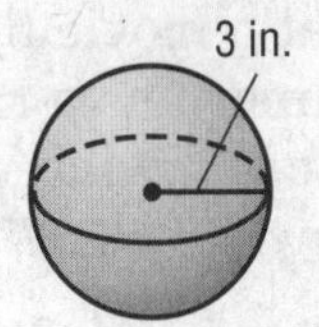

2. ________

3. A. 621.7 cm^3 C. 2,799.2 cm^3
   B. 932.6 cm^3 D. 3,419.5 cm^3

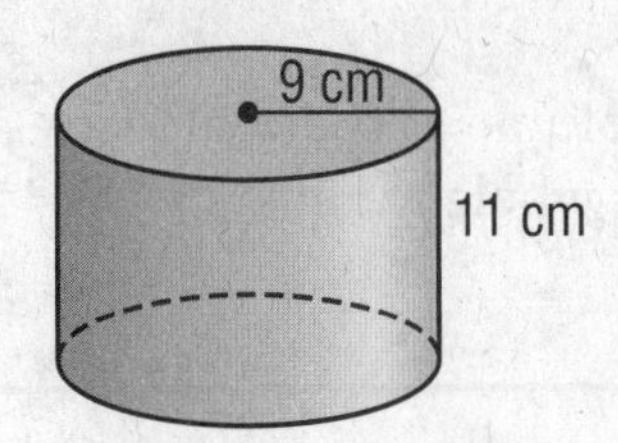

3. ________

4. F. 33.5 cm^3 H. 133.8 cm^3
   G. 44.6 cm^3 I. 150.2 cm^3

4. ________

5. What is the surface area of the cylinder? Round to the nearest tenth.
   A. 4,628.4 cm^2 C. 15,750.2 cm^2
   B. 10,781.9 cm^2 D. 21,553.0 cm^2

5. ________

6. What is the surface area of a cone with a radius of 6 meters and a slant height of 8 meters? Round to the nearest tenth.
   F. 263.9 m^2 H. 904.0 m^2
   G. 452.0 m^2 I. 1,055.2 m^2

6. ________

Test, Form 1B *(continued)*

SCORE ________

7. A cylindrical tea cup has a height of 7 centimeters, and its base has a radius of 3 centimeters. What is the volume of the tea cup to the nearest tenth?

A. 28.3 cm^3 **C.** 197.9 cm^3
B. 66.0 cm^3 **D.** 395.8 cm^3

7. ________

8. The storage tank shown at right is to be painted. What is the surface area to be painted to the nearest tenth? Assume that the bottom does not need painting.

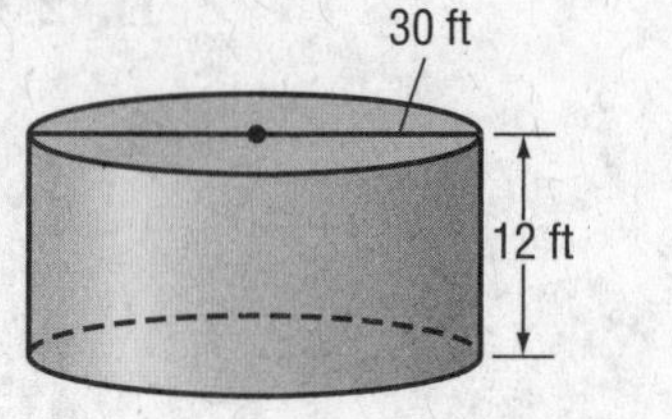

F. 1,130.4 ft^2 **H.** 2,544.7 ft^2
G. 1,837.8 ft^2 **I.** 7,771.5 ft^2

8. ________

9. The popcorn containers at a movie theater are in the shape of cones. Suppose a popcorn container has a radius of 6 inches and a slant height of 14 inches. What is the lateral area of the popcorn container rounded to the nearest inch?

A. 377 in^2 **C.** 264 in^2
B. 283 in^2 **D.** 113 in^2

9. ________

10. What is the volume of the composite shape to the nearest whole number?

F. 619 in^3
G. 366 in^3
H. 336 in^3
I. 283 in^3

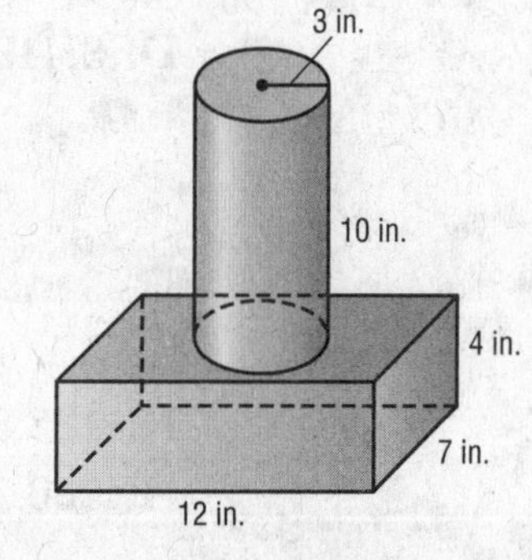

10. ________

11. The volume of a triangular prism is 42 cubic centimeters. What is the volume of a similar prism that is twice as large as the first prism?

A. 21 cm^3 **C.** 168 cm^3
B. 84 cm^3 **D.** 336 cm^3

11. ________

12. The surface area of a cone is 54 square inches. What is the surface area of a similar cone that is three times as large?

F. 1,458 in^2 **H.** 162 in^2
G. 486 in^2 **I.** 18 in^2

12. ________

NAME _______________ DATE _______________ PERIOD _______________

Test, Form 2A

SCORE _______________

Write the letter for the correct answer in the blank at the right of each question.

For Exercises 1–4, find the volume of each solid. Round to the nearest tenth if necessary.

1.

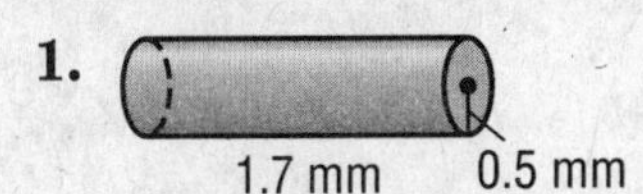

A. 6.9 mm^3 **C.** 1.3 mm^3

B. 6.1 mm^3 **D.** 0.4 mm^3

1. _______________

2. cylinder: diameter = 4 mm
height = 8 mm

F. 131.1 mm^3 **H.** 100.5 mm^3

G. 116.9 mm^3 **I.** 230.0 mm^3

2. _______________

3. **A.** 261.8 yd^3 **C.** 523.6 yd^3

B. 314.2 yd^3 **D.** 1,570.8 yd^3

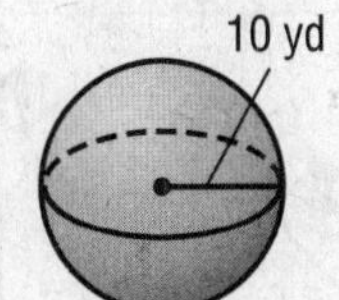

3. _______________

4. cone: diameter = 5.5 mm
height = 11.2 mm

F. 288.6 mm^3 **H.** 354.8 mm^3

G. 338.8 mm^3 **I.** 1,063.8 mm^3

4. _______________

For Exercises 5 and 6, find the surface area of each solid. Round to the nearest tenth if necessary.

5. **A.** 157.1 m^2 **C.** 383.3 m^2

B. 226.2 m^2 **D.** 565.4 m^2

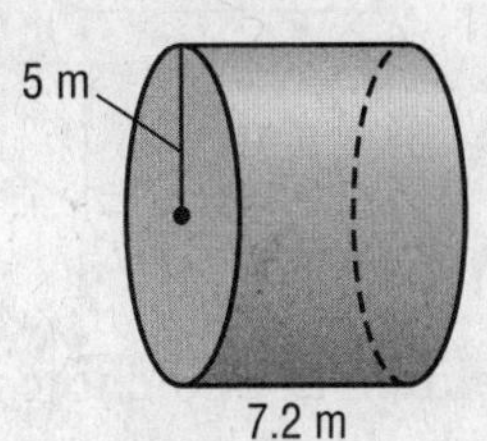

5. _______________

6. **F.** 1,230.9 in^2 **H.** 2,813.5 in^2

G. 2,198 in^2 **I.** 3,501.0 in^2

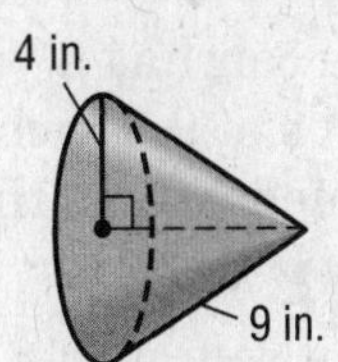

6. _______________

NAME ________________ DATE ____________ PERIOD ________

Test, Form 2A *(continued)*

SCORE ________

7. A kitchen sink has a volume of 1,800 cubic inches. What is the volume of a similar sink that is larger by a scale factor of 2?

A. 14,400 in^3 **C.** 7,200 in^3

B. 10,800 in^3 **D.** 3,600 in^3

7. ________

8. The surface area of a prism is 54 square inches. What is the surface area of a similar prism that is smaller by a scale $\frac{1}{3}$?

F. 2 in^2 **H.** 9 in^2

G. 6 in^2 **I.** 18 in^2

8. ________

9. Find the surface area of the composite shape. Round to the nearest tenth.

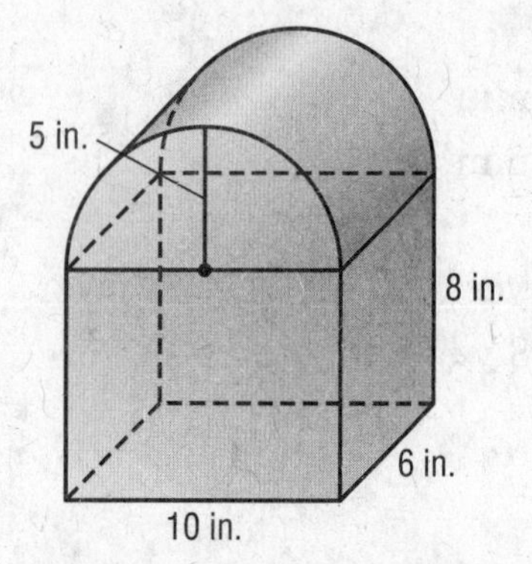

9. ________

10. A birdhouse is in the shape of the figure below. Suppose you wanted to fill the figure with birdseed. What is the volume of the figure below? Use the *solve a simpler problem* strategy.

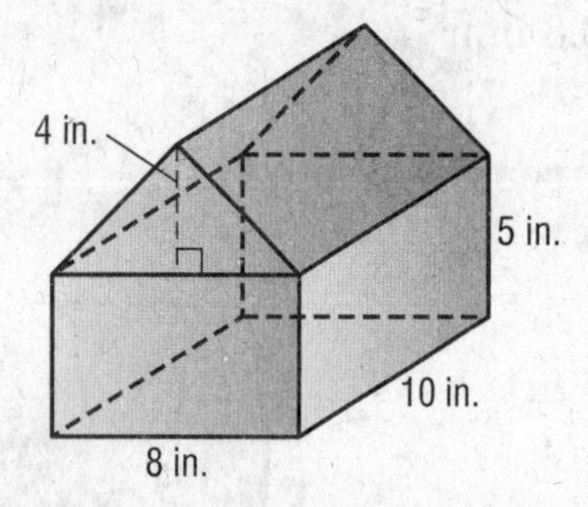

10. ________

11. A cone has a volume of 350 cubic meters. The area of the base is 70 square meters. What is the height of the cone?

11. ________

12. A cylindrical waste can has a volume of 5,667.7 cubic inches and its base has a radius of 9.5 inches. Find the height of the waste can. Round to the nearest tenth.

12. ________

13. A hemisphere has a radius of 8 centimeters. Find the volume of the hemisphere. Round to the nearest tenth.

13. ________

NAME ______________________ DATE ____________ PERIOD ________

Test, Form 2B

SCORE ________

Write the letter for the correct answer in the blank at the right of each question.

For Exercises 1–4, find the volume of each solid. Round to the nearest tenth if necessary.

1. A. 804.2 mm^3 C. 201.1 mm^3
   B. 268.1 mm^3 D. 134.0 mm^3

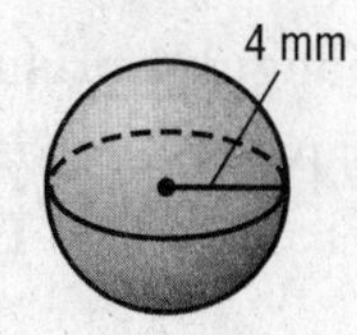

1. ________

2. cylinder: diameter = 6 mm
   height = 18 mm
   F. 169.6 mm^3 H. 508.9 mm^3
   G. 211.9 mm^3 I. 791.3 mm^3

2. ________

3. A. 84.8 cm^3 C. 269.6 cm^3
   B. 254.5 cm^3 D. 508.7 cm^3

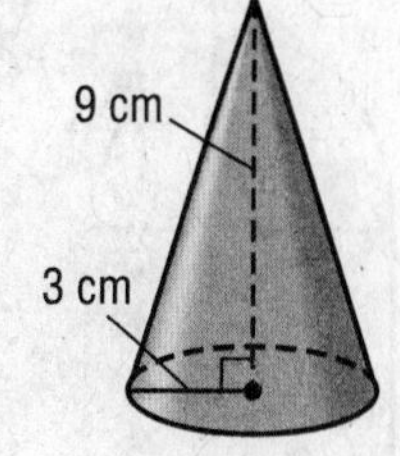

3. ________

4. cone: diameter = 5.5 in.
   height = 11.2 in.
   F. 376.9 in^3 H. 1,005.3 in^3
   G. 377.0 in^3 I. 3,015.9 in^3

4. ________

For Exercises 5 and 6, find the surface area of each solid. Round to the nearest tenth if necessary.

5. A. 353.4 m^2 C. 1,498.5 m^2
   B. 1,145.1 m^2 D. 4,294.2 m^2

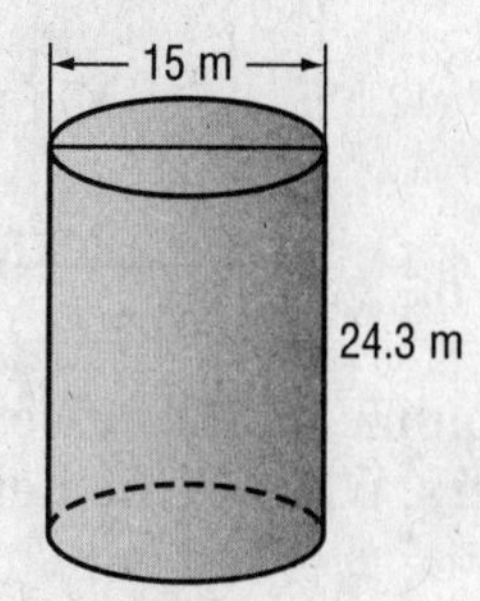

5. ________

6. F. 131.9 in^2 H. 301.4 in^2
   G. 263.8 in^2 I. 504 in^2

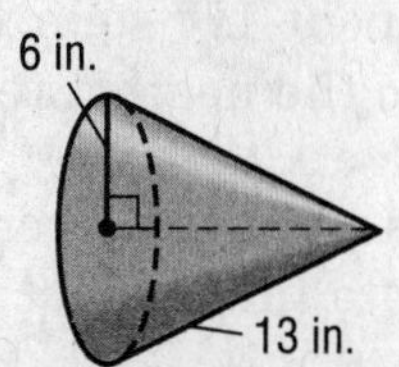

6. ________

Test, Form 2B *(continued)*

SCORE ________

7. A roof has a surface area of 4.8 square feet. What is the surface area of a similar roof that is larger by a scale factor of 3?

A. 1.6 ft^2 **C.** 43.2 ft^2
B. 14.4 ft^2 **D.** 129.6 ft^2

7. ________

8. The volume of a triangular prism is 84 cubic inches. What is the volume of a similar prism that is smaller by a scale $\frac{1}{2}$?

F. 10.5 in^3 **H.** 42 in^3
G. 21 in^3 **I.** 168 in^3

8. ________

9. Find the surface area of the composite shape. Round to the nearest tenth.

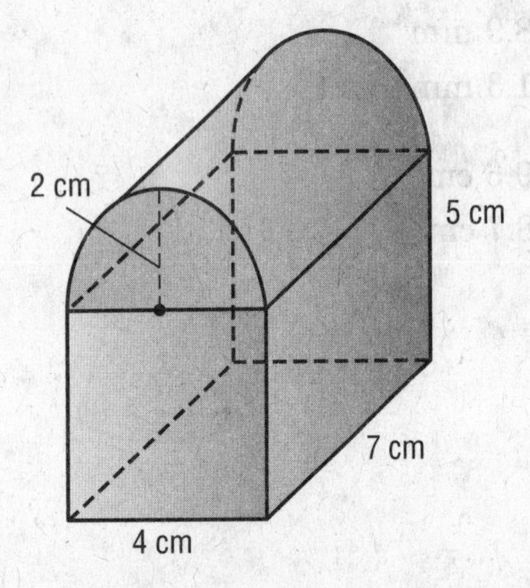

9. ________

10. Mrs. Powell is making a piñata for her son's birthday party. The piñata is shaped like the figure below. She wants to fill it with candy. What is the volume of the piñata? Use the *solve a simpler problem* strategy.

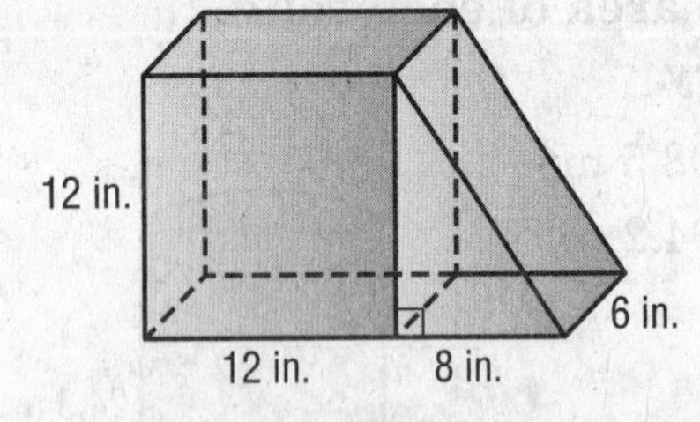

10. ________

11. A cone has a volume of 810 cubic meters. The area of the base is 90 square meters. What is the height of the cone?

11. 

12. A cylindrical waste can has a volume of 2,474 cubic inches and the area of its base is about 177 square inches. Find the height of the waste can. Round to the nearest tenth.

12. 

13. A hemisphere has a radius of 9 centimeters. Find the volume of the hemisphere. Round to the nearest tenth.

13. ________

NAME ______________________ DATE ____________ PERIOD ________

Test, Form 3A

SCORE ________

For Exercises 1–4, find the volume of each solid. Round to the nearest tenth if necessary.

1.

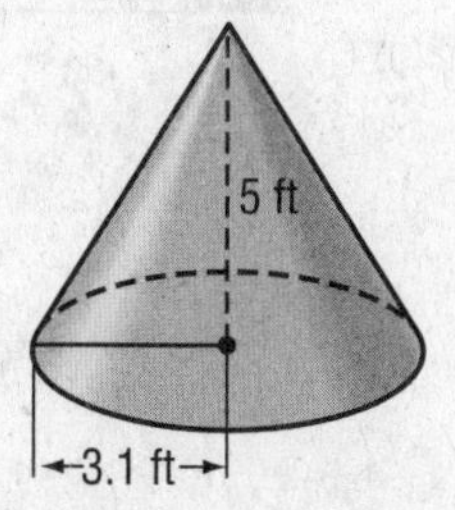

2.

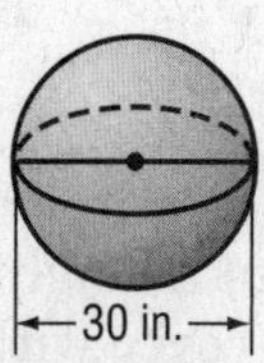

3. hemisphere: radius = 20 m

4. cylinder: radius = 3 m
height = 5.7 m

For Exercises 5–8, find the surface area of each solid. Round to the nearest tenth if necessary.

5.

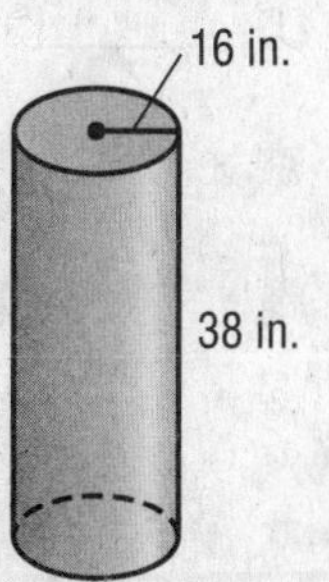

6.

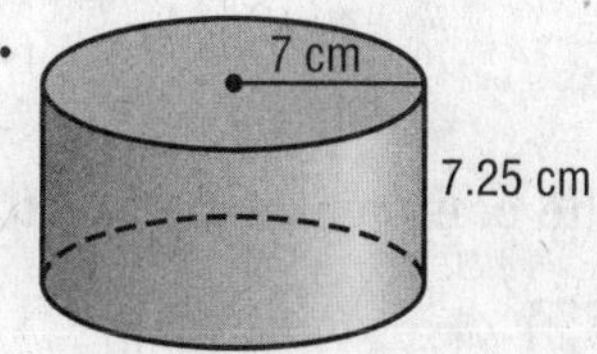

7. cone: diameter = 12 m
slant height = 8.4 m

8. cylinder: diameter = 7 cm
height = 5.1 cm

9. A beachball has a diameter of 12 inches. What is the volume to the nearest tenth?

10. The popcorn containers at a movie theater are in the shape of cones. Suppose a popcorn container has a diameter of 12 inches and a slant height of 20 inches. What is the lateral area of the popcorn container rounded to the nearest inch? Justify your answer.

11. The lateral area of a cone with a diameter of 10 centimeters is about 250.5 square centimeters. To the nearest tenth, what is the slant height of the cone?

1. ________
2. ________
3. ________
4. ________
5. ________
6. ________
7. ________
8. ________
9. ________
10. ________
11. ________

Test, Form 3A *(continued)*

SCORE ________

12. A container in the shape of a cone has a volume of 40 cubic units. Its base has an area of 15 square units. What is the height of the container?

12. ________

13. Marcos is buying paint to cover 10 cylindrical-shaped tables. Each table has a diameter of 2 feet and a height of 3 feet. How many square feet does Marcos need to cover with paint?

13. ________

14. Julie is making 8 cone-shaped party hats for her sister's birthday party from cardboard. Each party hat has a radius of 5 inches and a slant height of 6 inches. How much cardboard does Julie need? Round to the nearest tenth.

14. ________

15. Working separately, two bakers can make two wedding cakes in eight hours. Working at the same rate, how many wedding cakes can three bakers make in forty hours? Use the *solve a simpler problem* strategy.

15. ________

16. Find the volume of the composite shape. Round to the nearest tenth.

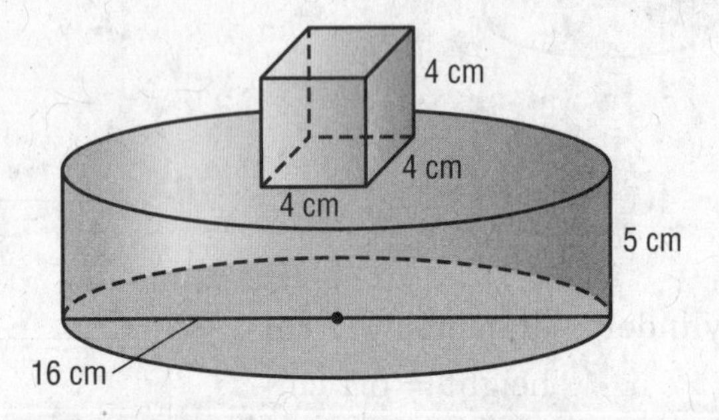

16. ________

17. The surface area of a pyramid is 327 square meters. What is the surface area of a similar pyramid that is smaller by a scale factor of $\frac{2}{3}$? Round to the nearest hundredth if necessary.

17. ________

18. Solid A is similar to Solid B. Solid B has a volume of 23,000 cubic meters. By what scale factor can you multiply every side of Solid A to get Solid B if the volume of Solid A is 23 cubic meters?

18. ________

19. A cylinder has a volume of 26 cubic inches. If all the dimensions are multiplied by 3.2, what would be the volume of the new cylinder? Round to the nearest hundredth if necessary?

19. ________

NAME ______________________ DATE ____________ PERIOD ________

Test, Form 3B

SCORE ________

For Exercises 1–4, find the volume of each solid. Round to the nearest tenth if necessary.

1.

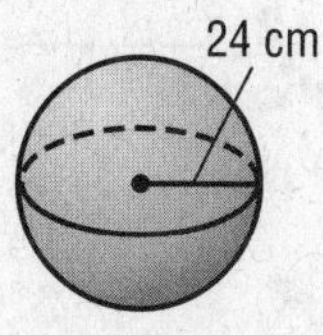

2.

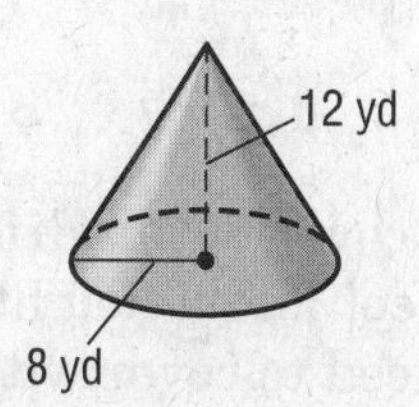

1. ________

2. ________

3. hemisphere: radius = 30 in.

4. cylinder: radius = 1.6 m
height = 1.3 m

3. ________

4. ________

For Exercises 5–8, find the surface area of each solid. Round to the nearest tenth if necessary.

5.

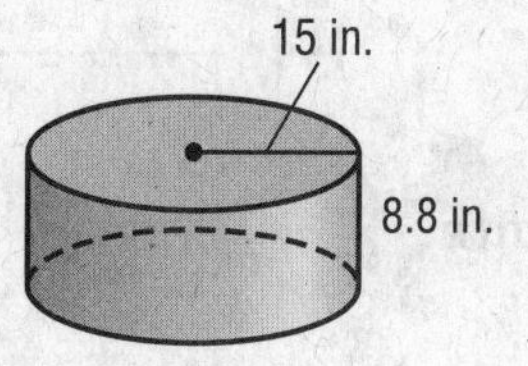

6.

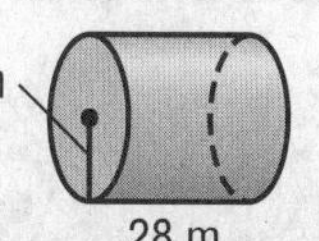

5. ________

6. ________

7. cone: diameter = 14 m
slant height = 8.4 m

8. cylinder: diameter = 7 cm
height = 5.1 cm

7. ________

8. ________

9. A beachball has a diameter of 14 inches. What is its volume to the nearest tenth?

9. ________

10. The popcorn containers at a movie theater are in the shape of cones. Suppose a popcorn container has a diameter of 10 inches and a slant height of 18 inches. What is the lateral area of the popcorn container rounded to the nearest inch? Justify your answer.

10. ________

11. The lateral area of a cone with a diameter of 15 yards is about 287.5 square yards. To the nearest tenth, what is the slant height of the cone?

11. ________

Test, Form 3B *(continued)*

SCORE ________

12. A container in the shape of a cone has a volume of 50 cubic units. Its base has an area of 20 square units. What is the height of the container? **12.** ________

13. Juanita is buying paint to cover 8 cylindrical-shaped tables. Each table has a diameter of 3 feet and a height of 2 feet. How many square feet does Juanita need to cover with paint? **13.** ________

14. Robert is making 5 cone-shaped party hats for his sister's birthday party from cardboard. Each party hat has a radius of 4 inches and a slant height of 7 inches. How much cardboard does Robert need? Round to the nearest tenth. **14.** ________

15. Working separately, six florists can make six floral arrangements in one half hour. Working at the same rate, how many flower arrangements can six florists make in 4 hours? Use the *solve a simpler problem* strategy. **15.** ________

16. Find the volume of the composite shape. Round to the nearest tenth.

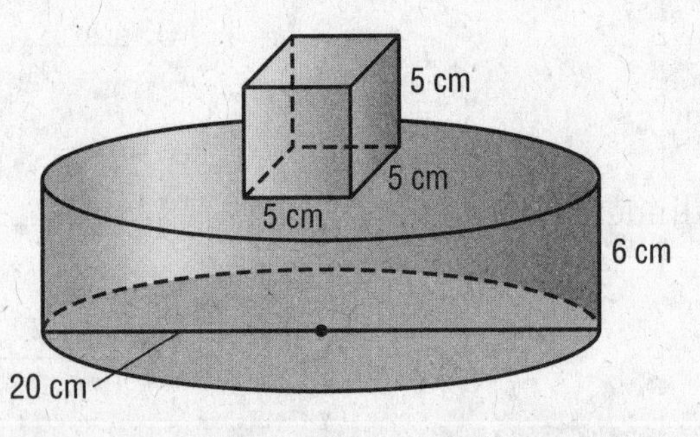

16. ________

17. The surface area of a cone is 830 square centimeters. What is the surface area of a similar cone that is smaller by a scale factor of $\frac{2}{5}$? **17.** ________

18. Solid A is similar to Solid B. Solid B has a volume of 65 cubic meters. By what scale factor can you multiply every side of Solid A to get Solid B the volume of Solid A is 65,000 cubic meters? **18.** ________

19. A cylinder has a volume of 46 cubic inches. If all the dimensions are multiplied by 2.5, what would be the volume of the new cylinder? **19.** ________

NAME ______________________ DATE ____________ PERIOD ________

Are You Ready?

Review

To find the average of a set of data, find the sum of the numbers. Then divide the sum by how many numbers are in the set. The resulting number is the average of the data in the set.

Example 1

Find the average for the data set.

17, 21, 13, 9, 15

To find the average, find the sum of the numbers. Then divide the sum by how many numbers are in the set.

$$\frac{17 + 21 + 13 + 9 + 15}{5} = \frac{75}{5}$$
$$= 15$$

Example 2

FISHING **Paula entered the local 4-day fishing tournament. Her catch for each of the days was 14, 6, 11, and 9 bass. What was the average number of bass caught per day?**

$$\frac{14 + 6 + 11 + 9}{4} = \frac{40}{4}$$
$$= 10$$

Exercises

Find the average for each data set. Round to the nearest tenth if necessary.

1. 26, 16, 11, 13, 24 — **1.** ____________

2. 51, 20, 32, 18, 45, 8 — **2.** ____________

3. 17, 39, 15, 42, 37, 61, 19, 40 — **3.** ____________

4. 62, 21, 18 — **4.** ____________

5. GYMNASTICS Clarence competes in the pommel horse event. His scores in the last four meets were 9.2, 8.6, 9.4, and 9.2. What is his average score? — **5.** ____________

NAME ______________________ DATE ____________ PERIOD ________

Are You Ready?

Practice

1. **SCHOOL** The results of a class survey about the number of miles students live from school is shown in the histogram. Describe the histogram. Then find the number of students who live less than 11 miles from school.

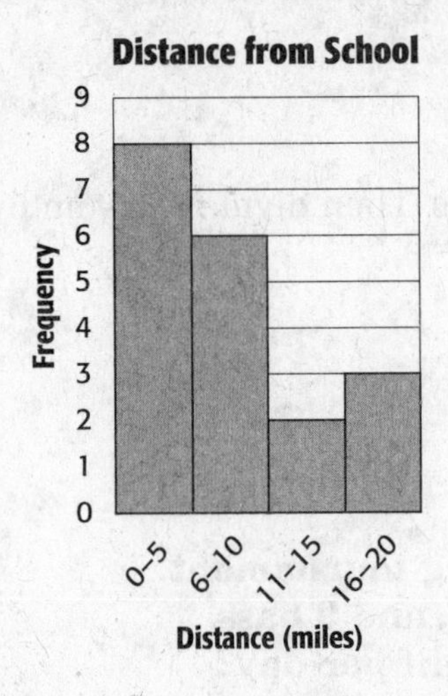

1. ______________

2. **FOOTBALL** The winning scores of recent Super Bowls games are shown in the histogram. Describe the histogram. Then find the number of games that had more than 40 points scored in it.

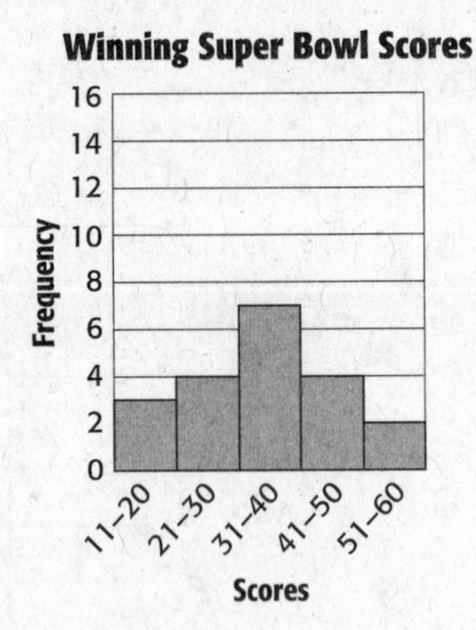

2. ______________

Find the average for each data set. Round to the nearest tenth if necessary.

3. 15, 10, 5, 8, 12, 4, 6, 4 — 3. ______________

4. 12, 29, 19, 22, 18, 8 — 4. ______________

5. 14, 27, 13, 21, 18, 32, 46, 59 — 5. ______________

6. 81, 19, 34, 17, 24, 32, 52, 17 — 6. ______________

7. **BOWLING** Jamal's bowling scores at this spring's bowl-a-thon were 113, 125, 155, and 119. What was his average score? — 7. ______________

NAME ______________________ DATE ____________ PERIOD ________

Are You Ready?

Apply

1. GOLF Sky is a member of her school's golf team. The table shows her scores from each of her matches. What was her average golf score? Round to the nearest tenth if necessary.

Sky's Golf Scores		
72	69	72
71	71	69
68	72	68

2. AUTUMN Karly owns a lawn service. During the month of October she averaged 42 customers the first week, 38 the second week, 55 the third week, and 65 the fourth week. How many customers did she average per week?

3. SALES The number of bags of popcorn sold at a certain movie theater for several days was 27, 17, 24, 38, 47, and 21. Find the average number of bags of popcorn sold during these days.

4. CUSTOMERS Big Mart keeps track of the number of customers entering the store in the first hour of each day. In the first four days of April, Big Mart had 215, 125, 118, and 214 customers in the first hour. Find the average number of customers Big Mart had in the first hour during those four days.

5. HEIGHT The histogram below shows the heights of students in a classroom. Describe the histogram. Then find the number of students who are more than 59 inches tall.

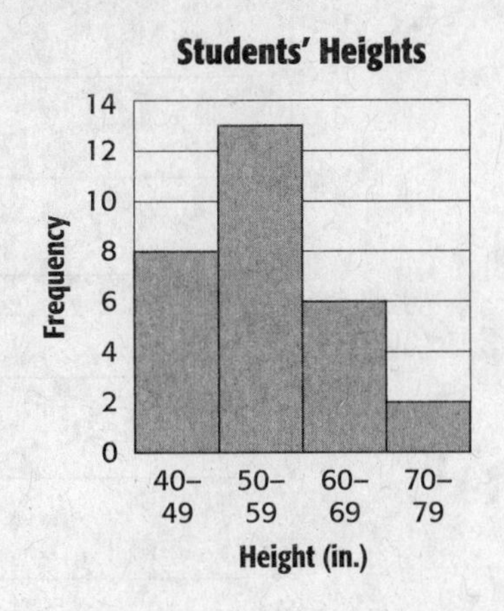

6. CHOIR Timmy belongs to the choir in his hometown. The histogram below shows the ages of the choir members. Describe the histogram. Then find the number of members who are less than 31 years old.

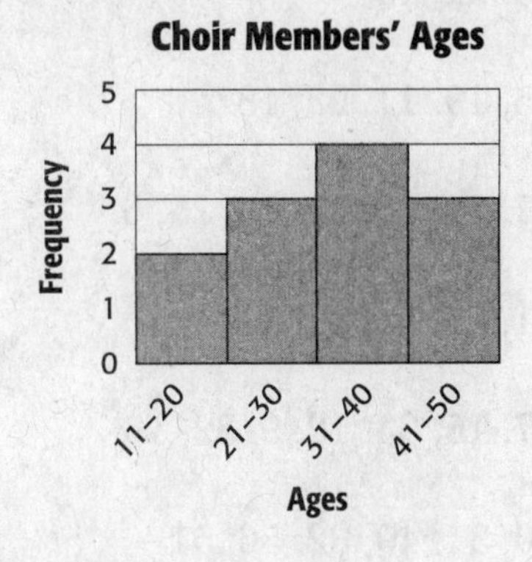

NAME ______________________ DATE ____________ PERIOD ________

Diagnostic Test

1. **ZOOS** The attendance at major U.S. zoos in a recent year is shown in the histogram. Describe the histogram. Then find the number of zoos who had an attendance of more than 1.9 million.

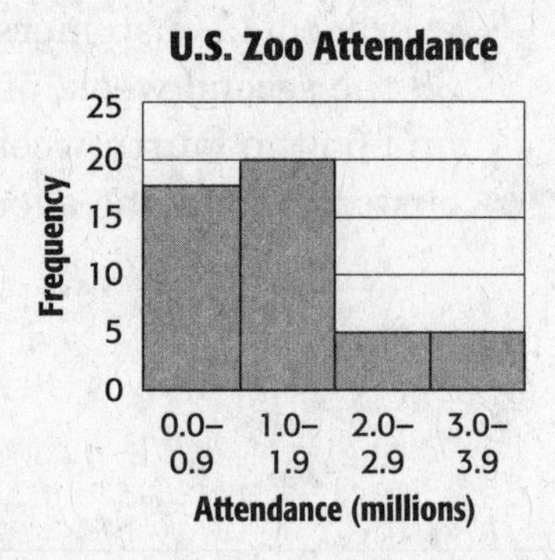

1. ____________

2. **BASKETBALL** The number of points a basketball team scored in a recent season is shown in the histogram. Describe the histogram. Then find the number of times the team scored more than 40 points.

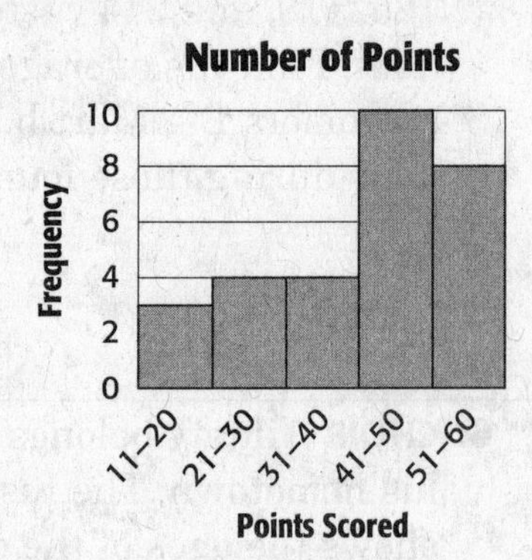

2. ____________

Find the average for each data set. Round to the nearest tenth if necessary.

3. 29, 45, 31, 15 — 3. ____________

4. 82, 75, 19, 11, 23, 18 — 4. ____________

5. 14, 27, 11, 19, 31, 19, 25, 6 — 5. ____________

6. 52, 31, 23, 12, 8, 14 — 6. ____________

7. 88, 27, 15, 63, 12, 9, 32, 15 — 7. ____________

8. 34, 40, 31, 49, 32, 56, 51 — 8. ____________

9. **TOURISTS** Clarine owns a sight-seeing boat in Florida. In each of the last five weeks she had 42, 59, 67, 51, and 91 customers. How many customers did she average per week? — 9. ____________

NAME ______________________ DATE ____________ PERIOD ________

Pretest

Find the mean, median, and mode for the given data set.

1. 62, 23, 14, 17, 62, 29, 31

1. ____________

2. ULTRA-MARATHON Dillon is training for the ultra-marathon. In the past five weeks he ran 21, 21, 25, 27, and 35 miles. What is the range for the number of miles he ran last week?

2. ____________

Find the five-number summary for the given data set.

3. 42, 13, 15, 29, 18, 71

3. ____________

Construct a box plot for the given data set.

4. 54, 62, 41, 37, 47, 63, 45, 39, 49

4. ____________

5. BASKETBALL The table below shows the number of points a team scored in their games. Find the five-number summary for the data. Then construct a box plot of the data.

Basketball Points	
24	28
27	35
44	38
46	

5. ____________

Construct a scatter plot of the data. Then draw a line that best seems to represent the data.

6.

x	1	3	4	5	6
y	1	2	4	6	5

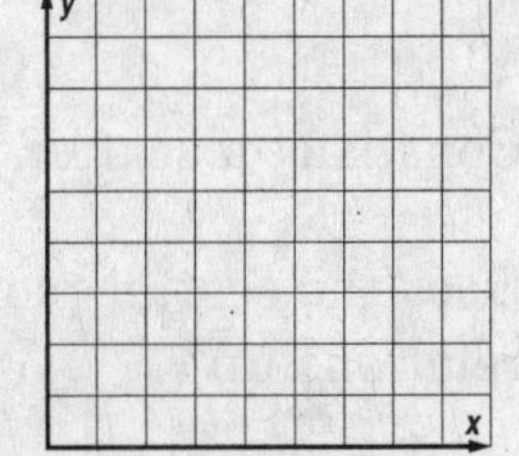

6.

NAME _______________ DATE _______________ PERIOD _______________

Chapter Quiz

1. **SURVEY** A survey was taken of local residents to determine their favorite sport to watch live. The survey results are shown on the graph. How many respondents preferred to watch either football or soccer?

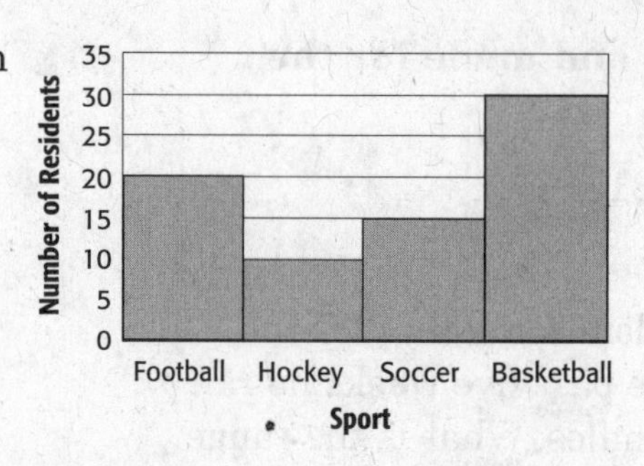

1. _______________

2. **WEATHER** Average temperatures for various months are given in the table below. Construct a scatter plot of the data. Draw a line that best seems to fit the data.

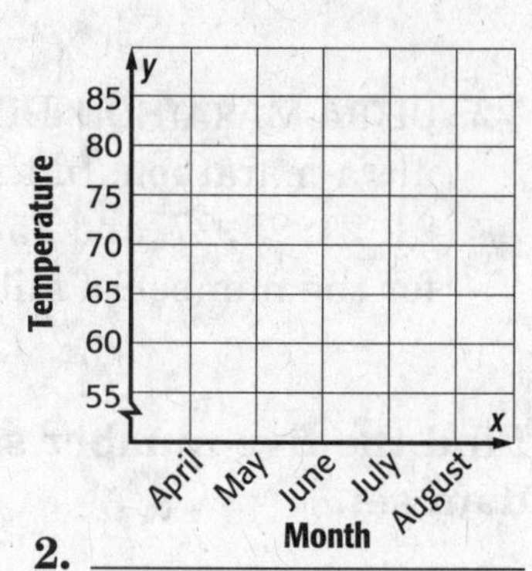

2. _______________

Month	1(April)	2(May)	3(June)	4(July)	5(Aug)
Average Temperature	55°	65°	72°	78°	81°

3. Write an equation for the line of fit that you have drawn.

3. _______________

4. Use the equation to make a conjecture about the average temperature in September.

4. _______________

For Exercises 5 and 6, use the following table.

Homework Grade (%)	10	92	80	32	60	40	15	50
Test Grade (%)	55	89	69	77	60	45	32	70

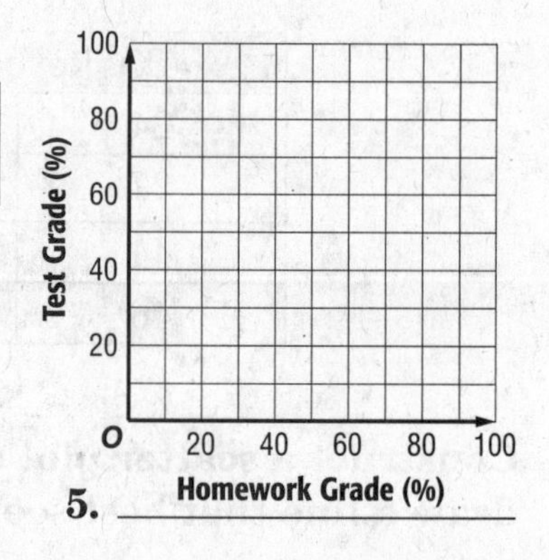

5. Construct a scatter plot of the data.

5. _______________

6. Does the scatter plot show a *positive, negative*, or *no* relationship?

6. _______________

7. Use the two-way table to determine the relative frequency of students who play a sport and like football to the total number of students plays a sport. Round to the nearest hundredth.

7. _______________

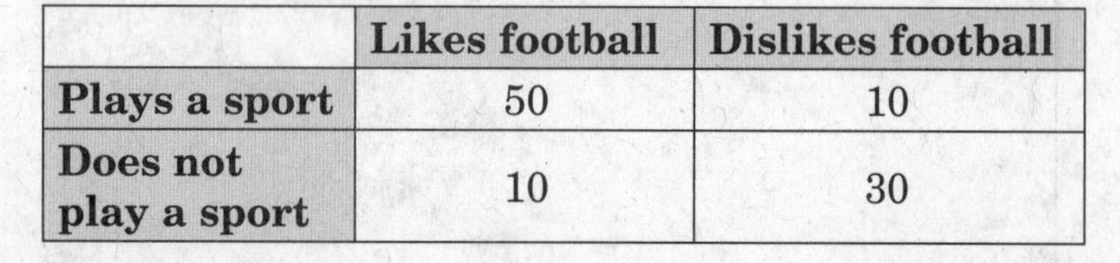

	Likes football	Dislikes football
Plays a sport	50	10
Does not play a sport	10	30

NAME ________________________ DATE ______________ PERIOD ________

Vocabulary Test

SCORE ________

bivariate data	relative frequency
distribution	scatter plot
five-number summary	standard deviation
line of best fit	symmetric
mean absolute deviation	two-way table
quantitative data	univariate data

Fill in the blank using a term found in the vocabulary list.

1. The ______________ of a data set is the average distance between each data value and the mean. **1.** ______________

2. A ______________ is a graph that shows the relationship, if any, between two sets of data. **2.** ______________

3. The ratio of the value of a subtotal to the value of the total is called ______________. **3.** ______________

4. The ______________ of a data set shows how the data deviates from the mean. **4.** ______________

5. A ______________ shows data that pertains to two categories. **5.** ______________

6. The ______________ of a data set shows the arrangement of data values. **6.** ______________

7. A ______________ is a line that is very close to most of the data points. **7.** ______________

Define each of the following in your own words.

8. symmetric **8.** ______________

9. quantitative data **9.** ______________

NAME ______________________ DATE ____________ PERIOD ________

Standardized Test Practice

Read each question. Then fill in the correct answer on the answer document provided by your teacher or on a sheet of paper.

1. The scatter plot below shows the cost of computer repairs in relation to the number of hours the repair takes.

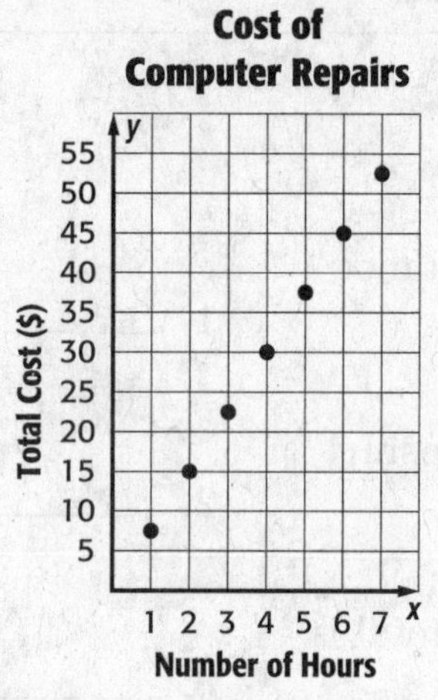

Based on the information in the scatter plot, which statement is a valid conclusion?

A. As the length of time increases, the cost of the repair increases.

B. As the length of time increases, the cost of the repair stays the same.

C. As the length of time decreases, the cost of the repair increases.

D. As the length of time increases, the cost of the repair decreases.

2. A store had daily sales of $15,696, $23,400, $19,080, $18,000, $23,400, $17,604, and $15,228 last week. Which data measure would make the sales last week appear the most profitable?

F. mean

G. median

H. mode

I. range

3. **GRIDDED RESPONSE** Erin jogged along the track around the outer edge of a park. She ran two miles along the one edge and then 3 miles along the other edge. She then cut across the park as shown by the dotted line. To the nearest tenth, how many miles was Erin's shortcut across the park?

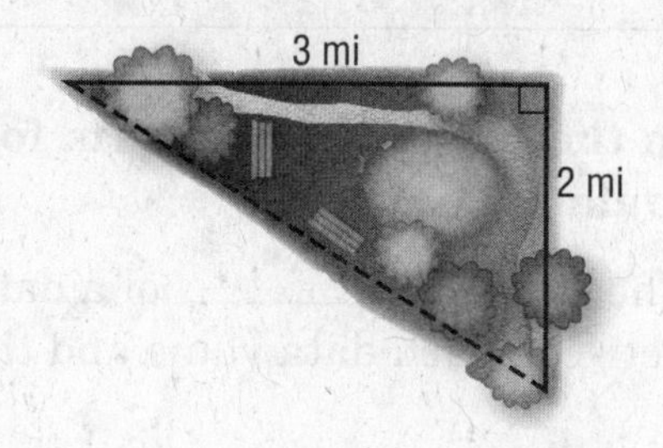

4. Rakim's French test scores are shown below.

86, 84, 80, 65, 90, 75, 88

Which measure of center would change the most if his lowest quiz score was dropped?

A. mean

B. median

C. mode

D. none of the above

5. **SHORT RESPONSE** Laurie wants to buy a new sweater that costs $45. The sweater is on sale for 25% off and 6.75% sales tax will be applied to the purchase. What will be the total cost of the sweater?

6. The moon is about 3.84×10^5 kilometers from Earth. Which of the following represents this number in standard notation?

F. 38,400,000 kilometers

G. 3,840,000 kilometers

H. 384,000 kilometers

I. 38,400 kilometers

7. An ice cream store surveyed 100 of its customers about their favorite flavor. The results are shown in the table.

Flavor	Frequency
Chocolate chip	40
Vanilla	15
Cookie dough	20
Chocolate	15
Other	10

If the store uses only these data to order ice cream, what conclusion can be drawn from the data?

A. More than half of each order should be chocolate chip and cookie dough ice cream.

B. Half of the order should be vanilla and chocolate ice cream.

C. Only chocolate, cookie dough, and vanilla ice cream should be ordered.

D. About one third of the order should be vanilla and chocolate chip ice cream.

8. **SHORT RESPONSE** The graph of rectangle $LMNP$ is shown below.

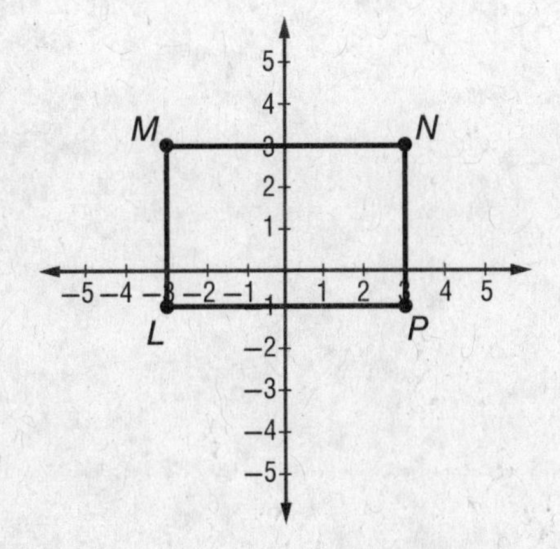

What is the area of rectangle $LMNP$?

9. Alisha's average math test score was 82. Which of the following students has the same average math test score as Alisha?

F. Jenny earned 492 points on 6 tests.

G. Frankie earned 352 points on 4 tests.

H. Benicio earned 468 points on 6 tests.

I. Dontonio earned 344 points on 4 tests.

10. **SHORT RESPONSE** A cone has an approximate volume of 83.8 cubic inches and a radius of 4 inches. What is the height of the cone? Round the nearest tenth.

11. **EXTENDED RESPONSE** The table shows the amount of time different students studied for an exam and the scores they received.

Student	Study Time	Test Score
Patrick	30 min	75
LaDonne	50 min	89
Marlena	1 hr 10 min	93
Jason	25 min	72
Joaquim	1 hour	91
Carla	45 min	83
Heather	1 hr 15 min	90

Part A Why is a scatter plot a good representation of the data?

Part B Graph the data. Do the data represent a *positive, negative,* or *no* relationship?

Part C Draw a line that seems to best represent the data. Then write an equation in slope-intercept form for the line of best fit.

Part D Use the equation to make a conjecture about the test score of a student that studied for 1 hour 30 minutes.

Student Recording Sheet

SCORE ________

Use this recording sheet with the Standardized Test Practice.

Fill in the correct answer. For gridded-response questions, write your answers in the boxes on the answer grid and fill in the bubbles to match your answers.

1. Ⓐ Ⓑ Ⓒ Ⓓ

2. Ⓕ Ⓖ Ⓗ Ⓘ

3.

4. Ⓐ Ⓑ Ⓒ Ⓓ

5. ______________

6. Ⓕ Ⓖ Ⓗ Ⓘ

7. Ⓐ Ⓑ Ⓒ Ⓓ

8. ______________

9. Ⓕ Ⓖ Ⓗ Ⓘ

10. ______________

Extended Response

Record your answers for Exercise 11 on the back of this paper.

NAME ______________________ DATE ______________ PERIOD ________

Extended-Response Test

SCORE ________

Demonstrate your knowledge by giving a clear, concise solution to each problem. Be sure to include all relevant drawings and justify your answers. You may show your solution in more than one way or investigate beyond the requirements of the problem. If necessary, record your answers on another piece of paper.

1. Zander has kept track of his family's gasoline usage for the past 10 weeks as shown in the table.

Week	1	2	3	4	5	6	7	8	9	10
Gallons	30.2	28.9	29.5	31.7	29.8	30.5	41.6	32.1	31.5	33.0

a. Explain how to find the mean of the data. Then find the mean.

b. Explain how to find the median of the data. Then find the median.

c. Explain why the data has no mode.

d. Explain how to find the range of the data. Then find the range.

e. Find the five-number summary of the data.

f. Explain how to identify outliers in a set of data. Then identify any outliers in the data above.

g. Draw a box plot for the data.

2. Antonio asked 50 people to donate to a charity. The table shows the number of people who donated each amount.

Donation Amount ($)	0	5	10	15	20	25
Number of People	13	8	11	7	7	4

a. Draw a scatter plot of the data.

b. Does the scatter plot show a *positive*, *negative*, or *no* relationship. Explain.

c. Draw a line of fit through the points (0, 13) and (25, 4) on your scatter plot in part **a**. Explain what is meant by a line of fit.

d. Use the slope formula to find the slope of the line of fit that you drew in part **c**. Show your work.

e. What is the y-intercept of the line of fit that you drew in part **c**? How do you know?

NAME ______________________ DATE ____________ PERIOD ________

Extended-Response Rubric

SCORE ________

Score	Description
4	A score of four is a response in which the student demonstrates a thorough understanding of the mathematics concepts and/or procedures embodied in the task. The student has responded correctly to the task, used mathematically sound procedures, and provided clear and complete explanations and interpretations. The response may contain minor flaws that do not detract from the demonstration of a thorough understanding.
3	A score of three is a response in which the student demonstrates an understanding of the mathematics concepts and/or procedures embodied in the task. The student's response to the task is essentially correct with the mathematical procedures used and the explanations and interpretations provided demonstrating an essential but less than thorough understanding. The response may contain minor flaws that reflect inattentive execution of mathematical procedures or indications of some misunderstanding of the underlying mathematics concepts and/or procedures.
2	A score of two indicates that the student has demonstrated only a partial understanding of the mathematics concepts and/or procedures embodied in the task. Although the student may have used the correct approach to obtaining a solution or may have provided a correct solution, the student's work lacks an essential understanding of the underlying mathematical concepts. The response contains errors related to misunderstanding important aspects of the task, misuse of mathematical procedures, or faulty interpretations of results.
1	A score of one indicates that the student has demonstrated a very limited understanding of the mathematics concepts and/or procedures embodied in the task. The student's response is incomplete and exhibits many flaws. Although the student's response has addressed some of the conditions of the task, the student reached an inadequate conclusion and/or provided reasoning that was faulty or incomplete. The response exhibits many flaws or may be incomplete.
0	A score of zero indicates that the student has provided no response at all, or a completely incorrect or uninterpretable response, or demonstrated insufficient understanding of the mathematics concepts and/or procedures embodied in the task. For example, a student may provide some work that is mathematically correct, but the work does not demonstrate even a rudimentary understanding of the primary focus of the task.

NAME ______________________ DATE ____________ PERIOD ________

Test, Form 1A

SCORE ________

Write the letter for the correct answer in the blank at the right of each question.

For Exercises 1–3, use the scatter plot shown at the right.

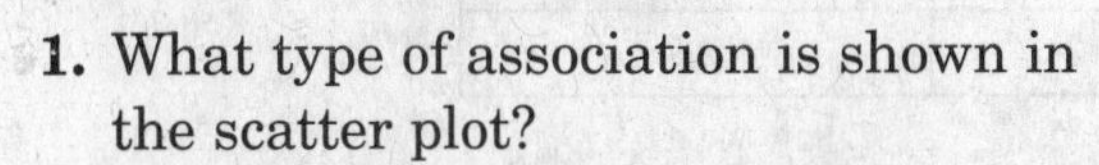

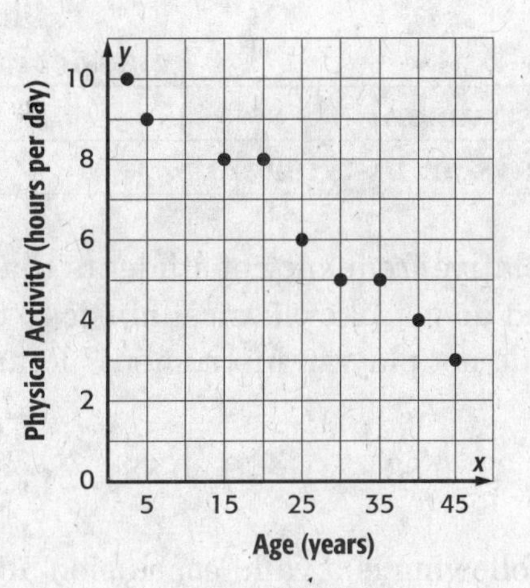

1. What type of association is shown in the scatter plot?

 A. a negative linear association
 B. a positive linear association
 C. a positive nonlinear association
 D. no association

1. ________

2. Which of the following statements is best supported by the scatter plot?

 F. As the age of a person increases, the time spent in physical activity increases.
 G. As the time spent in physical activity increases, a person's age increases.
 H. As the age of a person increases, the time spent in physical activity decreases.
 I. There is no relationship between the age of a person and his or her amount of physical activity.

2. ________

3. Which of the following is a reasonable estimate for the daily amount of physical activity for a person who is 50 years old?

 A. 6 h **C.** 4 h
 B. 5 h **D.** 2 h

3. ________

For Exercises 4 and 5, use the scatter plot shown at the right. The scatter plot shows the length of a metal spring when weights are attached.

4. Which of the following is the most reasonable equation for the line of best fit?

 F. $y = 5x + 9$ **H.** $y = -1.5x + 9$
 G. $y = -5x + 9$ **I.** $y = 1.5x + 9$

4. ________

5. Which of the following is the most reasonable estimate for the length of a spring when 20 weights are attached?

 A. 25 in. **C.** 54 in.
 B. 39 in. **D.** 62 in.

5. ________

Test, Form 1A *(continued)*

SCORE ________

For Exercises 6 and 7, use the two-way table shown below.

	Likes classical music	Dislikes classical music
Plays an instrument	15	2
Does not plays an instrument	3	25

6. What is the relative frequency of students that do not play an instrument and do not like classical music to the total number of students who do not play an instrument? Round to the nearest hundredth.

F. 0.10 **G.** 0.83 **H.** 0.88 **I.** 0.89

6. ________

7. Which of the following is a valid conclusion about the data?

A. Of the students that like classical music, most do not play a musical instrument.

B. Of the students that like classical music, most play a musical instrument.

C. There were a total of 27 students surveyed.

D. Most of the students surveyed like classical music.

7. ________

For Exercises 8 and 9, use the following data set.

2, 3, 3, 4, 5, 7, 8, 8, 8, 10, 10, 12

8. What are the first and third quartiles of the data?

F. 2, 12 **G.** 4.5, 8 **H.** 3, 10 **I.** 3.5, 9

8. ________

9. The standard deviation for the data is 3.23. Which of the following is within one standard deviation of the mean?

A. 2 **B.** 5 **C.** 10 **D.** 12

9. ________

10. The table given below shows the number of students who attended the Spanish Club meetings during the school year. To the nearest tenth, what is the mean absolute deviation of the data?

Spanish Club Attendance		
14	21	17
26	13	20

F. 18.5 **G.** 13.8 **H.** 3.8 **I.** 3.3

10. ________

11. If a data distribution is not symmetric, which should you use to describe the center?

A. mean **B.** mode **C.** median **D.** range

11. ________

NAME ______________________ DATE ____________ PERIOD ________

Test, Form 1B

SCORE ________

Write the letter for the correct answer in the blank at the right of each question.

For Exercises 1–3, use the scatter plot shown at the right.

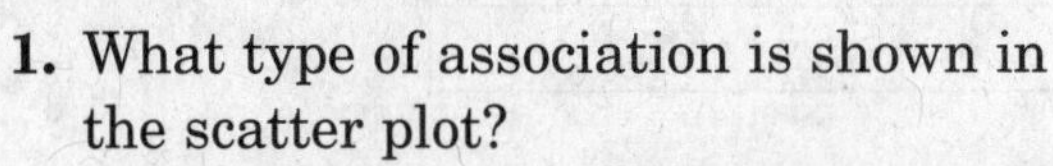

1. What type of association is shown in the scatter plot?

   A. a negative linear association
   B. a positive linear association
   C. a negative nonlinear association
   D. no association

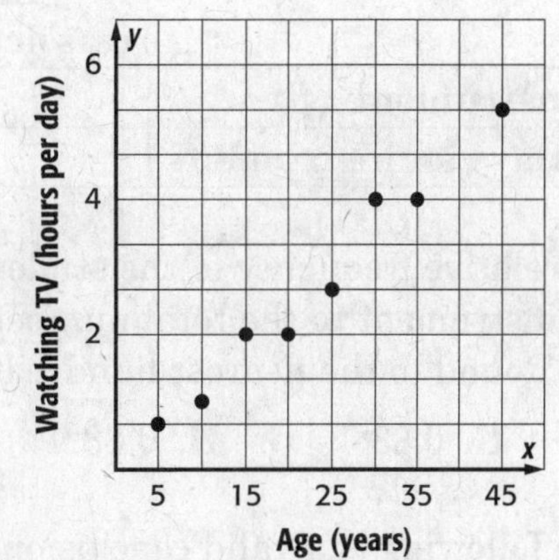

1. ________

2. Which of the following statements is best supported by the scatter plot?

   F. As the age of a person increases, the time spent watching TV increases.
   G. As the time spent watching TV increases, a person's age decreases.
   H. As the age of a person increases, the time spent watching TV decreases.
   I. There is no relationship between the age of a person and the amount of time spent watching TV.

2. ________

3. Which of the following is a reasonable estimate for the amount of TV watched per day for a person who is 50 years old?

   A. 6 h **B.** 5 h **C.** 4 h **D.** 2 h

3. ________

For Exercises 4 and 5, use the scatter plot shown at the right. The scatter plot shows Bryan's keyboarding speed after a number of weeks of keyboarding class.

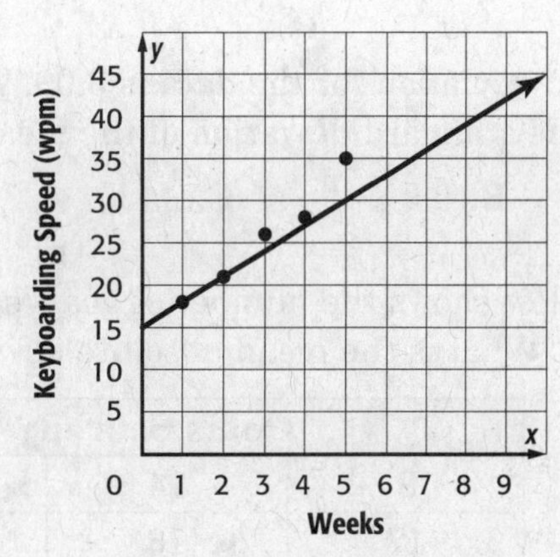

4. Which of the following is the most reasonable equation for the line of best fit?

   F. $y = 1.5x + 15$
   G. $y = -1.5x + 15$
   H. $y = -3x + 15$
   I. $y = 3x + 15$

4. ________

5. Which of the following is the most reasonable estimate for Bryan's keyboarding speed after 15 weeks?

   A. 50 wpm **B.** 55 wpm **C.** 60 wpm **D.** 65 wpm

5. ________

NAME ______________________________ DATE ______________ PERIOD ________

Test, Form 1B *(continued)*

SCORE ________

For Exercises 6 and 7, use the two-way table shown below.

	Likes classical music	Dislikes classical music
Plays an instrument	15	2
Does not plays an instrument	3	25

6. What is the relative frequency of the students who like classical music and play an instrument to the total number of students who play an instrument? Round to the nearest hundredth.

F. 0.12 **G.** 0.83 **H.** 0.88 **I.** 0.89 **6.** ________

7. Which of the following is a valid conclusion about the data?

A. Of the students that like classical music, most do not play a instrument.

B. Of the students that play an instrument, most do not like classical music.

C. There were a total of 45 students surveyed.

D. Most of the students surveyed play an instrument. **7.** ________

For Exercises 8 and 9, use the following data set.

6, 7, 7, 8, 9, 11, 12, 12, 12, 14, 14, 16

8. What are the first and third quartiles of the data?

F. 6, 16 **H.** 12, 11.5

G. 7.5, 13 **I.** 12, 14 **8.** ________

9. The standard deviation for the data is 3.09. Which of the following is *not* within one standard deviation of the mean?

A. 8 **B.** 9 **C.** 11 **D.** 16 **9.** ________

10. The table below shows the number of goals scored by Elain's field hockey team. What is the mean absolute deviation of the data?

Goals Scored		
8	14	11
17	16	24

F. 6 **G.** 5 **H.** 4 **I.** 1 **10.** ________

11. If a data distribution is symmetric, which should you use to describe the center?

A. mean **B.** mode **C.** median **D.** range **11.** ________

NAME ______________________ DATE ______________ PERIOD ________

Test, Form 2A

SCORE ________

Write the letter for the correct answer in the blank at the right of each question.

For Exercises 1–3, use the data in the table that shows the ages of people in a ceramics class at a community center.

Age of Class Members			
10	15	19	37
29	8	6	30
20	25	62	15

1. What is the mean absolute deviation?

A. 23 **C.** 11.33
B. 15.67 **D.** 9.6 **1.** ________

2. What is the median of the data?

F. 15 **G.** 19.5 **H.** 22.5 **I.** no median **2.** ________

3. What are the first and third quartiles of the data?

A. 6, 62 **B.** 15, 29 **C.** 10, 30 **D.** 12.5, 29.5 **3.** ________

4. The standard deviation of the ages of class members is 14.8. Which of the following best describes the ages that are within one standard deviation of the mean age?

F. 4.7 – 34.3 years **H.** 8.2 – 37.8 years
G. 6 – 20.8 years **I.** 47.2 – 62 years **4.** ________

For Exercises 5 and 6, use the scatter plot at the right that shows the number of assisted tackles for various players in one season.

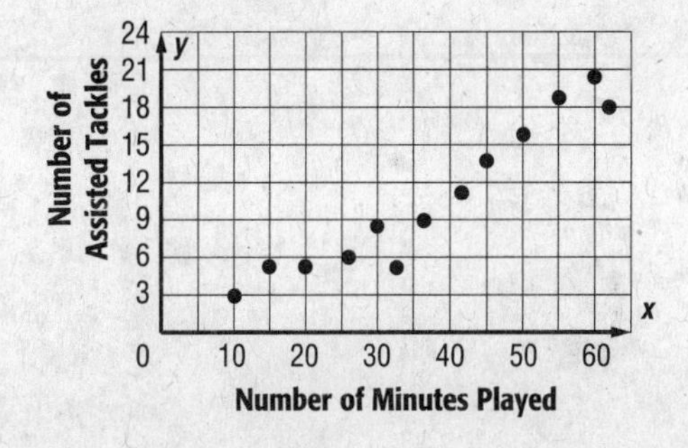

5. What type of association is shown in the scatter plot?

A. positive linear
B. negative linear
C. nonlinear
D. no association **5.** ________

6. Which of the following is a reasonable estimate for the number of assisted tackles for a player that played for 80 minutes?

F. 9 **G.** 18 **H.** 26 **I.** 40 **6.** ________

7. Which is appropriate to describe the spread of data if the data distribution is symmetric?

A. mean **C.** interquartile range
B. median **D.** mean absolute deviation **7.** ________

Test, Form 2A *(continued)*

SCORE ________

8. A teacher surveyed the students in the cafeteria and found that 35 males like fishing while 15 do not like fishing. There were 45 females surveyed and 24 of them dislike fishing.

a. Complete the two-way table summarizing the data.

8a, b.	Like Fishing	Dislike Fishing	Total
Male			
Female			
Total			

b. Find the relative frequencies of students by columns. Round to the nearest hundredth if necessary. Write the answer in the table.

c. Interpret the relative frequencies of students by columns.

8c. ________

For Exercises 9–13, use the data in the table below. The table shows the membership for a fitness center in the years 2003–2010.

Years Since 2002	1	2	3	4	5	6	7	8
Membership	75	150	125	200	175	300	250	350

9. Construct a scatter plot for the data.

9.

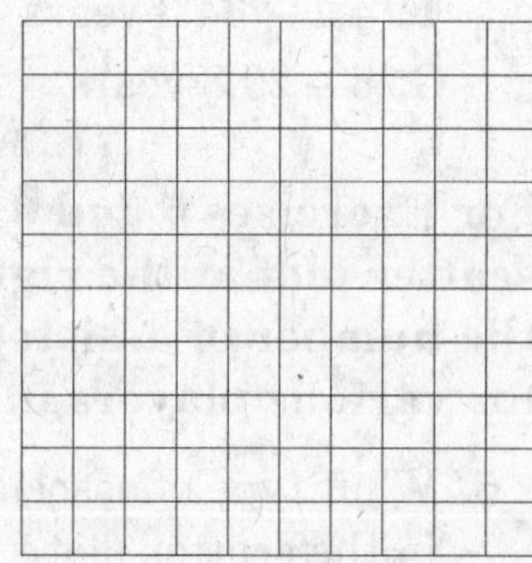

10. Draw and assess a line that seems to best represent the data on the scatter plot.

10. ________

11. Write an equation in slope-intercept form for the line of best fit that is drawn.

11. ________

12. Interpret the slope and y-intercept of the line of best fit.

12. ________

13. Use your equation from Exercise 11 to make a conjecture about the number of fitness center members in the year 2011.

13. ________

NAME ______________________ DATE ____________ PERIOD ________

Test, Form 2B

SCORE ________

Write the letter for the correct answer in the blank at the right of each question.

For Exercises 1–3, use the data in the table that shows number of kilometers run by people in a running club.

Distance Run (km)			
20	28	30	6
15	18	21	22
25	29	24	26

1. What is the mean absolute deviation?

A. 5 **C.** 15
B. 9.6 **D.** 22

1. ________

2. What is the median of the data?

F. 25 **G.** 23 **H.** 15.3 **I.** no median

2. ________

3. What are the first and third quartiles of the data?

A. 6, 28 **B.** 19, 27 **C.** 19, 28.5 **D.** 25.5, 27

3. ________

4. The standard deviation of the distance run is 6.5. Which of the following best describes the distances that are within one standard deviation of the mean distance run?

F. 14 – 26.5 km **H.** 15.5 – 28.5 km
G. 14 – 28.5 km **I.** 28.5 – 29 km

4. ________

For Exercises 5 and 6, use the scatter plot at the right that shows the semester score for students that missed some days of school.

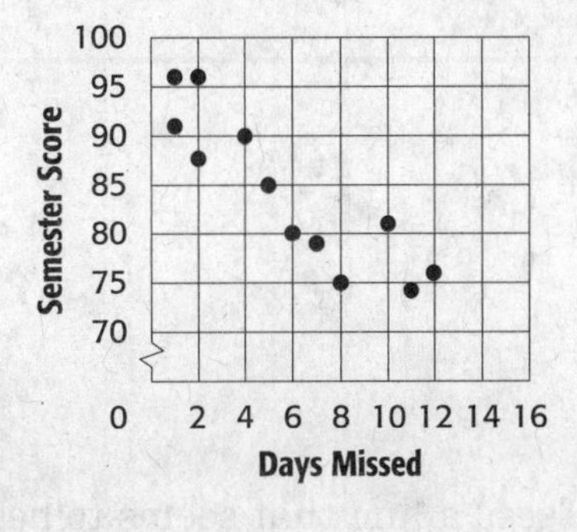

5. What type of association is shown in the scatter plot?

A. positive linear
B. negative linear
C. nonlinear
D. no association

5. ________

6. Which of the following is a reasonable estimate for the semester score for a student that missed 20 days of school?

F. 75 **G.** 73 **H.** 70 **I.** 60

6. ________

7. Which is appropriate to describe the spread of data if the data distribution is *not* symmetric?

A. mean **C.** interquartile range
B. median **D.** mean absolute deviation

7. ________

Test, Form 2B *(continued)*

SCORE _______________

8. A teacher surveyed the students in the cafeteria and found that 25 males like skiing while 10 do not like skiing. There were 50 females surveyed and 30 of them dislike skiing.

a. Complete the two-way table summarizing the data.

8a, b.	Like Skiing	Dislikes Skiing	Total
Male			
Female			
Total			

b. Find the relative frequencies of students by rows. Round to the nearest hundredth if necessary. Write the answer in the table.

c. Interpret the relative frequencies of students by rows.

8c. _______________

For Exercises 9–13, use the data in the table below. The table shows the membership for a savings club in the years 2005–2012.

Years Since 2004	1	2	3	4	5	6	7	8
Membership	120	100	110	100	90	75	85	65

9. Construct a scatter plot for the data.

9a.

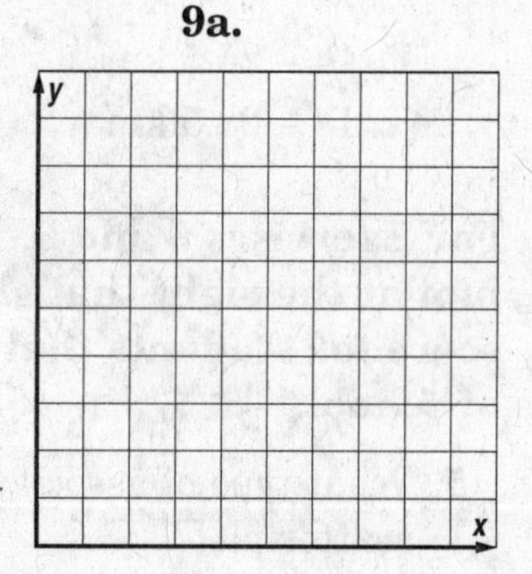

10. Draw and assess a line that seems to best represent the data on the scatter plot.

10. _______________

11. Write an equation in slope-intercept form for the line of best fit that is drawn.

11. _______________

12. Interpret the slope and y-intercept of the line of best fit.

12. _______________

13. Use your equation from Exercise 11 to make a conjecture about the number of savings club members in the year 2014.

13. _______________

NAME ________________________ DATE ____________ PERIOD ________

Test, Form 3A

SCORE ________

For Exercises 1–5, use the table below. The table shows the amount of television watched by a group of people.

Age (years)	Hours of TV Watched per Week
5	5
5	15
10	20
15	15
20	20
25	30
30	20
30	25
35	30
40	20

1.

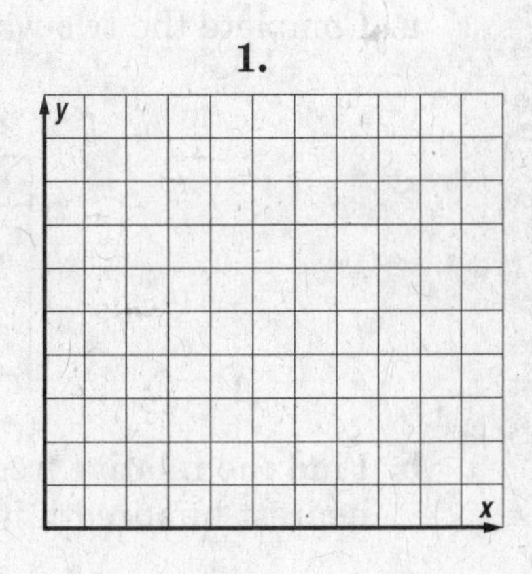

1. Construct a scatter plot of the data.

2. Interpret the scatter plot based on the shape of the distribution. **2.** ________

3. Draw and assess a line that seems to best represent the data on the scatter plot created for Exercise 1. **3.** ________

4. Write an equation in slope-intercept form for the line of best fit. **4.** ________

5. Use the line of best fit found in Exercise 4 to make a conjecture about the number of hours a 55-year-old would spend watching TV. **5.** ________

For Exercises 6 and 7, use the table below. The table shows the heights of plants in centimeters for a science fair project.

Plant Height (cm)					
16	16	29	20	33	17
24	31	25	33	23	19

6. ________

6. Find the five-number summary of the set of data.

7. Construct a box plot of the data. **7.**

NAME ______________________ DATE ____________ PERIOD ________

Test, Form 3A *(continued)*

SCORE ________

8. A teacher surveyed the students in the cafeteria and found that 20 males like soccer while 5 do not like soccer. There were 30 females surveyed and 6 of them do not like soccer.

a. Complete the two-way table summarizing the data.

8a, b.	Likes Soccer	Does Not Like Soccer	Total
Male			
Female			
Total			

b. Find the relative frequencies of students by rows. Round to the nearest hundredth if necessary. Write the answer in the table.

c. Interpret the relative frequencies of students by rows. **8c.** ____________

d. Does the data support the statement below? Justify your reasoning. **8d.** ____________

Girls do not like soccer.

For Exercises 9–12, use the table of quiz scores shown at the right.

Quiz Scores, Period 3					
25	13	16	30	27	22
19	22	15	28	27	29

9. Find the mean of the data. **9.** ____________

10. Find the mean absolute deviation for the data set. Round to the nearest tenth. **10.** ____________

11. Describe what the mean absolute deviation represents. **11.** ____________

12. The standard deviation of quiz scores is 5.6. Describe the quiz scores that are within one standard deviation of the mean. **12.** ____________

13. Explain what symmetric means with respect to a data distribution. **13.** ____________

NAME ______________________ DATE ______________ PERIOD ________

Test, Form 3B

SCORE ________

For Exercises 1–5, use the table below. The table shows the time spent listening to the radio by a group of people.

Age (years)	Hours Listening to the Radio (Weekly)
10	2
10	3
15	2
15	1
20	4
25	3
30	3
30	7
35	7
40	7

1.

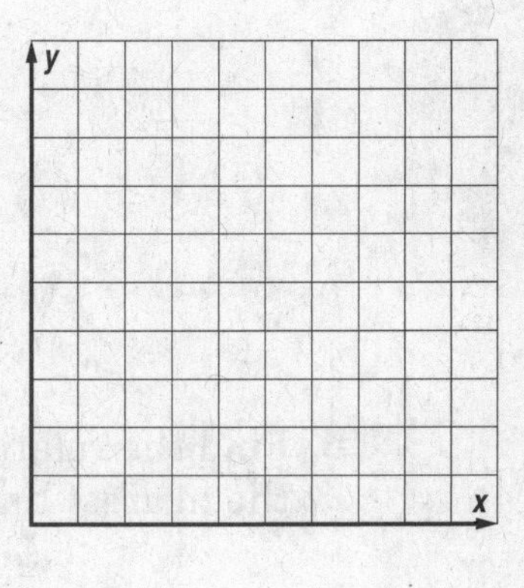

1. Construct a scatter plot of the data.

2. Interpret the scatter plot based on the shape of the distribution. 2. ________

3. Draw and assess a line that seems to best represent the data on the scatter plot created for Exercise 1. 3. ________

4. Write an equation in slope-intercept form for the line of best fit. 4. ________

5. Use the line of best fit found in Exercise 4 to make a conjecture about the number of hours a 60-year-old would spend listening to the radio. 5. ________

For Exercises 6 and 7, use the table below. The table shows the heights of children at a day care center.

Child's Height (cm)					
85	59	82	78	63	83
90	88	71	74	68	59

6. ________

6. Find the five-number summary of the set of data.

7. Construct a box plot for the data. 7.

NAME ______________________ DATE ____________ PERIOD ________

Test, Form 3B *(continued)*

SCORE ________

8. A teacher surveyed the students in the cafeteria and found that 30 males like bowling while 15 do not like bowling. There were 40 females surveyed and 20 of them dislike bowling. Construct a two-way table summarizing the data.

a. Complete the two-way table summarizing the data.

8a, b.	Likes Bowling	Does Not Like Bowling	Total
Male			
Female			
Total			

b. Find the relative frequencies of students by columns. Round to the nearest hundredth if necessary. Write the answer in the table.

c. Interpret the relative frequencies of students by columns. **8c.** __________

d. Does the data support the statement below? Justify your reasoning. **8d.** __________

Girls like bowling more than boys.

For Exercises 9–12, use the table of quiz scores shown at the right.

Quiz Scores, Period 3					
43	50	37	39	42	49
36	35	50	48	42	40

9. Find the mean of the data. Round to the nearest tenth. **9.** __________

10. Find the mean absolute deviation for the data set. Round to the nearest tenth. **10.** __________

11. Describe what the mean absolute deviation represents. **11.** __________

12. The standard deviation of quiz scores is 5.3. Describe the quiz scores that are within one standard deviation of the mean. **12.** __________

13. Explain what the distribution of a data set does. **13.** __________

NAME ______________________ DATE ____________ PERIOD ________

Course 3 Benchmark Test – First Quarter (Chapters 1–2)

1. The average distance from the Earth to the moon is about 384,000 kilometers. What is this number written in scientific notation?

   A. 384×10^5

   B. 384×10^3

   C. 3.84×10^6

   D. 3.84×10^5

2. **SHORT ANSWER** Marc is finding the product of the monomials $3c^2d^4$ and $-4c^3d$. His work is shown below. What error did he make?

   Marc

   $3c^2d^4(-4c^3d)$

   $= 3(-4)(c^2c^3)(d^4d)$

   $= -12c^6d^4$

3. Which point on the number line shows $\sqrt{45}$?

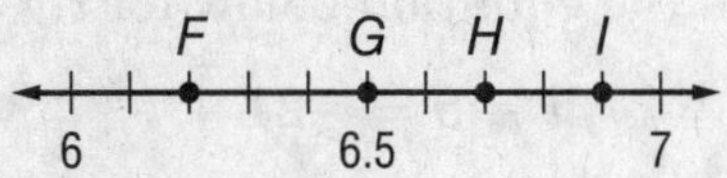

   F. point F

   G. point G

   H. point H

   I. point I

4. A moving company charges \$40 plus \$0.25 per mile to rent a van. Another company charges \$25 plus \$0.35 per mile to rent the same van. For what number of miles will the rental cost be the same for both companies?

   A. 150 miles

   B. 180 miles

   C. 260 miles

   D. 650 miles

5. A taxicab service charges \$3.75 plus \$0.40 per mile. Molly takes a taxicab from the hotel to the airport. If the total charge was \$10.95, which equation could be used to determine the number of miles from the hotel to the airport?

   F. $3.75m + 0.4 = 10.95$

   G. $3.75 + 0.4m = 10.95$

   H. $4.15m = 10.95$

   I. $3.35m = 10.95$

6. Which value is equivalent to 4^{-3}?

   A. -12

   B. -1

   C. $-\frac{1}{64}$

   D. $\frac{1}{64}$

Course 3 Benchmark Test – First Quarter *(continued)*

7. SHORT ANSWER The Venn diagram shows the real number system. Write the names of the missing sets of numbers.

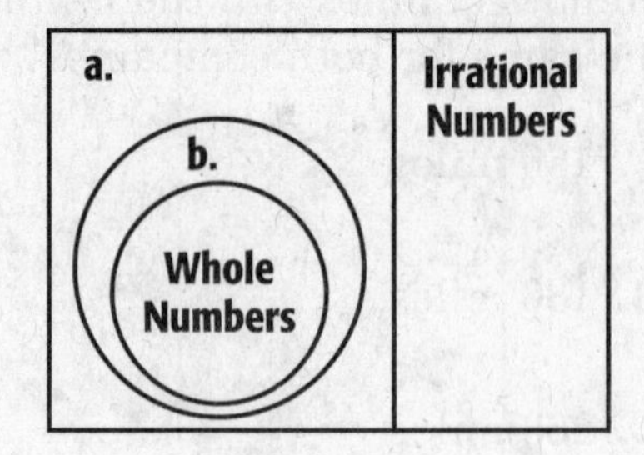

8. Which of the following does *not* represent a rational number?

F. -250

G. $\frac{11}{39}$

H. $\sqrt{60}$

I. $12.0\overline{982}$

9. The school marching band has 36 members. The band director wants to arrange the band members into a square formation. How many band members should be in each row?

A. 8

B. 6

C. 5

D. 4

10. Which expression is equivalent to the expression below?

$$a \cdot a \cdot a \cdot b \cdot a \cdot b \cdot b \cdot a \cdot b \cdot a$$

F. $a^6 b^4$

G. $a^{-6} b^{-4}$

H. $(ab)^{10}$

I. $(ab)^2$

11. What is the solution to the equation below?

$$-\frac{2}{3}p + \frac{1}{6} = \frac{7}{10}$$

A. $-\frac{13}{10}$

B. $-\frac{4}{5}$

C. $-\frac{26}{45}$

D. $-\frac{16}{45}$

12. Solve the equation below for t.

$$3t - 5 = -21 + t$$

F. -52

G. -32

H. -13

I. -8

Course 3 Benchmark Test – First Quarter *(continued)*

13. The distance from the Sun to Earth is about 1.5×10^{11} meters. Suppose light travels at a speed of 3×10^{8} meters per second. About how long does it take light from the Sun to reach Earth?

A. 4.5×10^{19} seconds

B. 1.503×10^{11} seconds

C. 5×10^{3} seconds

D. 5×10^{2} seconds

14. What is the value of b in the equation below?

$$4(b - 1) = 2b + 10$$

F. 4

G. 5.5

H. 7

I. 11.5

15. The table shows the populations of several states. What is the population of Ohio written in scientific notation?

State	Population
Georgia	9,400,000
Illinois	12,900,000
Ohio	11,500,000
California	36,900,000

A. 1.15×10^{-8}

B. 1.15×10^{-7}

C. 1.15×10^{7}

D. 1.15×10^{8}

16. Which of the expressions below is *not* equivalent to the other three?

F. 0.015625

G. 15.625%

H. 4^{-3}

I. $\frac{1}{64}$

17. **SHORT ANSWER** What is the result when the monomial $-5x^3y^2z$ is raised to the third power?

18. The area of a square living room is 169 square feet. What is the perimeter of the room?

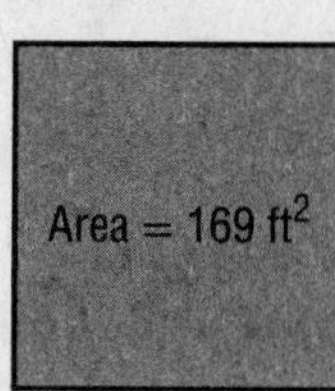

A. 13 ft

B. 17 ft

C. 52 ft

D. 68 ft

Course 3 Benchmark Test – First Quarter *(continued)*

19. Between which two integers does $\sqrt{88}$ lie on the number line?

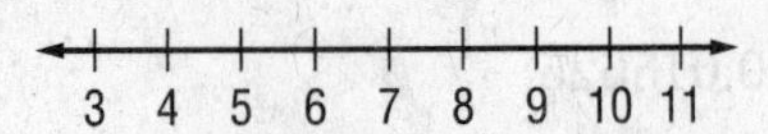

F. between 6 and 7

G. between 7 and 8

H. between 8 and 9

I. between 9 and 10

20. Which of the following symbols results in a true number sentence when placed in the blank?

$$\sqrt{12.96} \;\_\_\; 3\frac{3}{5}$$

A. =

B. >

C. <

D. ×

21. **SHORT ANSWER** The area of an equilateral triangle is given by the expression $\frac{s^2\sqrt{3}}{4}$, where s is the side length of the triangle. What is the area of triangle below? Round to the nearest tenth.

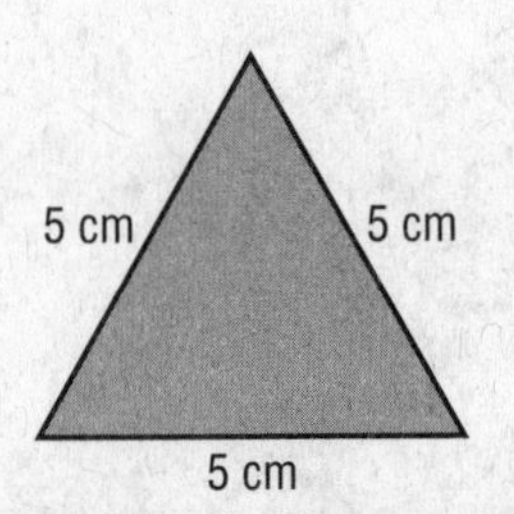

22. Which of the following numbers has the least absolute value?

F. 3.5×10^{-5}

G. 8.75×10^{-7}

H. 5.62×10^{3}

I. 1.002×10^{12}

23. Which equation shows the following relationship?

Seven less than four times a number is equal to 5.

A. $7 - 4n = 5$

B. $4n - 7 = 5$

C. $7n - 4 = 5$

D. $4 - 7n = 5$

24. Which equation is equivalent to the equation below?

$$5(n + 6) = 2(n - 3) + 4$$

F. $5n + 6 = 2n + 1$

G. $5n + 6 = 2n - 2$

H. $5n + 30 = 2n + 1$

I. $5n + 30 = 2n - 2$

NAME ______________________ DATE ____________ PERIOD ________

Course 3 Benchmark Test – First Quarter *(continued)*

25. **SHORT ANSWER** Juanita has saved $65 for vacation. She plans to save an additional $5 per week. How many weeks will it take for Juanita to save a total of $125? Write and solve an equation.

NAME ______________________ DATE ______________ PERIOD ________

Course 3 Benchmark Test – Second Quarter (Chapters 3–4)

1. The table shows how much Addison earns for working various numbers of hours at a part-time job.

Hours, x	Earnings ($), y
10	72.50
15	108.75
20	145.00

Which of the following describes the constant rate of change?

A. 5 hours per dollar

B. $5.00 per hour

C. 7.25 hours per dollar

D. $7.25 per hour

2. Let n represent the figure number in the pattern below.

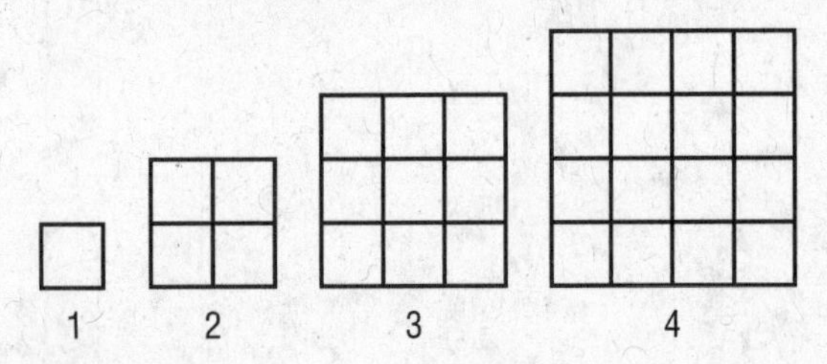

Which function represents the number of squares in each figure?

F. $f(n) = n^2$

G. $f(n) = 2n$

H. $f(n) = n^3$

I. $f(n) = 4n$

3. Which systems of linear equations has a solution of $(-2, 1)$?

A. $2x + 3y = -1$
$x - y = -3$

B. $2x + 3y = 1$
$x - y = 3$

C. $2x + 3y = -1$
$x - y = 3$

D. $2x + 3y = 1$
$x - y = -3$

4. What is the solution to the system of equations below?

$$3x - 2y = 7$$
$$-3x + 5y = 5$$

F. (3, 1)

G. (0, 1)

H. (5, 4)

I. no solution

5. **SHORT ANSWER** Missy walked around the school track to warm up. Then she ran several laps before walking to cool down. Sketch a graph to represent Missy's distance run over time.

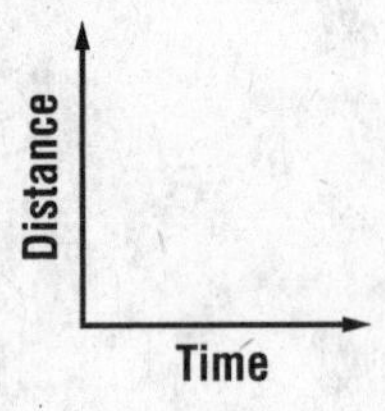

6. Which term describes the function shown below?

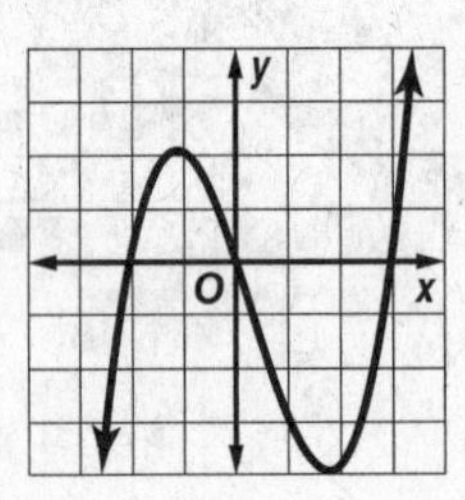

A. constant

B. linear

C. nonlinear

D. quadratic

NAME ______________________ DATE ____________ PERIOD ________

Course 3 Benchmark Test – Second Quarter *(continued)*

7. What is the equation of the quadratic function shown in the graph?

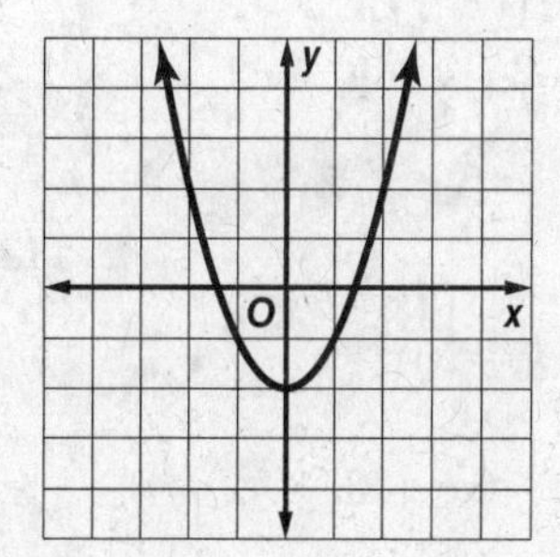

F $y = x^2 + 2$

G $y = x^2 - 2$

H $y = 2x^2$

I $y = \frac{1}{2}x^2$

8. **SHORT ANSWER** Find the x- and y-intercepts of the linear equation below.

$$4x - 5y = 20$$

9. What is the slope of the line that passes through $M(-6, 1)$ and $N(2, 5)$?

A 2

B $\frac{1}{2}$

C $-\frac{1}{2}$

D -2

10. What is the domain of the function shown in the table?

x	−4	−2	0	2	4
y	−3	7	5	0	−1

F. all real numbers

G. all even integers

H. {−3, −1, 0, 5, 7}

I. {−4, −2, 0, 2, 4}

11. What are the slope and y-intercept of the linear equation below?

$$y = -5x + 2$$

A. slope: 2, y-intercept: (0, −5)

B. slope: 2, y-intercept: (−5, 0)

C. slope: −5, y-intercept: (0, 2)

D. slope: −5, y-intercept: (2, 0)

12. A tank contains 550 gallons of water. When the valve is opened, the tank drains at a rate of 12 gallons per minute. Which function shows the relationship between the time t the valve is opened and the amount of water in the tank?

F. $A(t) = -12t + 550$

G. $A(t) = 12t + 550$

H. $A(t) = 12 + 550t$

I. $A(t) = -12 + 550t$

NAME ______________________ DATE ____________ PERIOD ________

Course 3 Benchmark Test – Second Quarter *(continued)*

13. Which relation is *not* a function?

A.

x	−2	0	2	4	6
y	3	3	3	3	3

B.

x	−3	0	2	−3	1
y	−5	4	2	0	−1

C.

x	1	2	3	4	5
y	1	2	3	4	5

D

x	−4	1	2	−3	4
y	0	3	−1	−2	3

14. What is the solution to the system of linear equations shown below?

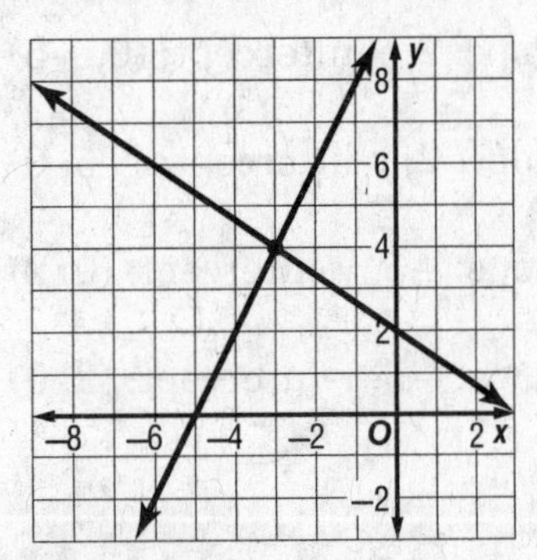

F. (4, −3)

G. (−4, 3)

H. (−3, 4)

I. (3, −4)

15. **SHORT ANSWER** What is the equation in slope-intercept form of the line that passes through (−2, 17) and (3, −13)?

16. Which linear function has the steepest slope?

A. $y = \frac{1}{2}x - 5$

B. $y = -\frac{2}{5}x + 3$

C. $y = 4x - 2$

D. $y = -6x + 1$

17. The table shows the cost of renting a van from a moving company for different numbers of miles driven.

Miles, *m*	Cost, *C*
50	$42.50
100	$65.00
150	$87.50
200	$110.00

Construct a function that relates the cost of renting a van to the number of miles driven.

F. $C(m) = 0.85m$

G. $C(m) = 0.85m + 10$

H. $C(m) = 0.45m$

I. $C(m) = 0.45m + 20$

18. Which two points form a line that has a slope of −3?

A. (−5, 3) and (2, 4)

B. (1, −6) and (−4, 9)

C. (−4, −3) and (5, 0)

D. (2, 8) and (−1, −1)

Course 3 Benchmark Test – Second Quarter *(continued)*

19. What are the x– and y–intercepts of the linear equation below?

$$6x - 2y = 12$$

F. (2, 0) and (0, −6)

G. (0, 2) and (−6, 0)

H. (−6, 0) and (2, 0)

I. (0, 2) and (0, −6)

20. The quadratic function $h(t) = -16t^2 + 120$ represents the height of an object in feet t seconds after when it falls from a height of 120 feet. What is the height of the object after 1.5 seconds?

A. 58 ft

B. 84 ft

C. 92 ft

D. 156 ft

21. **SHORT ANSWER** The table below shows the number of teams remaining in each round of a tournament. Is the number of teams a linear function of the number of rounds? Explain.

Round	Teams
1	32
2	16
3	8
4	4
5	2

22. What is the constant rate of change of the function represented in the table below?

x	y
−5	23
−1	7
3	−9
7	−25

F. 16

G. 4

H. −4

I. −16

23. The slope of a line is $-\frac{1}{5}$ and the y-intercept is (0, 6). What is the equation of the line in slope-intercept form?

A. $x + 5y = 30$

B. $x - 5y = 30$

C. $y = -\frac{1}{5}x - 6$

D. $y = -\frac{1}{5}x + 6$

24. Which of the following equations represents a horizontal line?

F. $y = x$

G. $y = -x + 1$

H. $y = -12$

I. $x = 5$

Course 3 Benchmark Test – Second Quarter *(continued)*

25. SHORT ANSWER The graph below shows the length of Michael's hair as a function of time. Describe the change in the length of Michael's hair over time.

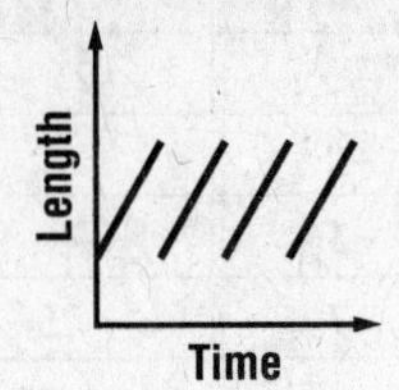

NAME ________________ DATE ________ PERIOD ________

Course 3 Benchmark Test – Third Quarter (Chapters 5–6)

1. **SHORT ANSWER** Alfonso leans a 20-foot long ladder against a wall with the base of the ladder 6 feet from the wall. How far up the wall does the ladder reach? Round to the nearest tenth if necessary.

2. What is the sum of the measures of the interior angles of a pentagon?

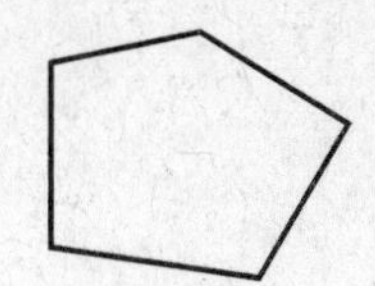

A. 900°

B. 720°

C. 540°

D. 450°

3. What is the distance between points A and B shown on the coordinate plane?

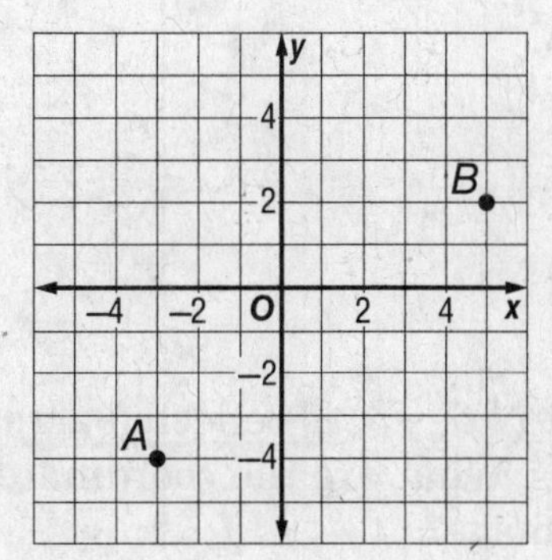

F. 8 units

G. 10 units

H. 12 units

I. 14 units

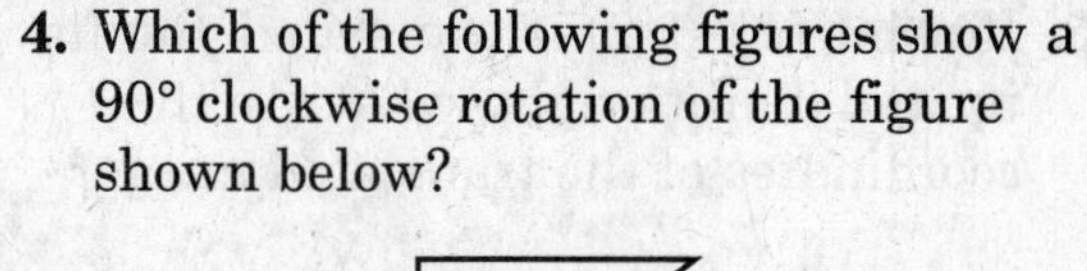

4. Which of the following figures show a 90° clockwise rotation of the figure shown below?

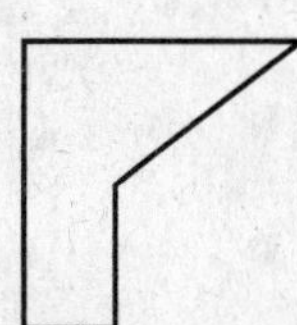

A.

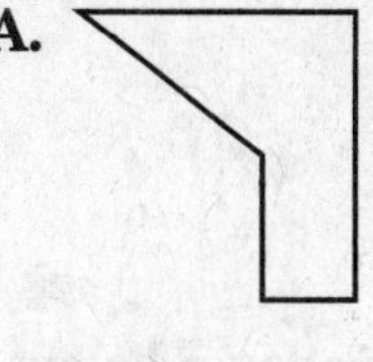

B.

C.

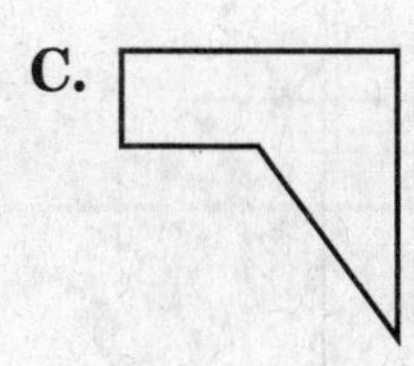

D.

Course 3 Benchmark Test – Third Quarter *(continued)*

5. If point $H(-6, 2)$ is translated 4 units up and 3 units right, what are the coordinates of the translated image?

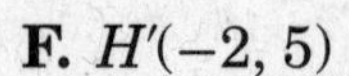

F. $H'(-2, 5)$

G. $H'(-3, 6)$

H. $H'(-9, -2)$

I. $H'(-9, 6)$

6. The dilation of $\overline{CD}$ is shown below. What is the scale factor of the dilation?

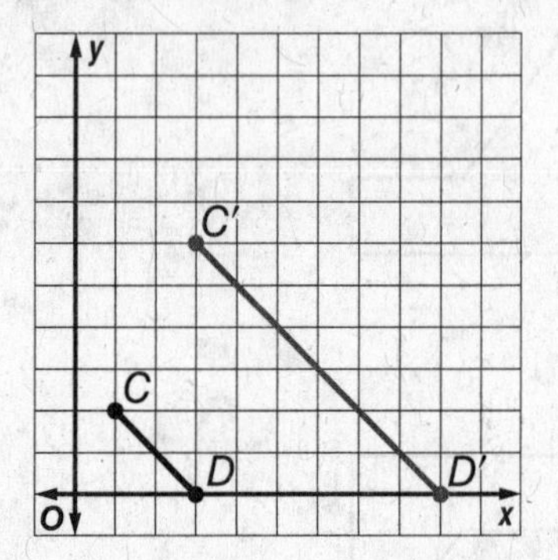

A. $\frac{1}{3}$

B. $\frac{1}{2}$

C. 2

D. 3

7. Which of the following terms describes two lines that intersect to form right angles?

F. parallel

G. perpendicular

H. skew

I. straight

8. What is the measure of angle 3?

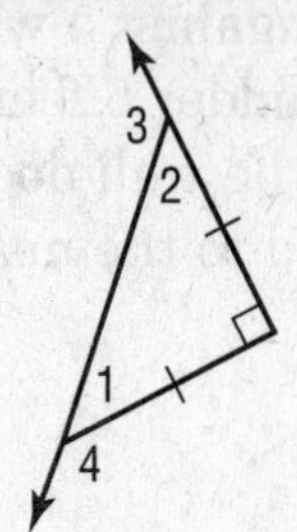

A. 45°

B. 90°

C. 135°

D. 225°

9. **SHORT ANSWER** Determine whether the following figure is a right triangle. Justify your answer.

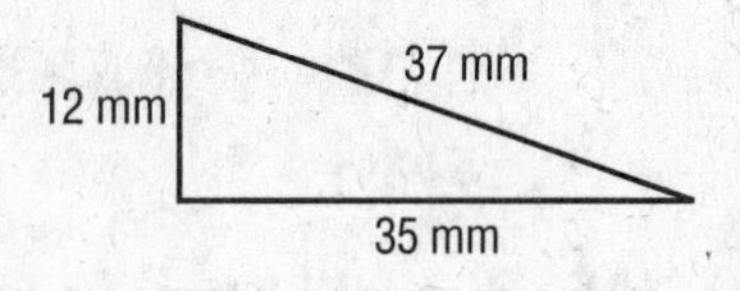

10. Point $N(6, -5)$ is reflected across the x-axis. What are the coordinates of the image?

F. $N'(-6, -5)$

G. $N'(-5, 6)$

H. $N'(5, -6)$

I. $N'(6, 5)$

Course 3 Benchmark Test – Third Quarter *(continued)*

11. Parallel lines l and m are intersected by transversal t as shown below. Which of the following angles are *not* congruent?

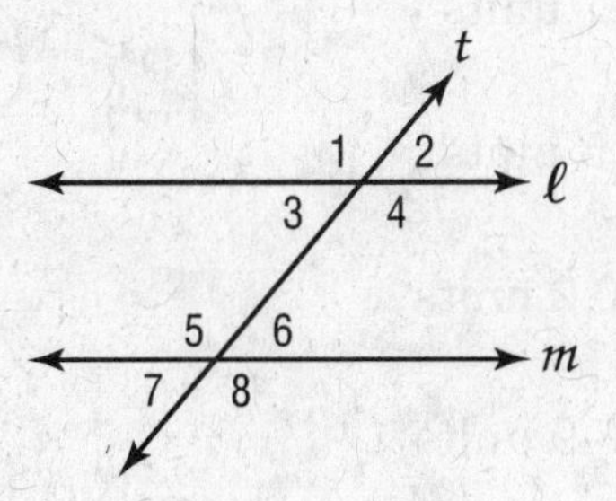

A. 1 and 2

B. 2 and 3

C. 3 and 6

D. 4 and 8

12. Suppose triangle RST shown on the coordinate grid is reflected across the y-axis. Which ordered pair does *not* represent a vertex of the reflected triangle?

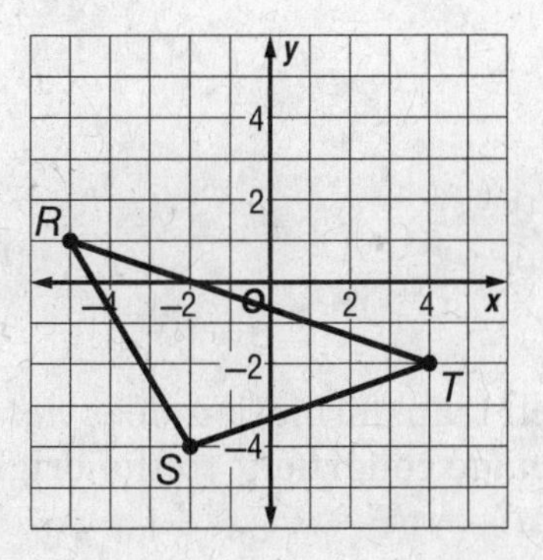

F. (5, 1)

G. (−4, −2)

H. (2, −4)

I. (−2, 4)

13. **SHORT ANSWER** Using the figure below, write a paragraph proof to show that $m\angle a = m\angle b = 45°$.

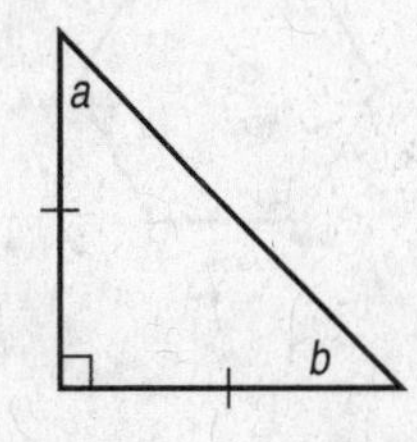

14. What is the approximate distance between points $W(-4, 1)$ and $Z(3, 7)$? Round to the nearest tenth.

A. 10.8 units

B. 9.2 units

C. 8.3 units

D. 6.1 units

15. What is the value of n in the triangle below?

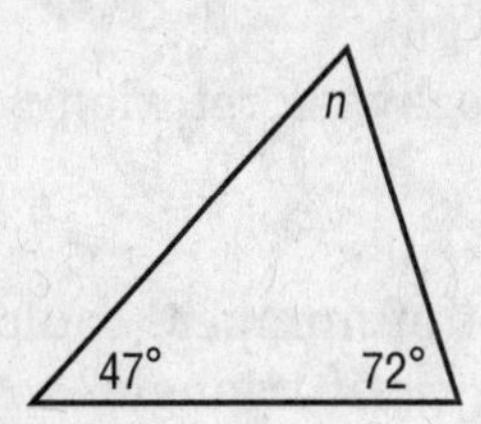

F. 47°

G. 51°

H. 61°

I. 72°

Course 3 Benchmark Test – Third Quarter *(continued)*

16. What is the measure of an interior angle of a regular hexagon?

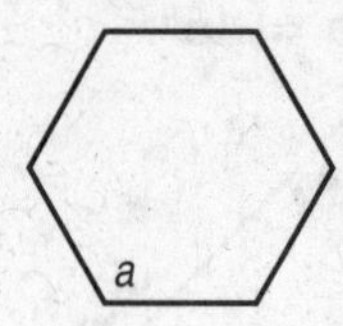

A. 120°

B. 135°

C. 720°

D. 810°

17. Which rotation best describes the transformation shown below?

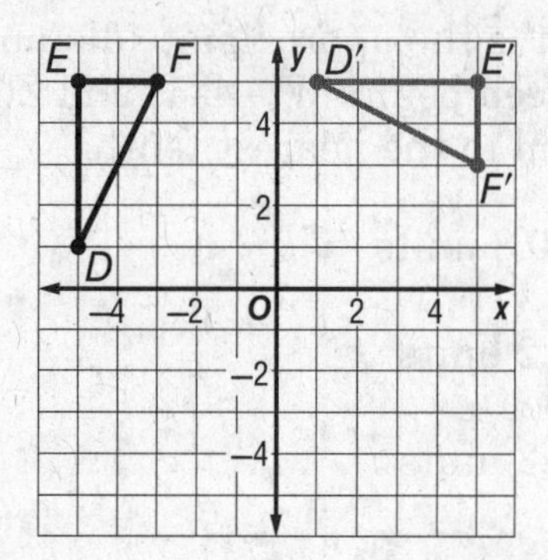

F. 90° counterclockwise rotation

G. 270° clockwise rotation

H. 180° rotation

I. 90° clockwise rotation

18. Which set of numbers could be the sides of a right triangle?

A. 6, 8, 12

B. 8, 15, 17

C. 4, 12, 16

D. 9, 11, 21

19. What is the approximate length of $\overline{NP}$ with endpoints $N(7, 3)$ and $P(-6, -2)$? Round to the nearest tenth.

F. 5.7 units

G. 6.5 units

H. 10.2 units

I. 13.9 units

20. **SHORT ANSWER** What is the length of the diagonal of a square with 8-foot sides? Round to the nearest tenth.

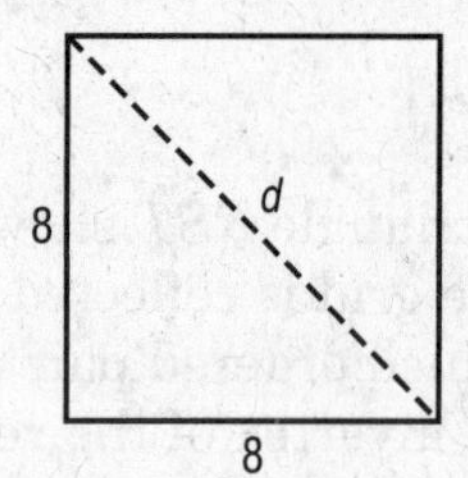

21. Which transformation does *not* result in an image congruent to the original figure?

A. translation

B. rotation

C. reflection

D. dilation

Course 3 Benchmark Test – Third Quarter *(continued)*

22. What is the value of x in the figure below?

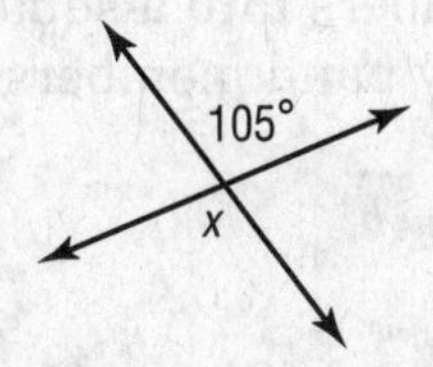

F. 75°

G. 85°

H. 105°

I. 115°

23. Mary enlarged a 4- by 6-inch photo to a 10- by 15-inch photo. What is the scale factor of the dilation?

A. 2

B. 2.5

C. 6

D. 9

24. The legs of a right triangle measure 7 units and 24 units. What is the measure of the hypotenuse? Round to the nearest tenth if necessary.

F. 17 units

G. 20.4 units

H. 23.0 units

I. 25 units

25. **SHORT ANSWER** Prove that triangle ABC is an isosceles triangle.

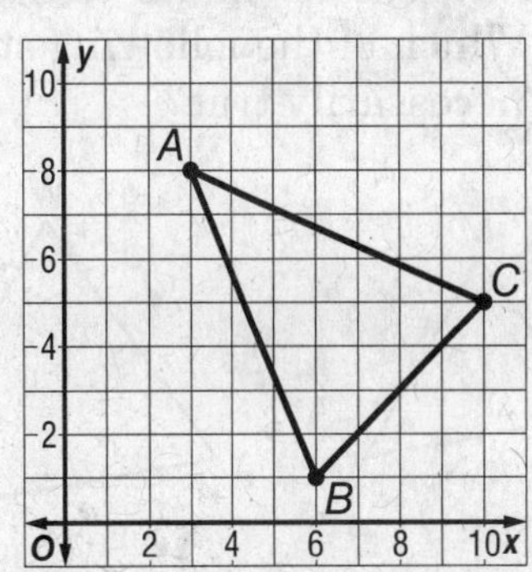

NAME ______________________ DATE ____________ PERIOD ________

Course 3 Benchmark Test – End of Year

1. The area of a figure is 64 square centimeters. Suppose the sides of the figure are doubled. What will be the new area of the similar figure?

A. 16 square centimeters

B. 32 square centimeters

C. 128 square centimeters

D. 256 square centimeters

2. Triangle *MNO* is similar to triangle *WXY*. Which of the following statements is not necessarily true?

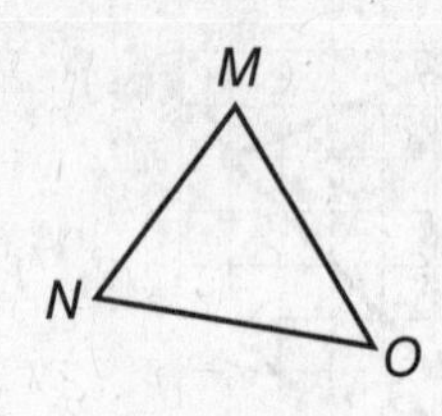

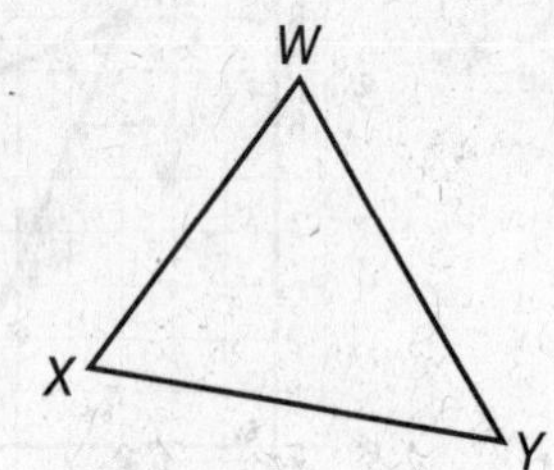

F. $\angle Y = \angle O$

G. $\frac{MO}{MN} = \frac{WX}{WY}$

H. $\angle N = \angle X$

I. $\frac{MN}{NO} = \frac{WX}{XY}$

3. **SHORT ANSWER** A moving company charges \$30 plus \$0.15 per mile to rent a moving van. Another company charges \$15 plus \$0.20 per mile to rent the same van. For how many miles will the cost be the same for the two companies? Write and solve an equation.

4. A marching band has 64 members. The band director wants to arrange the band members into a square formation. How many band members will be in each row?

A. 8

B. 7

C. 6

D. 5

5. Between which two integers does $\sqrt{42}$ lie on the number line?

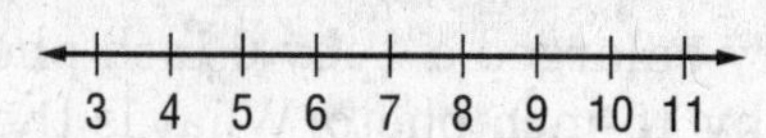

F. between 5 and 6

G. between 6 and 7

H. between 7 and 8

I. between 8 and 9

6. What are the slope and *y*-intercept of the linear equation below?

$$y = \frac{2}{3}x - 1$$

A. slope: $\frac{2}{3}$, *y*-intercept: $(0, -1)$

B. slope: $\frac{2}{3}$, *y*-intercept: $(-1, 0)$

C. slope: -1, *y*-intercept: $\left(0, \frac{2}{3}\right)$

D. slope: -1, *y*-intercept: $\left(\frac{2}{3}, 0\right)$

NAME ______________________ DATE ______________ PERIOD ________

Course 3 Benchmark Test – End of Year *(continued)*

7. What is the equation of the quadratic function shown in the graph?

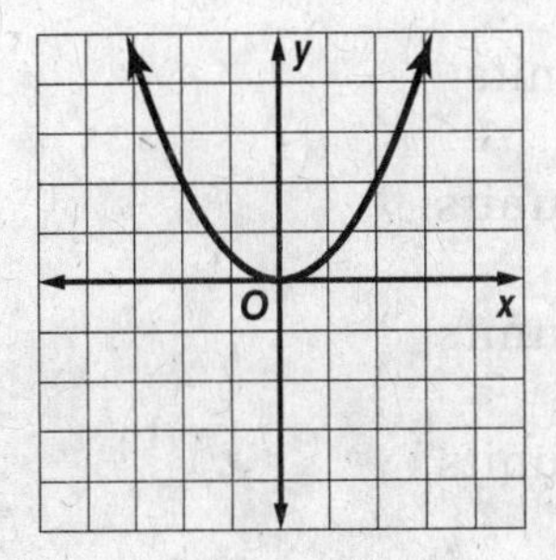

F. $y = x^2$

G. $y = -x^2$

H. $y = 2x^2$

I. $y = \frac{1}{2}x^2$

8. What is the volume of a sphere with a radius of 9 inches?

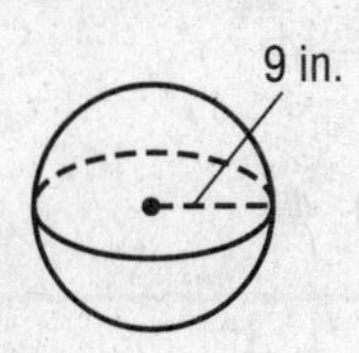

A. 1016π in^3

B. 972π in^3

C. 486π in^3

D. 324π in^3

9. What are the x- and y-intercepts of the linear equation below?

$$-5x + 3y = -15$$

F. (3, 0) and (0, –5)

G. (0, 3) and (–5, 0)

H. (–5, 0) and (3, 0)

I. (0, 3) and (0, –5)

10. SHORT ANSWER The two-way table shows the number of boys and girls in the school band and choir. Is there a greater percentage of girls in the school band or in the choir? Explain.

	Band	Choir
Boys	14	5
Girls	12	9

11. What is the sum of the measures of the interior angles of a hexagon?

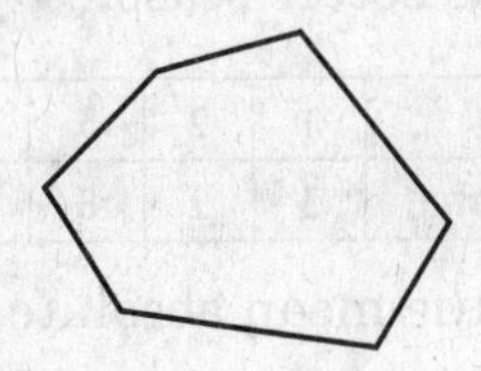

A. 540°

B. 720°

C. 900°

D. 1,080°

12. SHORT ANSWER Determine whether the following figure is a right triangle. Justify your answer.

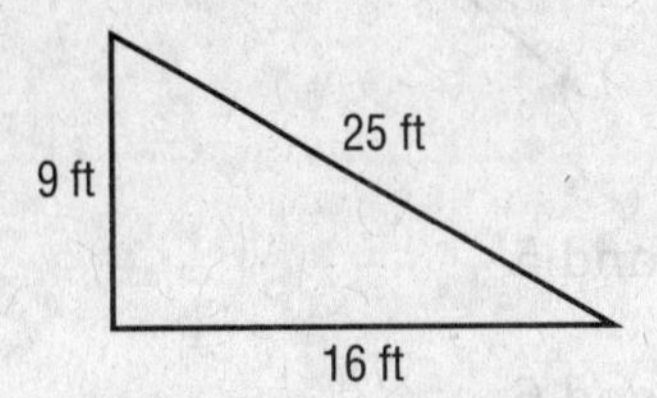

Course 3 Benchmark Test – End of Year *(continued)*

13. A soup can has a diameter of 8 centimeters and a height of 15 centimeters. About how much soup does the can hold? Use 3.14 for π. Round to the nearest tenth.

F. 376.8 cm^3

G. 753.6 cm^3

H. 1028.7 cm^3

I. 3014.4 cm^3

14. SHORT ANSWER The table shows the number of goals scored by the Cougars so far this soccer season.

Game	1	2	3	4	5
Goals Scored	3	2	6	5	4

What is the mean absolute deviation?

15. Parallel lines l and m are intersected by transversal t as shown below. Which of the following angles are alternate interior angles?

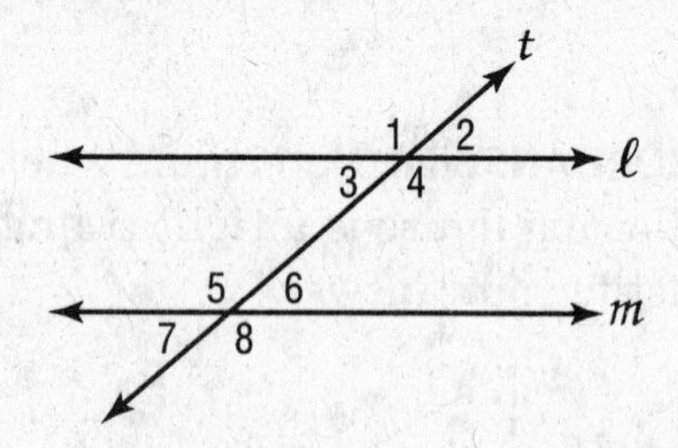

A. 1 and 5

B. 4 and 6

C. 2 and 7

D. 3 and 6

16. What is the distance between points $L(-5, 7)$ and $M(3, -8)$?

F. 9 units

G. 13 units

H. 15 units

I. 17 units

17. The slope of a line is -3 and the y-intercept is $(0, 4)$. What is the equation of the line in slope-intercept form?

A. $y = -\frac{1}{3}x + 4$

B. $y = \frac{1}{3}x - 4$

C. $y = 3x + 4$

D. $y = -3x + 4$

18. What is the value of n in the triangle below?

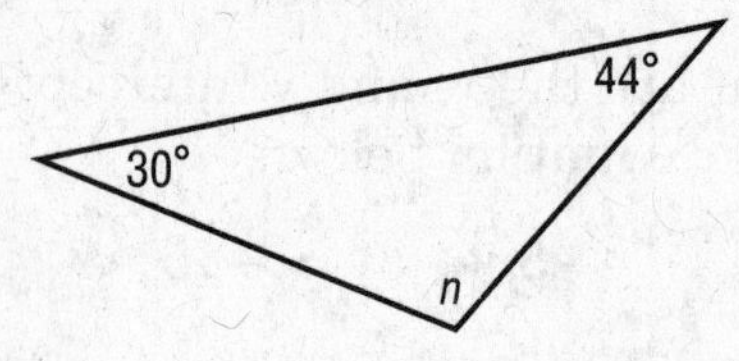

F. 68°

G. 74°

H. 96°

I. 106°

NAME ______________________ DATE ____________ PERIOD ________

Course 3 Benchmark Test – End of Year *(continued)*

19. Suppose the dimensions of a rectangular prism are enlarged by a factor of 3. By what scale factor will the volume of the prism be scaled?

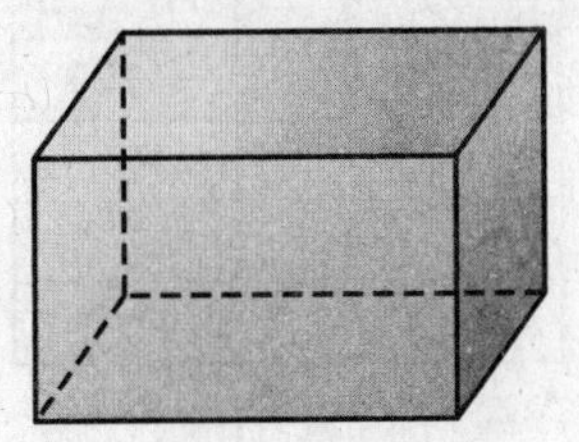

A. $\frac{1}{3}$

B. 3

C. 9

D. 27

20. What is the measure of an interior angle of a regular octagon?

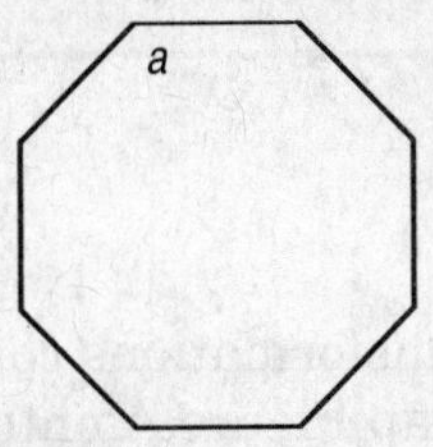

F. 1,080°

G. 720°

H. 540°

I. 135°

21. **SHORT ANSWER** What is the expression $(3x^2y^3)^3$ simplified?

22. Which equation is equivalent to $3x + 2y = -2$?

A. $y = -\frac{2}{3}x - 5$

B. $y = \frac{3}{2}x + 7$

C. $y = -\frac{3}{2}x - 1$

D. $y = \frac{2}{3}x + 4$

23. Which of the following symbols when placed in the blank results in a true number sentence?

$$1.7\overline{3} \_\_\_\_ \sqrt{3}$$

F. =

G. >

H. <

I. ×

24. What type of relationship is shown in the scatter plot below?

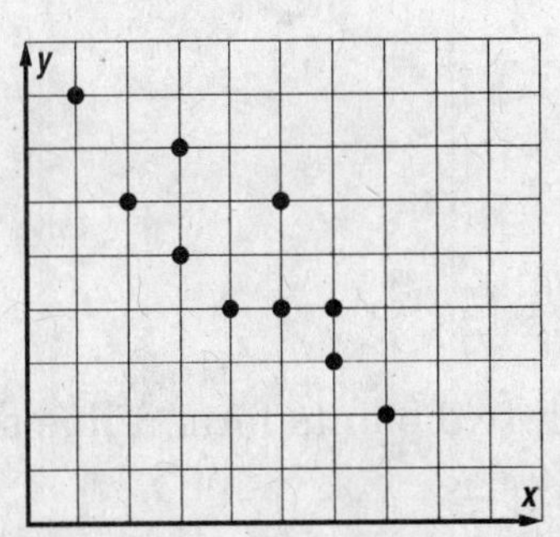

A. positive

B. negative

C. skewed

D. no relationship

Course 3 Benchmark Test – End of Year *(continued)*

25. About how much water can the paper drinking cup shown below hold? Use 3.14 for π. Round to the nearest tenth.

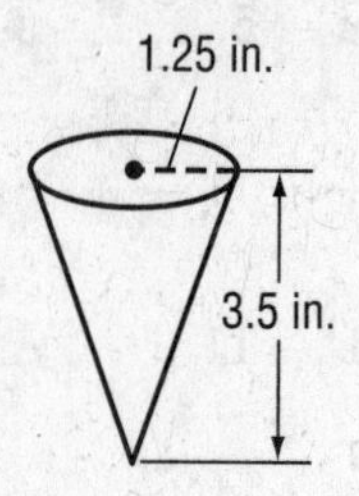

F. 17.2 cubic inches

G. 9.2 cubic inches

H. 5.7 cubic inches

I. 4.8 cubic inches

26. **SHORT ANSWER** Determine if the two figures below are congruent by using transformations. Explain your reasoning.

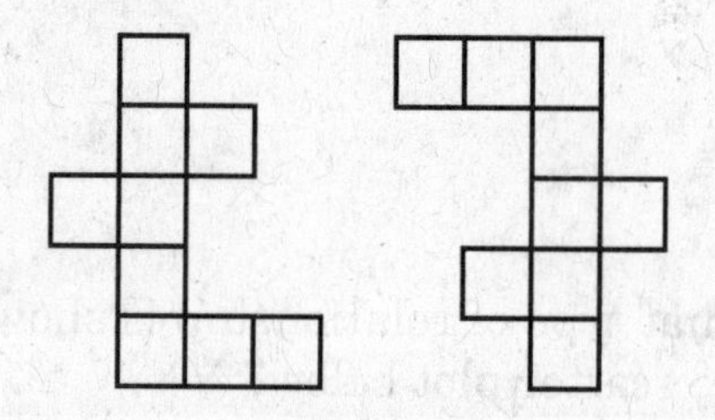

27. Which two points form a line that has a slope of $\frac{5}{2}$?

A. (3, 6) and (−1, −4)

B. (−4, 2) and (7, −1)

C. (−4, 7) and (−9, 5)

D. (3, −7) and (8, 4)

28. What is the constant rate of change of the function represented in the table below?

x	y
−6	−7
−3	−1
0	5
3	11

F. 2

G. 3

H. 5

I. 6

29. **SHORT ANSWER** What is the equation of the line that passes through (−6, −6) and (12, 9)?

30. Which transformations could have been used to map Figure A onto Figure B?

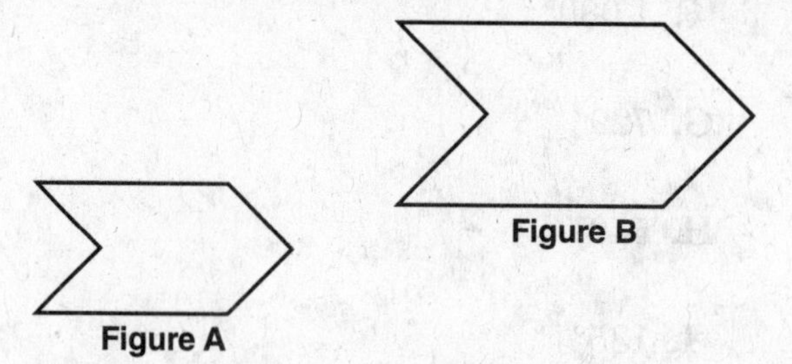

A. dilation, translation

B. dilation, reflection

C. reflection, rotation

D. translation, rotation

Course 3 Benchmark Test – End of Year *(continued)*

31. Katie is 5 feet tall. She casts a 3-foot long shadow at the same time that a flagpole casts an 18-foot long shadow.

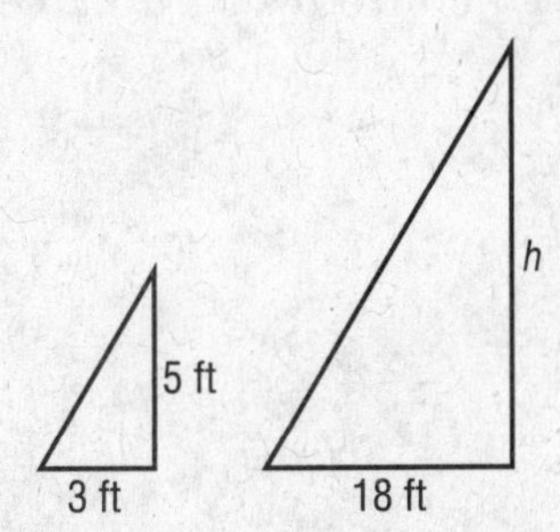

What is the height of the flagpole?

F. 10.8 ft

G. 22.4 ft

H. 28 ft

I. 30 ft

32. What is the approximate surface area of a cylinder with a height of 12 meters and a base radius of 2 meters? Use 3.14 for π. Round to the nearest tenth if necessary.

A. 242.1 m^2

B. 175.8 m^2

C. 150.7 m^2

D. 124.5 m^2

33. The distance from the Sun to Venus is about 1.08×10^{11} meters. If light travels at a speed of 3×10^8 meters per second, about how long does it take light from the sun to reach Venus?

F. 3.6×10^2 seconds

G. 4.2×10^2 seconds

H. 1.083×10^{11} seconds

I. 3.24×10^{19} seconds

34. Which of the following is equivalent to 2^{-4}?

A. -16

B. -8

C. $\frac{1}{32}$

D. $\frac{1}{16}$

35. What is the range of the function shown in the table?

x	−7	−5	−3	−1	1
y	4	6	1	−2	−3

F. all integers

G. all odd integers

H. $\{-3, -2, 1, 4, 6\}$

I. $\{-7, -5, -3, -1, 1\}$

36. **SHORT ANSWER** The area of a square patio is 225 square feet. What is the perimeter of the patio?

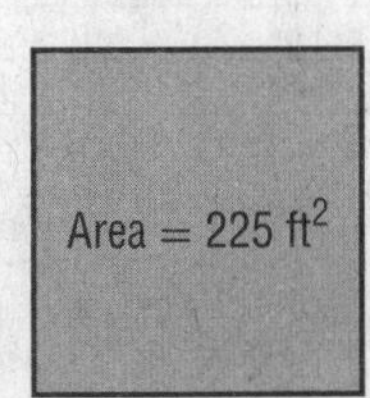

37. A cone has a height of 24 inches, a slant height of 25 inches, and a diameter of 14 inches. What is the surface area of the cone?

A. $1{,}176\pi$ in^2

B. 392π in^2

C. 224π in^2

D. 178π in^2

Course 3 Benchmark Test – End of Year *(continued)*

38. A hotel shuttle service charges \$7.50 plus \$0.85 per mile. A customer hires a shuttle, and the total charge is \$12.60. Which equation can be used to determine the number of miles from the hotel to the airport?

F. $0.85m + 7.5 = 12.6$

G. $7.5m + 0.85 = 12.6$

H. $8.35m = 12.6$

I. $6.65m = 12.6$

39. **SHORT ANSWER** What is the relationship between the slope of the line and the side lengths of the triangles?

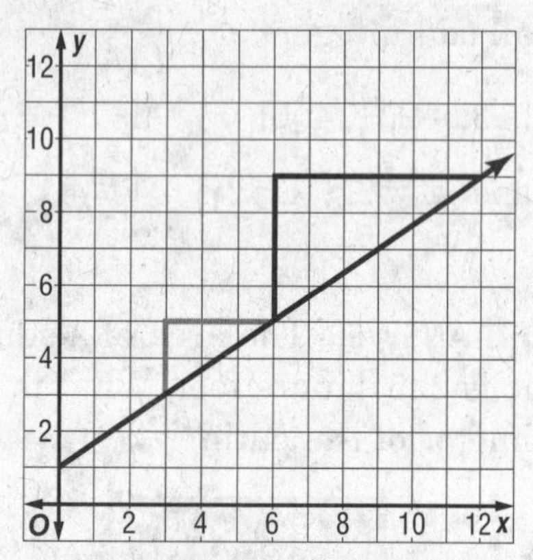

40. The population of the United States is about 3.1×10^8 people. What is this number written in standard form?

A. 3,100,000

B. 31,000,000

C. 310,000,000

D. 3,100,000,000

41. Which expression is equivalent to the expression below?

$$c \cdot c \cdot c \cdot c \cdot d \cdot d \cdot c \cdot d \cdot c \cdot c \cdot d$$

F. $(cd)^3$

G. $c^{-7}d^{-4}$

H. $(cd)^{11}$

I. c^7d^4

42. What is the solution to the system of linear equations shown below?

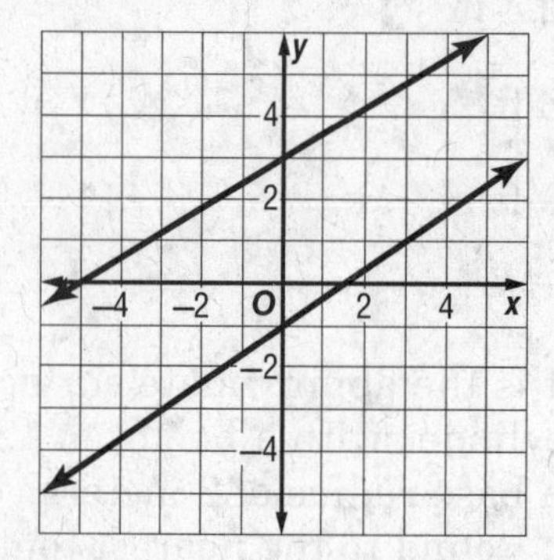

A. (0, 3)

B. (5, 6)

C. (−5, −4)

D. no solution

43. Jasmine determines figure $ABCD \cong$ figure $FGHI$. If $AB = 14$ meters, $BC = 11$ meters, $CD = 9$ meters, and $AD = 17$ meters, what is the length of $\overline{GH}$?

F. 9 m

G. 11 m

H. 14 m

I. 17 m

Course 3 Benchmark Test – End of Year *(continued)*

44. **SHORT ANSWER** Twenty years ago, Mr. Williams purchased a classic car for \$65,000. The table below shows the value of the car over time. Write an equation that represents the data.

Years from Purchase	Value (thousands)
0	\$65
5	\$67.5
10	\$70
15	\$72.5
20	\$75

What will be the value of the car when it has been 30 years since he purchased it?

45. What is the slope of the line that passes through points $R(0, 2)$ and $T(-3, -4)$?

A. 2

B. $\frac{1}{2}$

C. $-\frac{1}{2}$

D. -2

46. Robert has \$220 in his savings account. He plans to save an additional \$15 each week. Which function can Robert use to determine how much he will have saved s after m months?

F. $s(m) = 220m + 15$

G. $s(m) = 235m$

H. $s(m) = 15m + 220$

I. $s(m) = 15m$

47. What type of transformation is represented by the figures below?

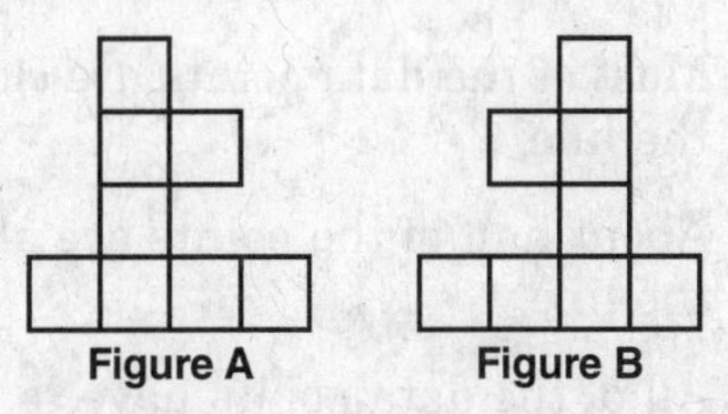

A. dilation

B. reflection

C. rotation

D. translation

48. Which of the following equations represents a vertical line?

F. $y = x$

G. $y = x + 10$

H. $y = 4$

I. $x = 5$

49. Which series of transformations can be used to prove that triangle RST is similar to triangle LMN?

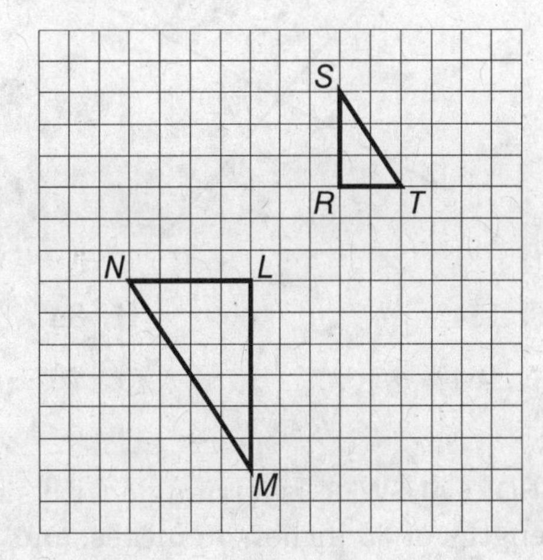

A. reflection, dilation

B. 90° rotation, dilation

C. translation, dilation

D. 180° rotation, dilation

Course 3 Benchmark Test – End of Year *(continued)*

50. Which of the following statements about a line of best fit is *not* true?

F. Most of the data points are close to the line.

G. About half of the points are above the line.

H. All of the data points have to be on the line.

I. The line can be used to make conjectures.

51. The endpoints of $\overline{AR}$ are $A(8, -2)$ and $R(-4, 1)$. What is the length of $\overline{AR}$? Round to the nearest tenth.

A. 12.4 units

B. 11.2 units

C. 7.5 units

D. 4.0 units

52. What is the value of x in the figure below?

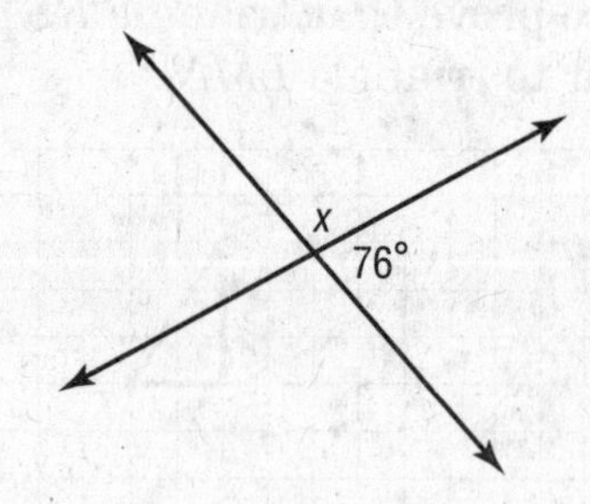

F. 114° **H.** 86°

G. 104° **I.** 76°

53. **SHORT ANSWER** Is a triangle with side lengths of 33 inches, 56 inches, and 65 inches a right triangle? Explain your reasoning.

54. Which set lists the values below from least to greatest?

$$3^{-2}, \sqrt{3}, 1.3 \times 10^{-1}, \frac{1}{3}$$

A $\left\{\sqrt{3}, \frac{1}{3}, 1.3 \times 10^{-1}, 3^{-2}\right\}$

B $\left\{\sqrt{3}, 1.3 \times 10^{-1}, \frac{1}{3}, 3^{-2}\right\}$

C $\left\{3^{-2}, 1.3 \times 10^{-1}, \frac{1}{3}, \sqrt{3}\right\}$

D $\left\{3^{-2}, \frac{1}{3}, 1.3 \times 10^{-1}, \sqrt{3}\right\}$

55. **SHORT ANSWER** The table below shows the prices of digital cameras at an electronics store. Summarize the data.

Prices of Digital Cameras ($)					
75	115	95	105	115	95
100	100	70	80	105	75
120	95	115	175	105	110

56. What is the value of v in the equation below?

$$3(2v + 1) = -15(5v + 16)$$

F. $\frac{13}{81}$ **H.** -2

G. $\frac{5}{27}$ **I.** -3

57. What is the solution to the equation below?

$$0.4p + 0.1 = 1.15$$

A. 3.125 **C.** 0.5

B. 2.625 **D.** 0.42

NAME ______________________ DATE ____________ PERIOD ________

Course 3 Benchmark Test – End of Year *(continued)*

58. Solve the system of equations below.

$$7x + 6y = -10$$
$$-2x + y = 11$$

F. $(-4, 3)$

G. $(-5, 1)$

H. $(7, 9)$

I. no solution

59. The quadratic function $h(t) = -16t^2 + 90$ represents the height, in feet, of an object t seconds after it begins falling from a height of 90 feet. What is the height of the object after 2 seconds?

A. 22 ft

B. 26 ft

C. 58 ft

D. 154 ft

60. Let n represent the figure number in the pattern below.

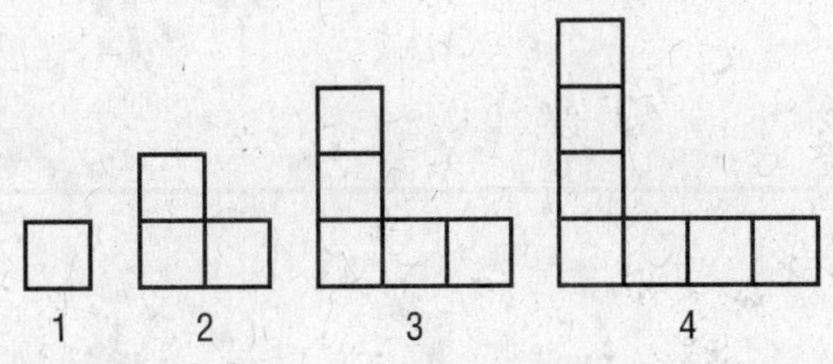

Which function represents the number of squares used to create each figure?

F. $f(n) = n^2$

G. $f(n) = n^2 - 1$

H. $f(n) = 2n - 1$

I. $f(n) = 2n + 1$

61. By what factor would you need to multiply the dimensions of a polygon in order for the resulting image to have a perimeter that is equal to $\frac{1}{4}$ the original perimeter?

A. $\frac{1}{4}$

B. $\frac{1}{2}$

C. 2

D. 4

62. A rectangular-shaped school courtyard has a length of 280 feet and a width of 150 feet wide. What is the approximate length of a diagonal of the courtyard to the nearest tenth?

F. 430.0 ft

G. 395.4 ft

H. 317.6 ft

I. 295.1 ft

63. **SHORT ANSWER** Does the data in the table represent a linear or nonlinear function? Explain your reasoning.

x	y
−7	−37
−2	−7
1	11
5	35
7	47

64. What is the scale factor of the dilated figure shown below?

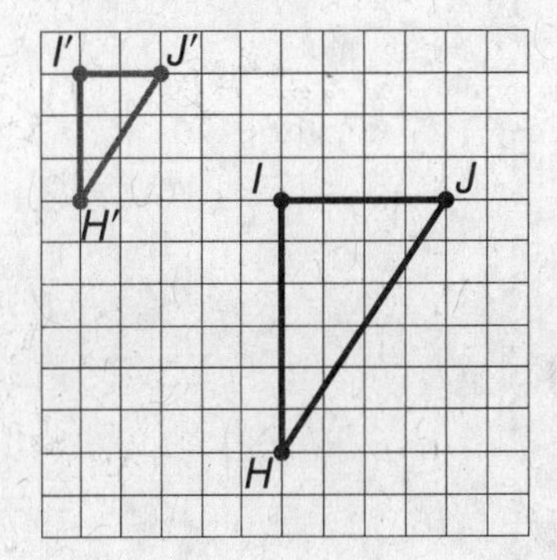

A. 0.25

B. 0.5

C. 2

D. 4

65. Point $A(-7, -3)$ is reflected across the y-axis. What are the coordinates of the image?

F. $A'(3, -7)$

G. $A'(-7, 3)$

H. $A'(-3, -7)$

I. $A'(7, -3)$

Chapter 1 Answer Key

Are You Ready?—Review
Page 1

1. 729
2. 144
3. 3,136
4. 4,608
5. 125
6. 6,561
7. 200
8. 576
9. 576 ft^2
10. $10,125

Are You Ready?—Practice
Page 2

1. 343
2. 32,768
3. −4,500
4. 576
5. 2,916
6. 2,187 calls
7. 6,125 ft^2
8. $-1 \times 2 \times 2 \times 2 \times 2 \times 2 \times 3$
9. $2 \times 3 \times 7$
10. $2 \times 2 \times 2 \times 2 \times 3 \times 3$
11. 1×17
12. $2 \times 3 \times 3 \times 3$
13. $-1 \times 2 \times 5 \times 11$
14. $2 \times 2 \times 2 \times 2$
15. $2 \times 2 \times 3 \times 13$
16. Mr. Smith's: $2 \times 2 \times 2 \times 2 \times 7$
    Ms. Gonzalez's: $2 \times 7 \times 7$
    Mrs. Johnson's: 1×83
    Mr. Cyzdin's: $2 \times 2 \times 2 \times 3 \times 3$

Answers

Chapter 1 Answer Key

Are You Ready?—Apply
Page 3

1. **DISTANCE** The distance from Eli's house to his grandparents' house is $4 \cdot 3 \cdot 4 \cdot 3 \cdot 3$ miles. How many miles away is the grandparents' house? **432 mi**

2. **TEMPERATURE** The table shows the length and width of Florida at its most distant points. Find the prime factorization of each number.

Measurement	Distance (mi)
Length	500
Width	160

Length: $2 \times 2 \times 5 \times 5 \times 5$
Width: $2 \times 2 \times 2 \times 2 \times 2 \times 5$

3. **DOGS** The table shows the weights of the dogs in a local dog show. Find the prime factorization of each number.

Dog	Weights (lb)
Irish Setter	68
Jack Russell Terrier	15
Great Dane	114
Beagle	28

Irish Setter: $2 \times 2 \times 17$
Jack Russell Terrier: 3×5
Great Dane: $2 \times 3 \times 19$
Beagle: $2 \times 2 \times 7$

4. **HAY** Mr. Day feeds the cows on his farm $2 \cdot 7 \cdot 5 \cdot 5 \cdot 5$ pounds of hay per week. How many pounds of hay do they eat per week? **1,750 pounds**

5. **NURSERY** A nursery owner wants to build a new greenhouse that will have $5 \cdot 2 \cdot 5 \cdot 5 \cdot 5$ square feet. How many square feet is the greenhouse? **1,250 ft^2**

6. **FUNDRAISER** The table shows the amount of money each student raised for the school fundraiser. Find the prime factorization of each number.

Student	Money Raised
Dorsey	\$125
Danica	\$88
Jazzra	\$96
Allen	\$150

Dorsey: $5 \times 5 \times 5$
Danica: $2 \times 2 \times 2 \times 11$
Jazzra: $2 \times 2 \times 2 \times 2 \times 2 \times 3$
Allen: $2 \times 3 \times 5 \times 5$

Chapter 1 Answer Key

Diagnostic Test
Page 4

1. 27
2. 7,776
3. −10,125
4. 392
5. 675
6. 4,096 calls
7. 310 mi
8. $-1 \times 2 \times 41$
9. 2×19
10. $2 \times 2 \times 2 \times 2 \times 13$
11. 1×11
12. $2 \times 2 \times 3 \times 5$
13. $-1 \times 2 \times 3 \times 13$
14. $2 \times 2 \times 5$
15. $2 \times 2 \times 31$
16. Oksana: 1×59
    Silvia: 2×31
    Joseph: $2 \times 2 \times 2 \times 2 \times 2 \times 2$
    Diego: $2 \times 3 \times 11$

Pretest
Page 5

1. 8^3
2. 4^4
3. −125
4. 729
5. 8^6
6. 7^6
7. $\frac{1}{9^4}$
8. $\frac{1}{6^3}$
9. 36,800
10. 0.00007924
11. −6
12. $\frac{4}{5}$
13. 3
14. 8

Answers

Chapter 1 Answer Key

Chapter Quiz
Page 6

1. 7.2
2. $4\frac{5}{8}$
3. $-\frac{5}{9}$
4. $4^2 \cdot p^2$
5. $x^4 \cdot y^2$
6. 160
7. 9^8
8. $5x$
9. $5m^2n$
10. 1,296
11. 15,625
12. $-125q^6p^3$
13. $25

Vocabulary Test
Page 7

1. exponent
2. base
3. Scientific notation
4. square root
5. power
6. radical sign
7. perfect squares
8. irrational
9. A monomial is a number, a variable, or a product of a number and one or more variables.
10. The set of rational numbers and the set of irrational numbers together make up the set of real numbers.

Chapter 1 Answer Key

Student Recording Sheet, Page 10

Use this recording sheet with the Standardized Test Practice pages.

Fill in the correct answer. For gridded–response questions, write your answers in the boxes on the answer grid and fill in the bubbles to match your answers.

1. Ⓐ ● Ⓒ Ⓓ

2. 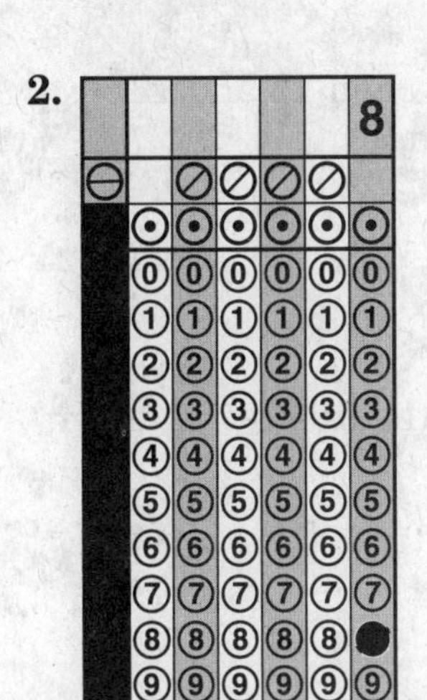

3. Ⓕ Ⓖ ● Ⓘ

4. 60 units

5. 20

6. $5m^7$ ft

7. ● Ⓑ Ⓒ Ⓓ

8. Ⓕ Ⓖ Ⓗ ●

9. Ⓐ Ⓑ Ⓒ ●

10. Ⓕ ● Ⓗ Ⓘ

11. Ⓐ Ⓑ ● Ⓓ

12. Ⓕ Ⓖ ● Ⓘ

Extended Response

Record your answers for Exercise 13 on the back of this paper.

Part A Find the volume of the container, which is $9 \times 9 \times 9$. Find the volume of each block which is $3 \times 3 \times 3$. Divide the volume of the container by the volume of the block.

Part B $\frac{9 \times 9 \times 9}{3 \times 3 \times 3} = 27$

Part C 27 blocks

Answers

Chapter 1 Answer Key

Extended-Response Test, Page 11
Sample Answers

In addition to the scoring rubric, the following sample answers may be used as guidance in evaluating extended response assessment items.

1. a. Sample answer: a number you multiply by itself to get the original number.

b. 2; 6 and -6

c. Sample answer: Since $12^2 = 144$ and $13^2 = 169$, $\sqrt{150}$ will be between 12 and 13. It will be closer to 12 since 150 is closer to 144 than it is to 169.

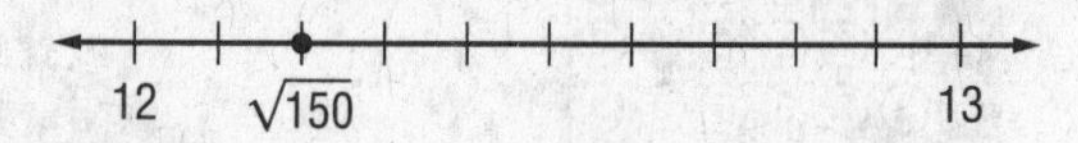

d. Sample answer: a number you use as a factor three times to get the original number.

e. Sample answer: I know that $\sqrt[3]{216} = 6$ and that $220 > 216$. So, $\sqrt[3]{220} > 6$.

2. a. Mercury: 3.6×10^7 miles
Earth: 9.3×10^7 miles
Neptune: 2.799×10^9 miles

b. 57,000,000 miles, 5.7×10^7 miles

c. 2,706,000,000 miles, 2.706×10^9 miles

3. a. The real number system is made up of the sets of rational numbers and irrational numbers. Together, they make up all the points of a number line.

b. Rational nubers are all numbers that can be expressed in the form $\frac{a}{b}$ where a and b are integers and $b \neq 0$. Rational numbers include whole numbers, such as 5, and integers, such as -3. An irrational number is a number that cannot be expressed as $\frac{a}{b}$ where a and b are integers and $b \neq 0$. An example of an irrational number is $\sqrt{2}$.

Chapter 1 Answer Key

Test, Form 1A
Page 13

1. B
2. F
3. D
4. G
5. A
6. H
7. D
8. H
9. D
10. H
11. D

Test, Form 1A *(continued)*
Page 14

12. F
13. B
14. G
15. C
16. H
17. B
18. F
19. A
20. F

Answers

Chapter 1 Answer Key

Test, Form 1B
Page 15

1. B
2. F
3. A
4. H
5. C
6. G
7. A
8. I
9. D
10. H
11. B

Test, Form 1B ***(continued)***
Page 16

12. G
13. B
14. H
15. A
16. F
17. C
18. F
19. D
20. G

Chapter 1 Answer Key

Test, Form 2A
Page 17

1. A
2. I
3. D
4. G
5. D
6. H
7. B
8. G
9. C

Test, Form 2A *(continued)*
Page 18

10. H
11. B
12. $200
13. 19
14. 21 ft
15. See students' work; Sample answers:
Whole: 5
Integer: −9
Rational: $\frac{2}{3}$
Irrational: $\sqrt{5}$
16. 6
17. 5
18. 20 or −20

Answers

Chapter 1 Answer Key

Test, Form 2B
Page 19

1. D
2. F
3. C
4. H
5. B
6. I
7. A
8. I
9. C

Test, Form 2B ***(continued)***
Page 20

10. F
11. C
12. $187.50
13. 3
14. 65 ft
15. See students' work; Sample answers:

Not Whole: -4

Not Integer: $\frac{1}{7}$

Not Rational: $\sqrt{3}$

Not Irrational: -2.1

16. 10
17. 3
18. 30 or -30

Chapter 1 Answer Key

Test, Form 3A
Page 21

1. 43

2. n^4

3. $12x^3y^4$

4. u^{24}

5. $-\frac{7}{c^8}$

6. $4g^4h^6$

7. $2\frac{9}{50}$

8. $\frac{1}{7^5}$

9. 5

10. 3×10^{-4}

11. 0.000307

Test, Form 3A *(continued)*
Page 22

12. 3.84×10^{-2}

13. 30

14. 21,000.0568

15. 9

16. 30 cm

17. 8; since $8^3 = 512$ and $512 > 510$

18. $\pi, \sqrt{13}$

19. 4.09, 4.509, $\frac{229}{50}$, $\sqrt{21}$

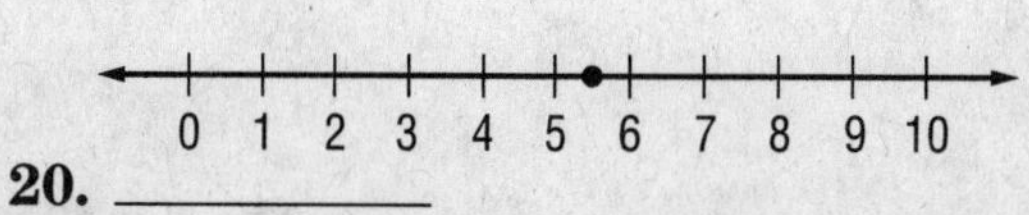

20.

Answers

Chapter 1 Answer Key

Test, Form 3B
Page 23

1. 17

2. p^2

3. $-10x^6y^3$

4. p^{20}

5. $\frac{-4}{d^{15}}$

6. $9m^8n^2$

7. $5\frac{31}{50}$

8. $\frac{1}{a^6}$

9. 4

10. 1.6×10^{-4}

11. 201,000

Test, Form 3B *(continued)*
Page 24

12. 4.73×10^{-5}

13. 100

14. 780.000361

15. 8

16. 24 in.

17. $\sqrt[3]{345}$; since $7^3 = 343$ and $345 > 343$

18. $\pi, \sqrt{19}$

19. 5.09, $\frac{537}{100}$, $\sqrt{29}$, 5.4

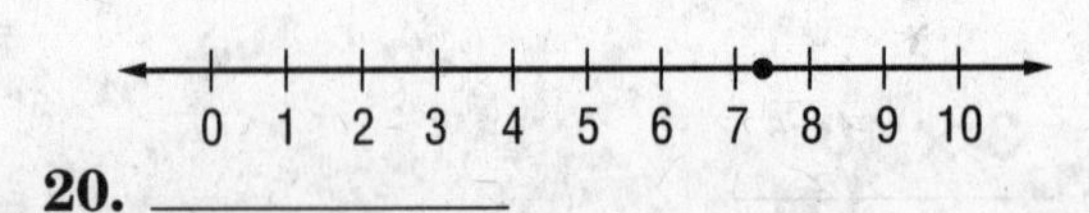

20.

Chapter 2 Answer Key

Are You Ready?—Review
Page 25

1. false
2. true
3. true
4. false
5. false
6. true

7. −80°F; on a number line −80 is to the left of 134.

Are You Ready?—Practice
Page 26

1. −9
2. 40
3. −240
4. 68

5. false
6. false

7. −5

8. $65

Answers

Chapter 2 Answer Key

Are You Ready?—Apply
Page 27

1. **TOURS** Fred and Mary's Ice Cream Company gives tours of their facilities on Mondays and Wednesdays. If 45 people went on the Monday tour and 77 people went on the Wednesday tour, how many people toured the facilities this week? **122 people**

2. **MOVIES** Mr. Ramiro wants to take his 6 adult children to the movies. If one adult ticket cost $9, how much will it cost Mr. Ramiro to take himself and his children to the movies? **$63**

3. **ELEVATION** Death Valley, located in the United States, is 282 feet below sea level. The Sea of Galilee, located in Israel, is 682 feet below sea level. How much further below the lowest point in Death Valley is the lowest point in the Sea of Galilee? **400 ft**

4. **ANGLES** Two angles are said to be supplementary when the sum of their measures is 180°. Find the measure of the supplementary angle to each of the given angle measures.

Angle Measure	Supplementary Measure
35°	**145°**
144°	**36°**
83°	**97°**

5. **POPULATION** In a small town, $\frac{3}{4}$ of the population is older than 55. In a larger town, $\frac{2}{3}$ of the population is older than 55. Which town has the greater fraction of its population over 55, the smaller or the larger town? **smaller**

6. **AVERAGING** Gabe recorded the low temperature on Saturday to be −13° and on Sunday to be −7°. What was the average low temperature over the weekend? **−10°**

Chapter 2 Answer Key

Diagnostic Test
Page 28

1. 50
2. –84
3. –19
4. 21
5. $153
6. 30 mi
7. Melina
8. 1
9. –4
10. –15
11. 40

Pretest
Page 29

1. –14
2. –42
3. 16
4. –30
5. $x + 42 = 78$; 36 students
6. Commutative Property of Addition
7. Associative Property of Multiplication
8. Distributive Property
9. 3
10. 2

Answers

Chapter 2 Answer Key

Chapter Quiz
Page 30

1. 42
2. 54
3. 46
4. 7.5
5. $-\frac{3}{2}$ or $-1\frac{1}{2}$
6. 4
7. 1
8. $92
9. $2x + 5 = 17$
10. $\frac{x}{3} - 4 = 20$
11. h = number of hours; $45 + 15h = 120$; $h = 5$ hours
12. m = amount paid each month; $12m + 125 = 725$; $m = \$50$

Vocabulary Test
Page 31

1. Properties
2. multiplicative inverses
3. two-step equation
4. coefficient
5. Sample answer: Another word for empty set, it is when an equation has no solution.
6. Sample answer: An equation that is true for every value of the variable.

Chapter 2 Answer Key

Student Recording Sheet, Page 34

Use this recording sheet with the Standardized Test Practice pages.

Fill in the correct answer. For gridded-response questions, write your answers in the boxes on the answer grid and fill in the bubbles to match your answers.

1. ● Ⓑ Ⓒ Ⓓ
2. $n + 6 = 23$
3. Ⓕ Ⓖ Ⓗ ●
4. Ⓐ Ⓑ ● Ⓓ
5. $18m^7n^3$
6. | 4 | . | 4 | | | |
   |---|---|---|---|---|---|
7. Ⓕ ● Ⓗ Ⓘ
8. 3
9. Ⓐ Ⓑ Ⓒ ●
10. Ⓕ Ⓖ ● Ⓘ
11. 3
12. −4
13. Ⓕ Ⓖ ● Ⓘ
14. 5^4
15. Ⓐ ● Ⓒ Ⓓ
16. ● Ⓖ Ⓗ Ⓘ

17a. $21y + 14$

17b. 3

Extended Response

Record your answers for Exercise 17 on the back of this paper.

Answers

Chapter 2 Answer Key

Page 35, Extended-Response Test
Sample Answers

In addition to the scoring rubric, the following sample answers may be used as guidance in evaluating open-ended assessment items.

1. a. length: $(4.5x - 18)$ ft or $(3.5x - 4)$ ft;
width: $2x - 2$

b. $(13x - 40)$ ft or $(11x - 12)$ ft or $(12x - 26)$ ft

c. 14

2. a. $x(2x + 2) = (2x^2 + 2x)$ ft^2

b. $2x^2 + 2x$

3. a. length: 30 ft; width; 14 ft

b. 420 ft^2

4. 420 ft^2

5. yes; Sample answer: A solution to
$2x^2 + 2x = 420$ is 14.

Chapter 2 Answer Key

Test, Form 1A
Page 37

1. D
2. H
3. B
4. G
5. A
6. H
7. C
8. H

Test, Form 1A ***(continued)***
Page 38

9. D
10. H
11. A
12. G
13. C
14. G
15. B
16. H
17. C

Answers

Chapter 2 Answer Key

Test, Form 1B
Page 39

1. C
2. H
3. B
4. G
5. B
6. H
7. C
8. G

Test, Form 1B ***(continued)***
Page 40

9. C
10. H
11. B
12. G
13. A
14. I
15. C
16. F
17. B

Chapter 2 Answer Key

Test, Form 2A
Page 41

1. C
2. F
3. B
4. I
5. D
6. I
7. B
8. H
9. C

Test, Form 2A *(continued)*
Page 42

10. I
11. C
12. G
13. B
14. F
15. B
16. 2
17. 52 points
18. 12 ft; 8 ft

Answers

Chapter 2 Answer Key

Test, Form 2B
Page 43

1. C
2. I
3. D
4. F
5. A
6. H
7. B
8. G
9. D

Test, Form 2B ***(continued)***
Page 44

10. G
11. C
12. H
13. D
14. H
15. B
16. –3
17. 12 times
18. 40 ft; 15 ft

Chapter 2 Answer Key

Test, Form 3A
Page 45

1. s = Susan's age; $2s - 5 = 51$
2. $2b + 56.8 = 180.4$; $61.8g$
3. $3t + 15 = 90$; $25
4. -0.5
5. $\frac{1}{10}$
6. 5
7. 2
8. $-7\frac{1}{5}$
9. -2.2
10. 36 movies

Test, Form 3A *(continued)*
Page 46

11. 8
12. 22
13. all real numbers
14. 3
15. null set
16. 12 points
17. 15 bulbs

18a. p = cost of popcorn $2p + 4(p - 3.50) = 32.50$

18b. $7.75

Answers

Chapter 2 Answer Key

Test, Form 3B
Page 47

1. c = number of Bobbie's comic books; $2c + 9 = 95$

2. $4d + 18.75 = 26.55$; $1.95

3. $2g + 30 = 202$; 86 points

4. −13

5. $\frac{1}{2}$

6. 6

7. 6

8. 9

9. −3.5

10. 48 movies

Test, Form 3B ***(continued)***
Page 48

11. 4

12. −12

13. null set

14. −1

15. all real numbers

16. 1 hit

17. 20 fish

18a. t = cost of lunch; $5t + 3(t - 5.25) = 60.25$

18b. $9.50

Chapter 3 Answer Key

Are You Ready?—Review
Page 49

1. −20
2. 1
3. −6
4. −12
5. 13
6. −3
7. 12
8. 0
9. 17°F

Are You Ready?—Practice
Page 50

1. 3
2. −5
3. −11
4. 26
5. 180°
6. 103 yd
7. $\frac{1}{2}$
8. $\frac{1}{5}$
9. −2
10. −1
11. $\frac{3}{5}$
12. −2

Answers

Chapter 3 Answer Key

Are You Ready?—Apply
Page 51

1. **DEBT** Clare had $110 in her savings. She wanted to buy an MP3 player for $118. What is the difference between how much money Clare had and how much she needed? **–$8**

2. **TEMPERATURE** How much warmer is it when the temperature is 17°F than when it is –11°F? **28°F**

3. **KITES** Dexter is flying his kite. Initially he flies his kite at an altitude of 157 feet. The kite then descends 35 feet. What is the height of the kite? **122 ft**

4. **HIKING** Neddy and Louisa were hiking when they came across the following sign along the trail. What is the difference in the elevations of White's Peak and Rock Falls? **210 ft**

Mountain	Elevation
White's Peak	720 feet
Little Meadow	612 feet
Rock Falls	510 feet

5. **SCUBA DIVING** Guadalupe is scuba diving. Four minutes ago, she was at –3 feet, but now is at –15 feet. What is the difference in the two depths? **–12 ft**

6. **RABBITS** Clint and his brother raise rabbits. Clint has 11 rabbits and his brother has 24. What is the difference between the number of rabbits Clint has and the number his brother has? **–13 rabbits**

Chapter 3 Answer Key

Diagnostic Test
Page 52

1. 9
2. –13
3. –2
4. –3
5. 1,405 ft
6. 16°F
7. –1
8. –2
9. 1
10. –1
11. 6

Pretest
Page 53

1. linear; 2
2. –1
3. $4.50
4. –2, 7
5. x-intercept: 5
y-intercept: –2

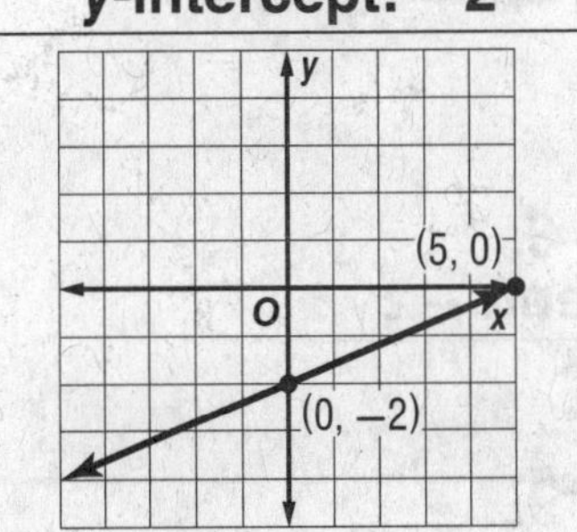

6.

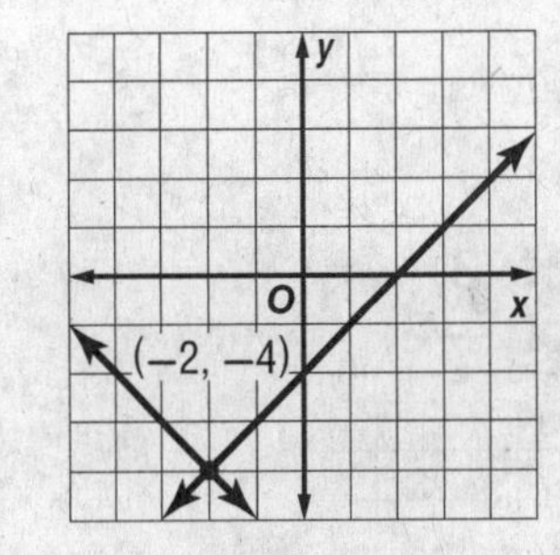

7. (2, –6)

Answers

Chapter 3 Answer Key

Chapter Quiz
Page 54

1. linear; \$25/hour

2. $-\frac{3}{4}$

3. \$46.25

4. slope: −3; *y*-intercept: −2

5. slope: −5; *y*-intercept: 7

6. *x*-intercept: 3, *y*-intercept: −2;

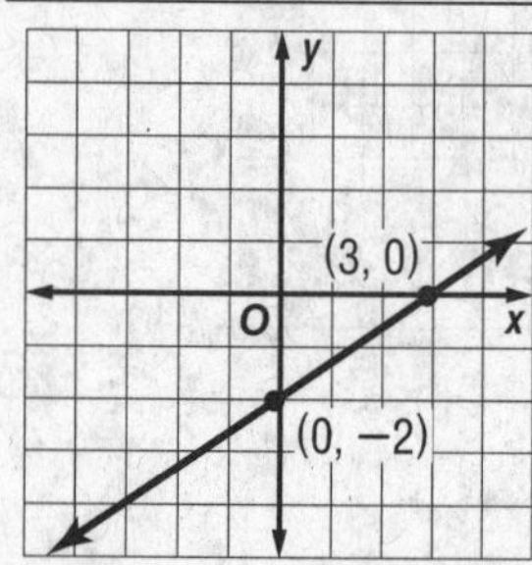

Vocabulary Test
Page 55

1. false
2. true
3. false
4. false
5. true
6. true
7. true
8. false
9. The horizontal change between two points.
10. The *y*-coordinate of point where the graph crosses the *y*-axis.

Chapter 3 Answer Key

Student Recording Sheet, Page 58

Use this recording sheet with the Standardized Test Practice pages.

Fill in the correct answer. For gridded-response questions, write your answers in the boxes on the answer grid and fill in the bubbles to match your answers.

1. ● Ⓑ Ⓒ Ⓓ

2. Ⓕ ● Ⓗ Ⓘ

3. 40.95

4. Ⓐ Ⓑ Ⓒ ●

5. slope: $\frac{1}{2}$; x-intercept: 6; y-intercept: −3

6. Ⓕ ● Ⓗ Ⓘ

7.

8. Ⓐ ● Ⓒ Ⓓ

9.

10. Ⓕ Ⓖ ● Ⓘ

11. Ⓐ ● Ⓒ Ⓓ

Part A $15x + 9y = 90$

Part B The x-intercept is at (6, 0). This means 6 adult tickets could be purchased for \$90. The y-intercept is at (0, 10). This means 10 children's tickets could be purchased for \$90.

Part C $y = -\frac{5}{3}x + 10$

Part D

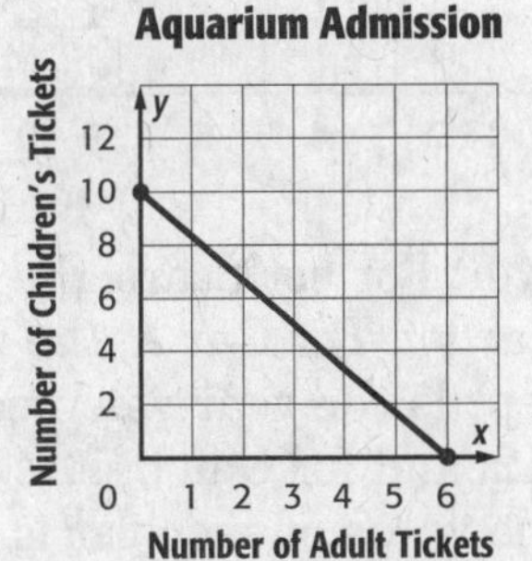

Part E $\frac{-5}{3}$; the rate of change between children's tickets and adults' tickets purchased

Extended Response

Record your answers for Exercise 12 on the back of this paper.

Answers

Chapter 3 Answer Key

Extended-Response Test, Page 59
Sample Answers

In addition to the scoring rubric, the following sample answers may be used as guidance in evaluating extended response assessment items.

1. $24,000; $22,500

2.

Car Value at the End of the Indicated Year	Car A	Car B
1st	$22,000	$21,000
2nd	$20,000	$19,500
3rd	$18,000	$18,000
4th	$16,000	$16,500
5th	$14,000	$15,000

3. –$2,000 per year, –$1,500 per year

4. $v = 24{,}000 - 2{,}000t$; $v = 22{,}500 - 1{,}500t$

5.

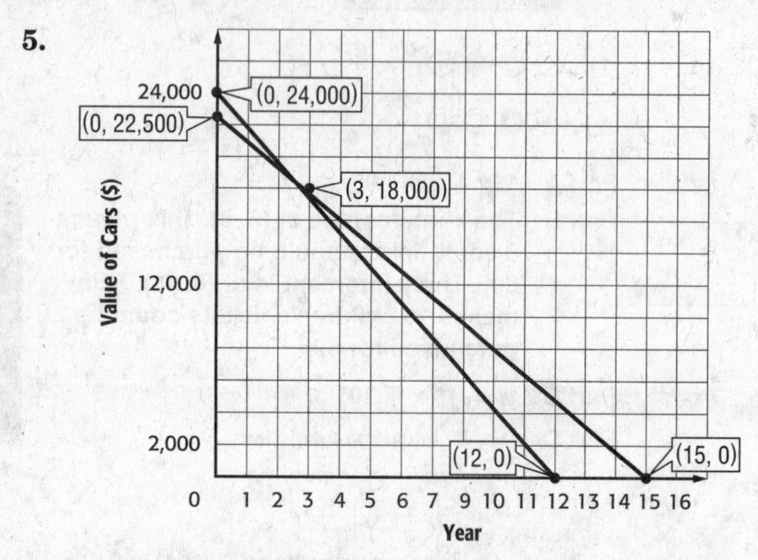

6. Sample answer: For both cars, the y-intercept is the value of the car after driving it off of the car lot. For Car A, the value is $24,000, and for Car B, the value is $22,500. The x-intercept is the year when each car has no value. In the Case of Car A, it is at the end of year 12, and for Car B, it is at the end of year 15. The point of intersection is the year when the value of each car is the same, or the end of the 3rd year with a value of $18,000.

Chapter 3 Answer Key

Test, Form 1A
Page 61

1. C
2. F
3. B
4. I
5. A
6. I

Test, Form 1A ***(continued)***
Page 62

7. C
8. F
9. C
10. G
11. D
12. I
13. C

Answers

Chapter 3 Answer Key

Test, Form 1B
Page 63

1. D
2. F
3. C
4. F
5. C
6. F

Test, Form 1B ***(continued)***
Page 64

7. B
8. G
9. C
10. G
11. D
12. H
13. D

Chapter 3 Answer Key

Test, Form 2A
Page 65

1. C
2. H
3. C
4. G
5. B
6. I
7. D

Test, Form 2A *(continued)*
Page 66

8. F
9. B
10. F
11. 3 tables seating 4 people, 4 tables seating 6 people
12. slope: her savings each week; *y*-intercept: the amount already saved.
13. (4, –5)

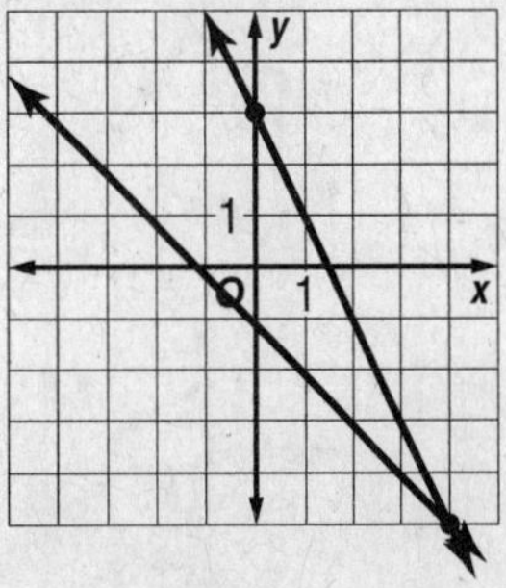

Chapter 3 Answer Key

Test, Form 2B
Page 67

1. C
2. F
3. A
4. H
5. B
6. I
7. D

Test, Form 2B ***(continued)***
Page 68

8. F
9. B
10. H
11. 6 boats seating 3 people, 7 boats seating 5 people
12. slope: amount he saves each week; *y*-intercept: the amount already saved
13. (−1, 1)

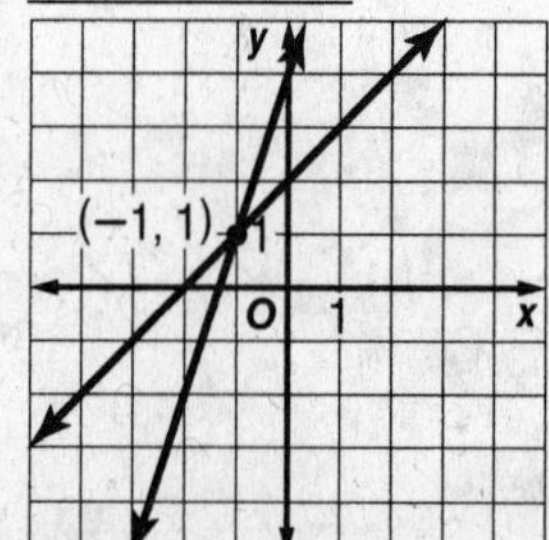

Chapter 3 Answer Key

Test, Form 3A
Page 69

1. 6 oranges, 2 bottles of juice

2. $\frac{22}{5}$

3. $y = 4x$; 26 pictures

4. Store A; Store A: \$3.75 per bottle; Store B: \$3.50 per bottle \$3.75 > \$3.50

5. slope: 8, y-intercept: -12

6. $y = -2x - 3$

7. slope: rate of descent, y-intercept: starting height of the bird

Test, Form 3A *(continued)*
Page 70

8. x-intercept: 4, y-intercept: 10

9a.

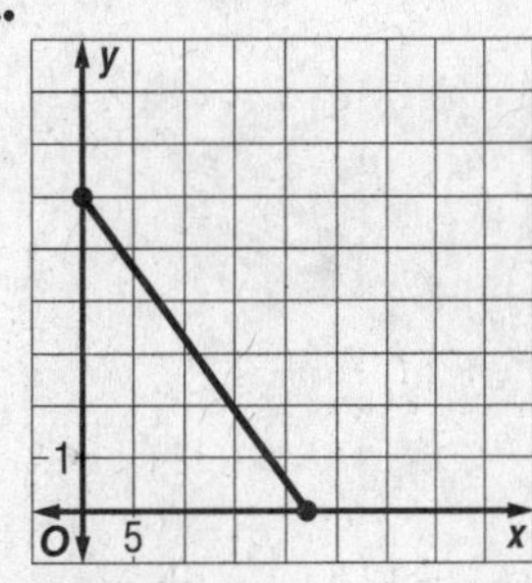

9b. Lakasha brought 21 hats and 0 coats; Lakasha brought 0 hats and 6 coats

10. (2, 4)

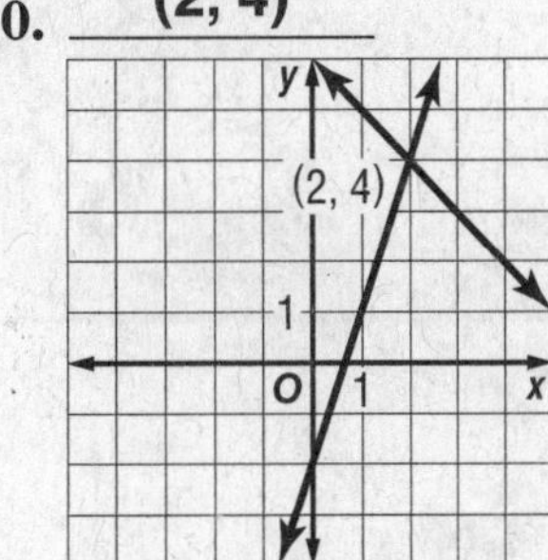

11a. $c + t = 20$, $c = t + 6$

11b. 13 car owners, 7 truck owners

12a. $s + h = 32$, $s = 3h$

12b. $s = 24$, $h = 8$

12c. Isaiah bought 24 soft pieces of candy and 8 hard pieces of candy.

Chapter 3 Answer Key

Test, Form 3B
Page 71

1. 24 bagels, 8 containers of cream cheese

2. $\frac{13}{3}$

3. $y = 5x$; $12\frac{1}{2}$ doz

4. Store B; Store A: \$3.50 per tube; Store B: \$4 per tube; \$4 > \$3.50

5. slope: –7, y-intercept: 3

6. $y = 3x - 1$

7. slope: rate of climb, y-intercept: starting height of bird

Test, Form 3B ***(continued)***
Page 72

8. x-intercept: –6
y-intercept: –9

9a.

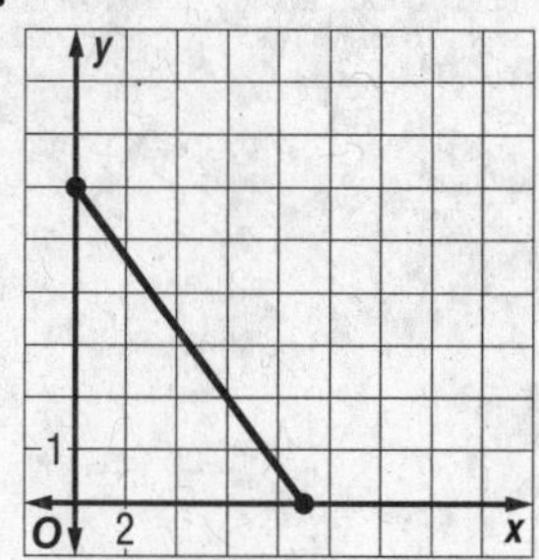

9b. Amy bought 9 streamers and no balloons, Amy bought no streamers and 6 balloons.

10. (1, –1),

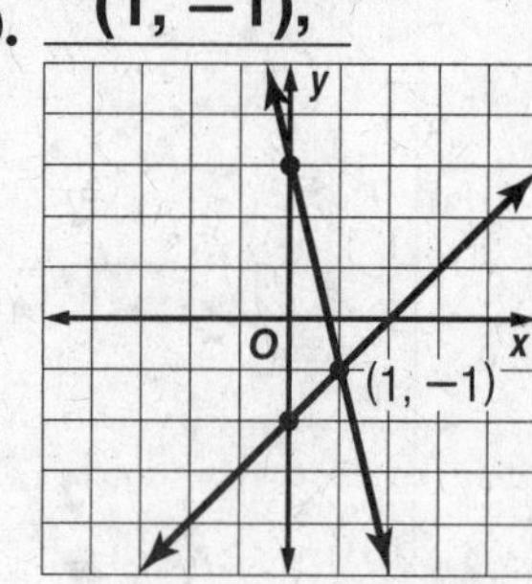

11a. $r + l = 19$, $r = l + 5$

11b. 12 right handed, 7 left handed

12a. $r + b = 35$, $r = 4b$

12b. $r = 28$, $b = 7$

12c. 28 red pieces and 7 black pieces of licorice

Chapter 4 Answer Key

Are You Ready?—Review
Page 73

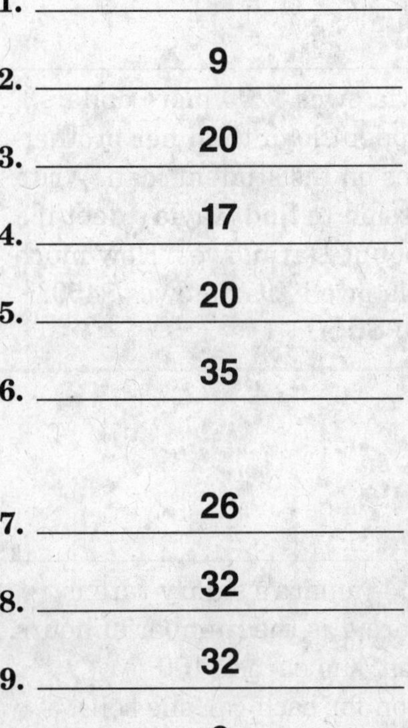

1. 22
2. 9
3. 20
4. 17
5. 20
6. 35
7. 26
8. 32
9. 32
10. 9

Are You Ready?—Practice
Page 74

1. (0, 4)
2. (2, 2)
3. (6, 1)
4. (3, 5)

5–10.

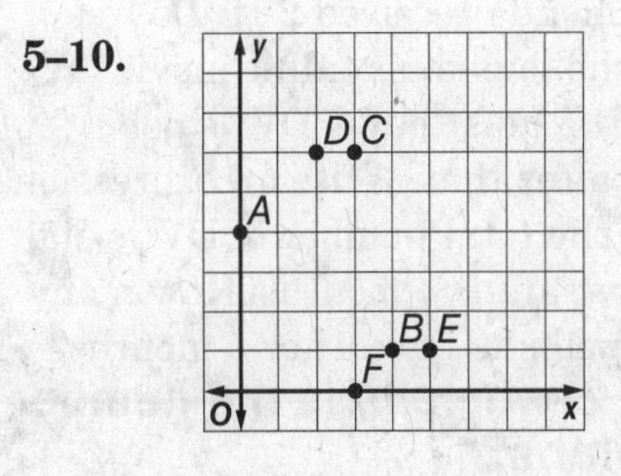

11. 72
12. 26
13. 14
14. 14
15. 12

Answers

Chapter 4 Answer Key

Are You Ready?—Apply
Page 75

1. **WALKING** From his cabin, Rodney walked 2 miles north and 4 miles east, where he rested. If the origin represents the cabin, graph the point on the coordinate grid representing Rodney's resting point.

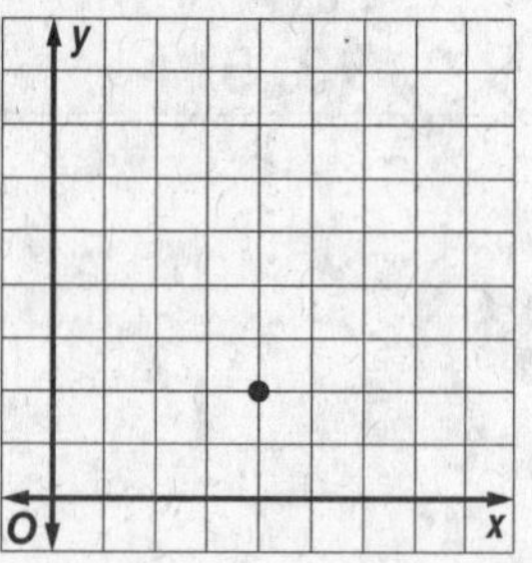

2. **PROFIT** The monthly profit of a flute store is $150x - 1{,}180$, where x represents the number of units sold. Find the monthly profit if the store sells 1,120 flutes. **$166,820**

3. **DVDs** Shonda received 2 DVDs free when she joined an online movie store. She must buy 3 DVDs per month after that. Write an expression to find the total number of DVDs she will have after m months. How many DVDs will she have after 5 months?
$3m + 2$, where m is the number of months; 17

4. **DEBT** Kayla owes $200 more dollars on her student loan than her brother Darin does on his student loan. Write an expression to find Kayla's debt if z is the amount Darin owes. How much does Kayla owe if Darin owes $450?
$z + 200$; $650

5. **CELL PHONE BILL** Jonathan's cell phone bill before taxes is $\$49.99 + 0.10m$, where m is the number of minutes over 400. What is Jonathan's bill if he used 550 minutes? **$64.99**

6. **CAR SALES** Samara's salary rate is $\$17h$, where h is the number of hours. In addition, she earns $100 commission for each car she sells. How much did she earn last week if she worked 40 hours and sold three cars? **$980**

Chapter 4 Answer Key

Diagnostic Test
Page 76

1. (0, 5)
2. (3, 3)
3. (5, 1)
4. (6, 5)
5.

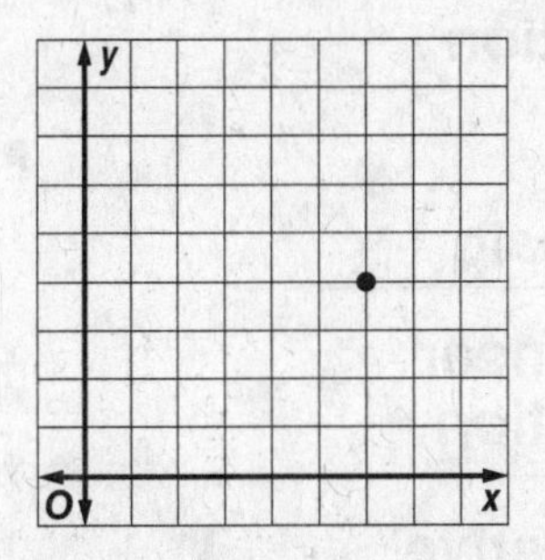

6. 24
7. 27
8. $10\frac{2}{3}$
9. 21
10. 16
11. $2,800

Pretest
Page 77

1. 10
2. domain: −1, 3, 4, 5; range: −3, −2, 0, 1
3. $2n + 1$; 11, 13, 15

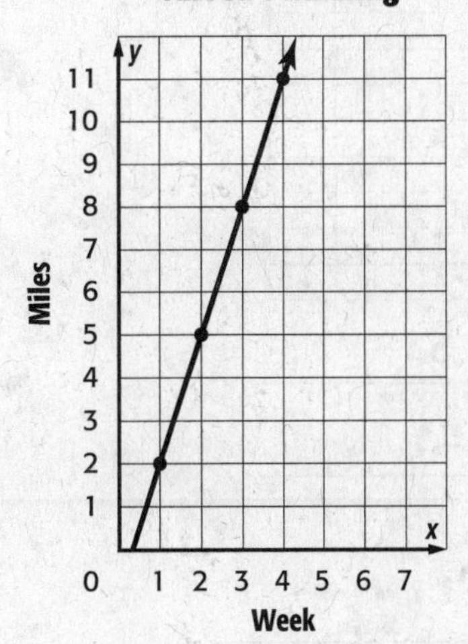

4.

5a. $f(n) = 8.00 + 0.10n$, where n is the number of text messages

5b. $12.60

Number of Texts	Cost ($)
1	8.10
2	8.20
3	8.30
4	8.40

5c.

Answers

Chapter 4 Answer Key

Chapter Quiz
Page 78

1. $m = 8.5h$

2.

Hours, h	Money, m
4	\$34.00
5	\$42.50
6	\$51.00
7	\$59.50

3.

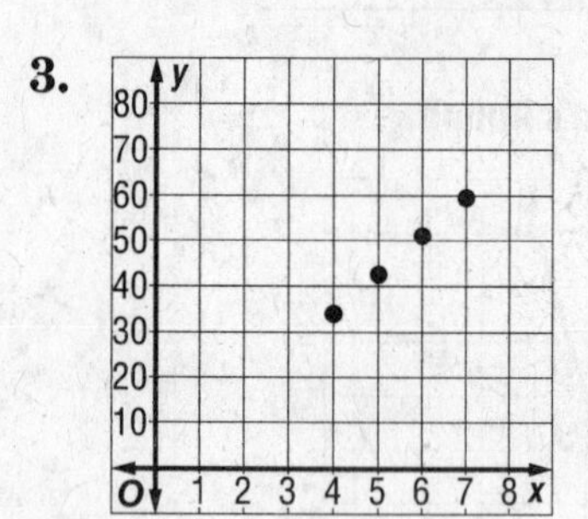

4. 5

5. −1

6.

x	$x + 4$	$f(x)$
−3	−3 + 4	1
−1	−1 + 4	3
2	2 + 4	6
11	11 + 4	15

7–8.

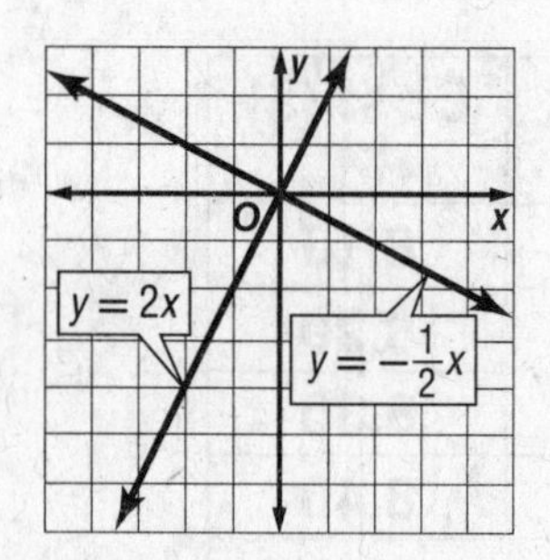

Vocabulary Test
Page 79

1. function table
2. relation
3. domain
4. nonlinear function
5. qualitative graph
6. range
7. discrete data
8. function
9. Sample answer: It is the variable for the range because it depends on the domain.
10. Sample answer: A *quadratic function* is a function in which the greatest power of the variable is 2.

Chapter 4 Answer Key

Student Recording Sheet, Page 82

Use this recording sheet with the Standardized Test Practice pages.

Fill in the correct answer. For gridded-response questions, write your answers in the boxes on the answer grid and fill in the bubbles to match your answers.

1. Ⓐ ● Ⓒ Ⓓ
2. Ⓕ Ⓖ Ⓗ ●
3. {−3, −1, 2, 4}
4. Ⓐ Ⓑ Ⓒ ●
5. Ⓕ Ⓖ ● Ⓘ
6. Ⓐ ● Ⓒ Ⓓ
7. Ⓕ Ⓖ Ⓗ ●
8. $0.05n$
9. Ⓐ Ⓑ Ⓒ ●
10. Ⓕ Ⓖ ● Ⓘ
11. ● Ⓑ Ⓒ Ⓓ

Extended Response

Record your answers for Exercise 12 on the back of this paper.

12. Part A

Number of DVDs	Total Cost ($)
1	5
2	7
3	9
4	11
5	13
6	15
7	17
8	19
9	21
10	23

12. Part B

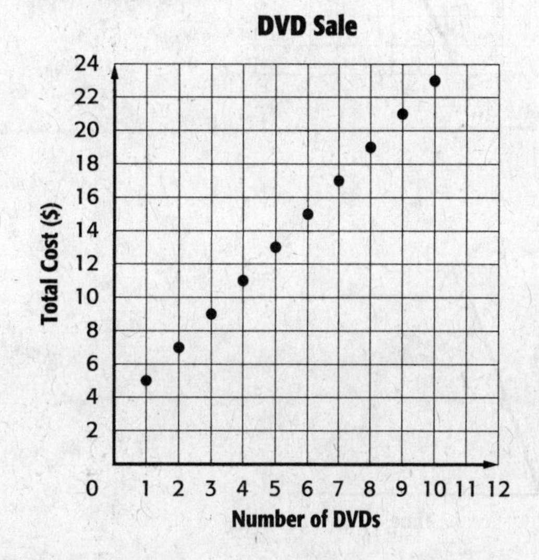

12. Part C $c = 2d + 3$

12. Part D $33

Answers

Chapter 4 Answer Key

Extended-Response Test, Page 83
Sample Answers

In addition to the scoring rubric, the following sample answers may be used as guidance in evaluating extended response assessment items.

1. a.

x	$2x - 3$	y
−2	2(−2) − 3	−7
−1	2(−1) − 3	−5
0	2(0) − 3	−3
1	2(1) − 3	−1

b. Linear; as the x values go up by 1, the y values go up by 2 each time. Since the rate of change is constant, the function is linear.

c. The graph is a straight line.

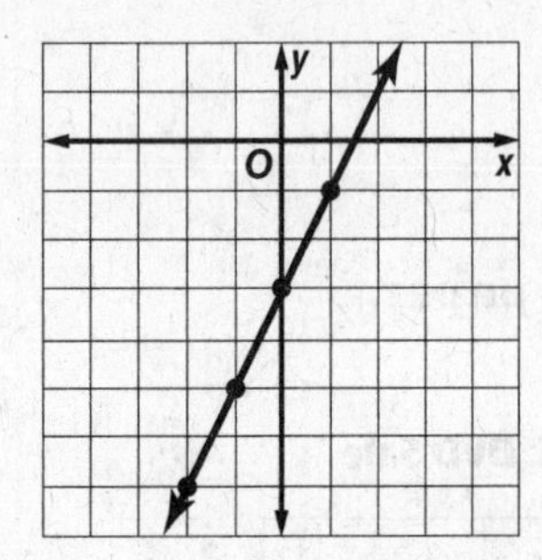

2.

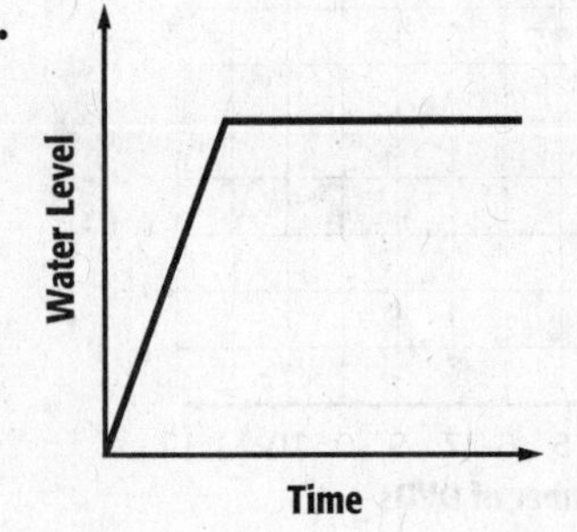

Chapter 4 Answer Key

Test, Form 1A
Page 85

1. C

2. G

3. C

4. F

5. D

6. H

Test, Form 1A ***(continued)***
Page 86

7. B

8. G

9. A

10. F

11. B

12. I

Chapter 4 Answer Key

Test, Form 1B
Page 87

1. A
2. F
3. C
4. I
5. A
6. I

Test, Form 1B *(continued)*
Page 88

7. D
8. F
9. C
10. I
11. A
12. G

Chapter 4 Answer Key

Test, Form 2A
Page 89

1. D

2. G

3. C

4. F

5. D

6. G

Test, Form 2A ***(continued)***
Page 90

7. C

8. domain: {−4, 0, 1, 3}; range: {2, 3, 4}

9.

x	$f(x)$
−2	**5**
0	**1**
1	**−1**
2	**−3**

10. $c = 4.5p$

11. continuous; You can have a fraction of a pound when you weigh cantaloupes.

12.

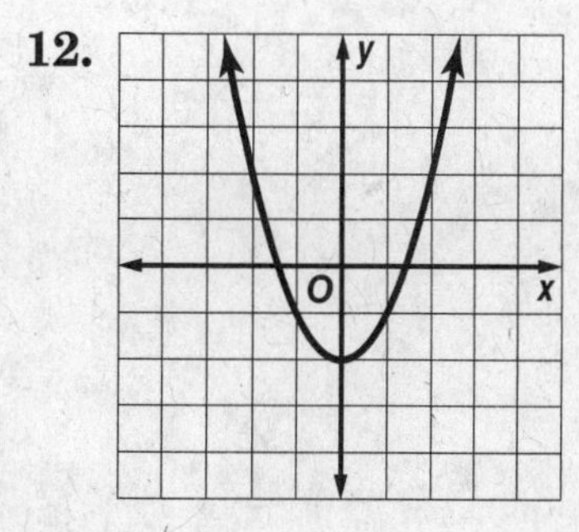

13.

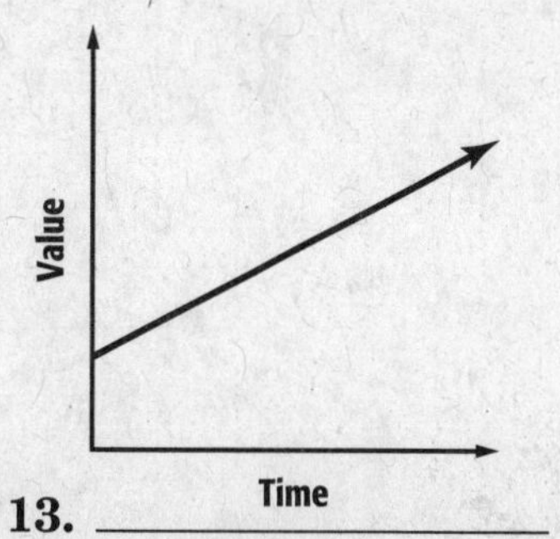

Chapter 4 Answer Key

Test, Form 2B
Page 91

1. A

2. H

3. C

4. G

5. D

6. F

Test, Form 2B ***(continued)***
Page 92

7. D

8. domain: {0, 1, 3, 4}; range: {–3, –1, 2, 4}

9.

x	$f(x)$
–2	**–4**
–1	**–1**
0	**2**
1	**5**

10. $c = 5.3p$

11. continuous; You can have a fraction of a pound when you weigh bacon.

12.

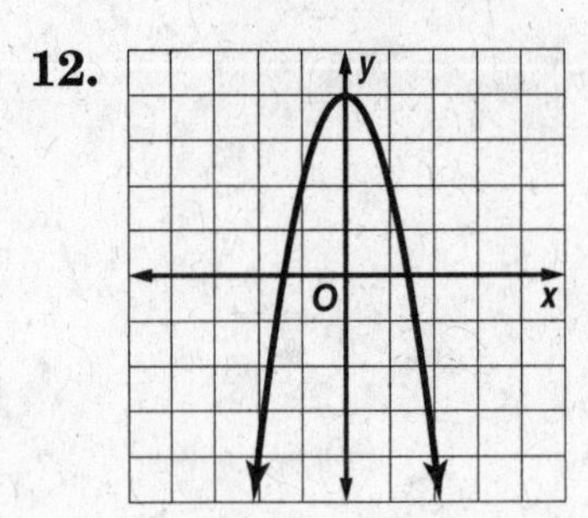

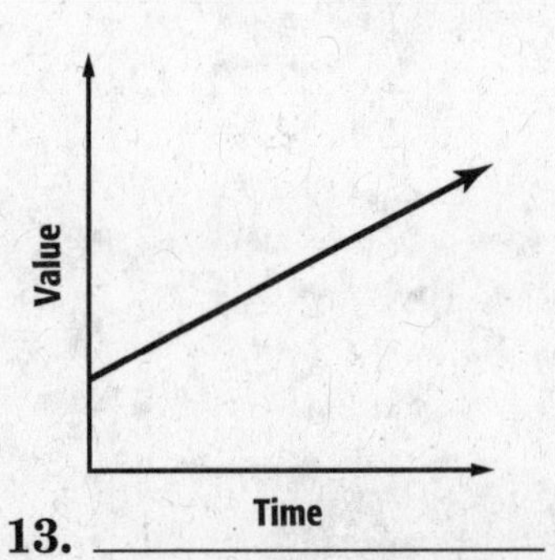

13.

Chapter 4 Answer Key

Test, Form 3A
Page 93

1. $y = 1.5x$

2.

Number of Customers, x	Cost of Coffee, y
5	7.50
10	15.00
15	22.50
20	30.00

3. 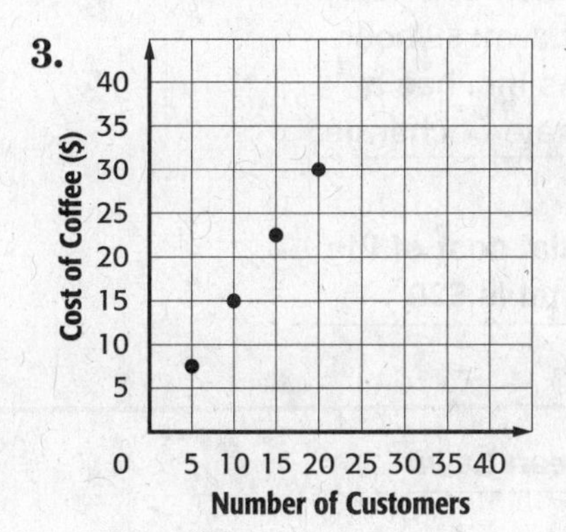

4. domain: {5, 10, 15, 20}; range: {7.5, 15, 22.5, 30}

5. discrete; There can only be a whole number amount of customers.

6. −18

7. −55

Test, Form 3A ***(continued)***
Page 94

8. nonlinear; If you graph the function, the ordered pairs do not lie on a straight line.

9. Carpets Inc.: $20/day; Clark Cleaners: $25/day. Clark Cleaners has a greater rate of change.

10. The initial cost of the rental is $10.

11. Carpets Inc.

12. 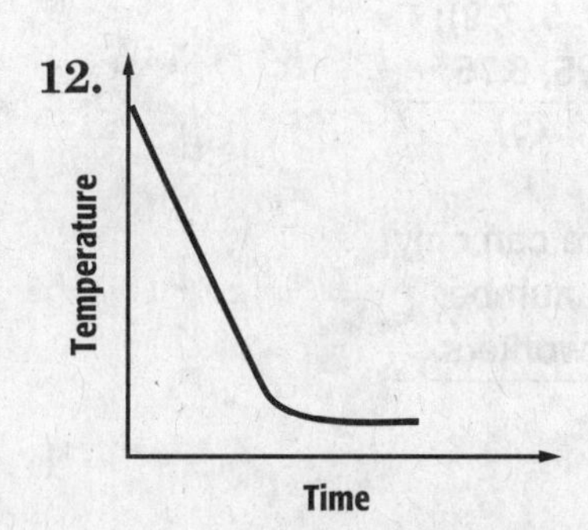

13. 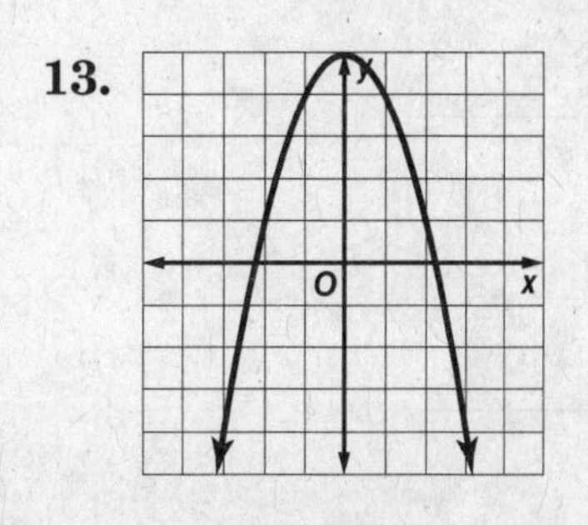

Answers

Chapter 4 Answer Key

Test, Form 3B
Page 95

1. $y = 1.75x$

2.

Number of Workers, x	Cost of Bagels, y
3	5.25
5	8.75
7	12.25
9	15.75

3.

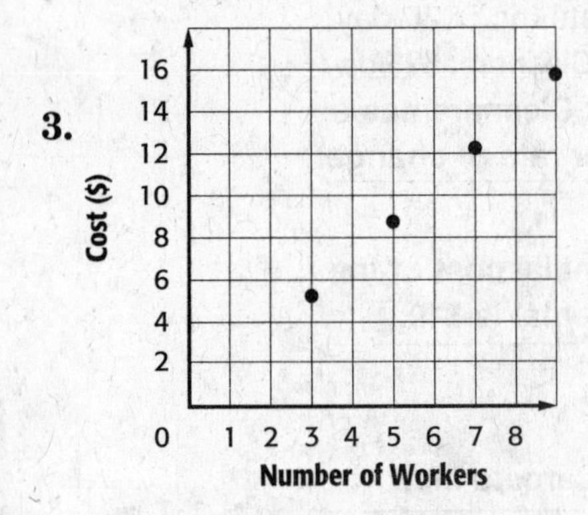

4. domain: {3, 5, 7, 9}; range: {5.25, 8.75, 12.25, 15.75}

5. discrete; There can only be a whole number amount of workers.

6. -19

7. -37

Test, Form 3B ***(continued)***
Page 96

8. nonlinear; If you graph the function, the ordered pairs do not lie on a straight line.

9. Lawns Inc.: $10/hour; Green Lawn: $9/hour. Lawns Inc. has a greater rate of change.

10. The initial cost of the rental is $20.

11. Green Lawn

12.

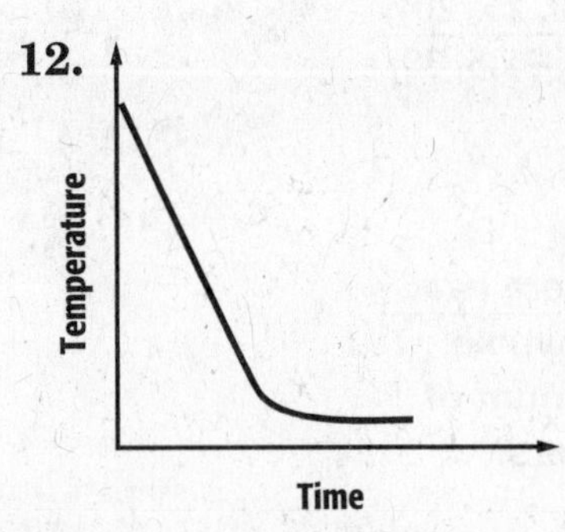

13.

Chapter 5 Answer Key

Are You Ready?—Review
Page 97

1–4. 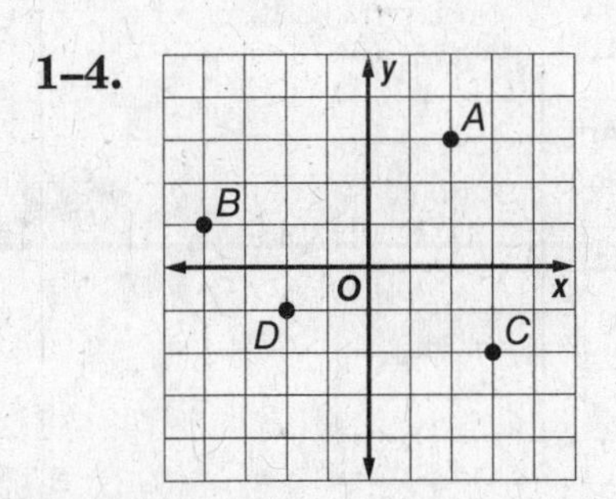

5–8. 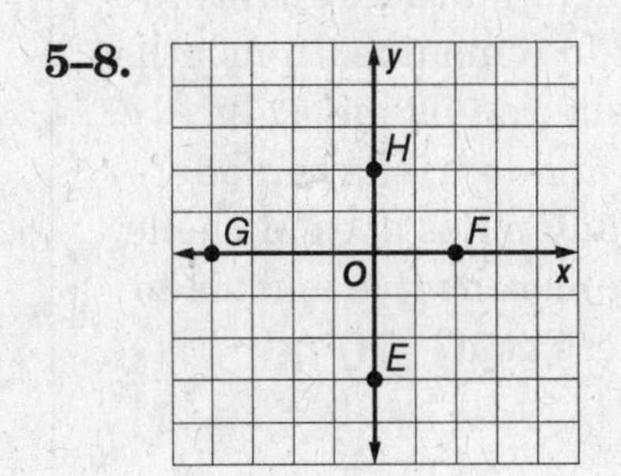

Are You Ready?—Practice
Page 98

1. 56
2. 76
3. 99
4. 71
5. $57

6–10. 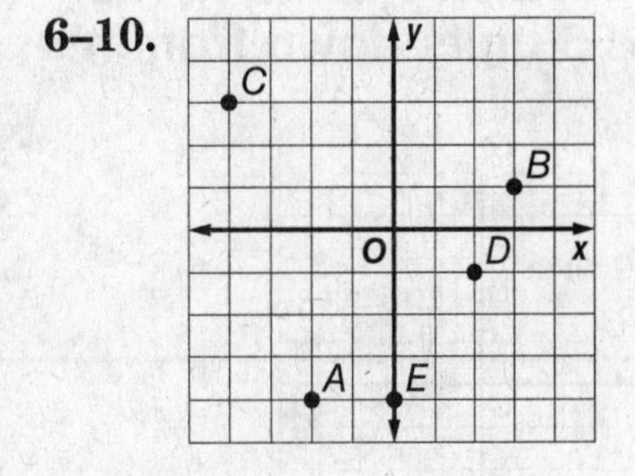

11. Sample answer: Lilac Park is located 3 units to the left and 2 units up from the school.

Chapter 5 Answer Key

Are You Ready?—Apply
Page 99

1. **TRAVEL** The total distance that Fred, Amy, and Leah need to drive is 180 miles. Fred will drive 65 miles and Leah will drive 45 miles. How many miles will Amy need to drive?
70 mi

2. **WEDDING** There were a total of 180 guests at a wedding. Of those guests, 37 ordered fish and 60 ordered chicken. If the remainder of the guests ordered steak, how many ordered steak? **83 guests**

3. **MAPPING** Conner's work place is located at the point (1, –3) on the grid below. Describe the location of his work place with respect to his home.
Sample answer: Conner's work place is located 3 units to the right and 5 units down from his home.

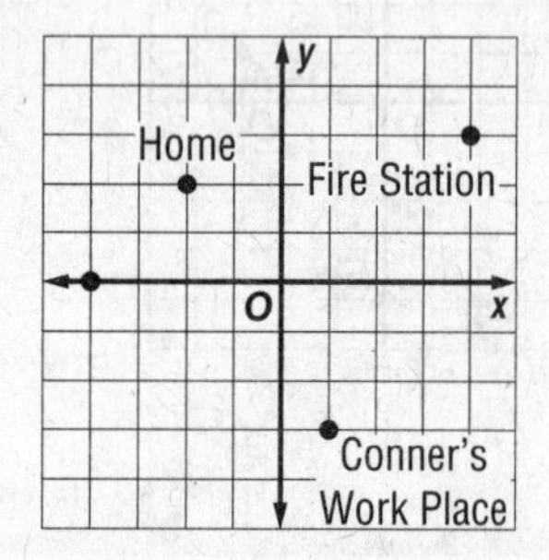

4. **MAPPING** Vikram's martial arts school is located at (–2, 4) on the grid below. Describe the location of his school with respect to his home. **Sample answer: Vikram's school is located 5 units to the left and 3 units up from home.**

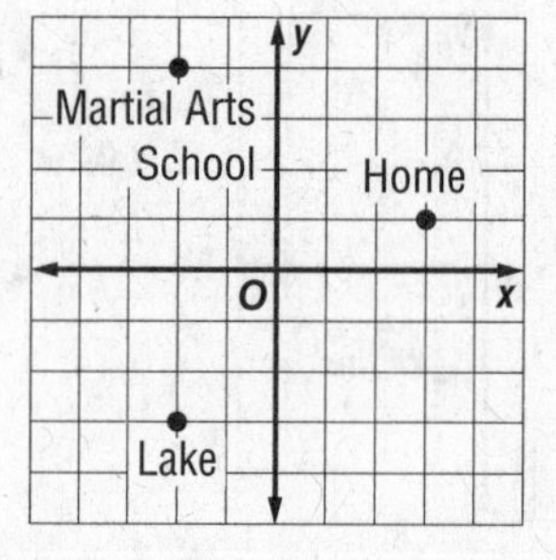

5. **MONEY** Dinner and admission tickets to the aquarium costs $180 for three adults. If two people each pay $70, how much will the third person have to pay? **$40**

6. **RUNNING** Michelle's fitness goal is to run 180 miles in 6 months. In months one and two, she ran 52 miles. In months three and four, she ran 58 miles. How many miles does Michelle need to run in months five and six to reach her fitness goal? **70 mi**

Chapter 5 Answer Key

Diagnostic Test
Page 100

1. 127
2. 84
3. 63
4. 69
5. 101 pounds

6–10.

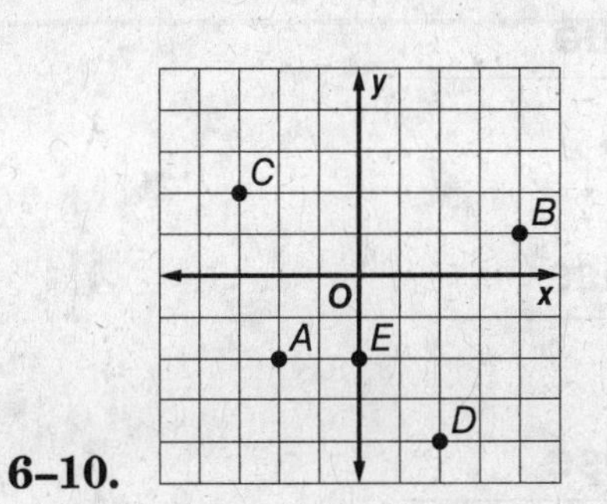

11. Sample answer: The bowling alley is located 5 units to the right and 4 units down from the police station.

Pretest
Page 101

1. 110°
2. 70°
3. 48°, 90°
4. 720°
5. $3.1^2 + 6.4^2 = c^2$; 7.1 in.
6. 9.2 units

Answers

Chapter 5 Answer Key

Chapter Quiz
Page 102

1. adjacent, supplementary
2. 100°
3. 100°
4. 38°
5. 45°
6. 65°
7. 63
8. 45°
9. 12
10. Sample answer. The sum of the measures of a right triangle is 180°. You know that a right angle measures 90°. So, 180 − 90 − 25 = 65.

Vocabulary Test
Page 103

1. true
2. false
3. true
4. false
5. false
6. true
7. Sample answer: a polygon where all of the sides and angles are equal to each other.
8. Sample answer: angles of a polygon formed by two adjacent sides.

Chapter 5 Answer Key

Student Recording Sheet, Page 106

Use this recording sheet with the Standardized Test Practice pages.

Fill in the correct answer. For gridded-response questions, write your answers in the boxes on the answer grid and fill in the bubbles to match your answers.

1. Ⓐ Ⓑ ● Ⓓ
2. 80
3. ● Ⓖ Ⓗ Ⓘ
4. Ⓐ ● Ⓒ Ⓓ
5. $74.92
6. ● Ⓖ Ⓗ Ⓘ
7. Ⓐ Ⓑ ● Ⓓ
8. Ⓕ ● Ⓗ Ⓘ
9. 4×10^{12}
10. Ⓐ ● Ⓒ Ⓓ
11. ● Ⓖ Ⓗ Ⓘ
12. Ⓐ Ⓑ Ⓒ ●
13. Ⓕ ● Ⓗ Ⓘ

14a. (–2, –4)

14b. (3, 4)

14c. 9.4 units

14d. 141 m

Extended Response

Record your answers for Exercise 14 on the back of this paper.

Answers

Chapter 5 Answer Key

Extended-Response Test, Page 107
Sample Answers

In addition to the scoring rubric, the following sample answers may be used as guidance in evaluating extended response assessment items.

1. a. If two angles have the same measure, they are congruent.

b. $\angle 1 \cong \angle 3$, $\angle 5 \cong \angle 7$, $\angle 1 \cong \angle 7$, $\angle 2 \cong \angle 4$, $\angle 6 \cong \angle 8$, $\angle 4 \cong \angle 6$, $\angle 8 \cong \angle 2$, $\angle 3 \cong \angle 5$, $\angle 4 \cong \angle 8$, $\angle 1 \cong \angle 5$, $\angle 2 \cong \angle 6$, $\angle 3 \cong \angle 7$; $m\angle 1 = 104°$, $m\angle 3 = 104°$, $m\angle 4 = 76°$, $m\angle 5 = 104°$, $m\angle 6 = 76°$, $m\angle 7 = 104°$, $m\angle 8 = 76°$

2. a.

The sides of a polygon are all line segments with no curves. A polygon is simple and closed.

b. A regular polygon has all sides congruent and all angles congruent.

3. a. The area of the square on leg $\overline{AC}$, 4, plus the area of the square on leg $\overline{BC}$, 4, equals the area of the square on the hypotenuse, 8.

b. Sample answer: How far is it from the top of a 16-foot pole to a point on the ground 12 feet from the bottom of the pole?

c. Sample answer: The pole makes a right angle with the ground. Thus a right triangle ABC is formed.

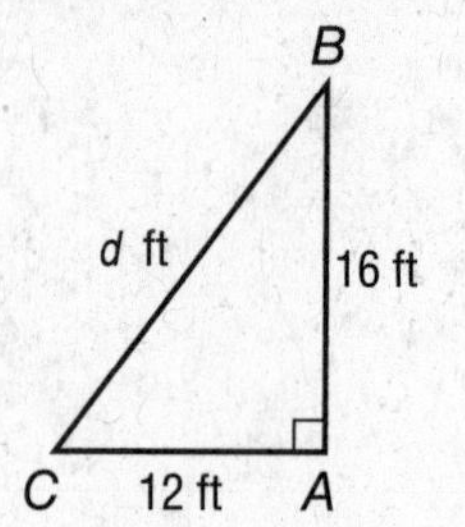

By the Pythagorean Theorem,
$d^2 = 16^2 + 12^2 = 400$. Thus, $d = 20$ or $d = -20$, but -20 is rejected since length cannot be negative.

Chapter 5 Answer Key

Test, Form 1A
Page 109

1. D

2. H

3. A

4. H

5. D

Test, Form 1A ***(continued)***
Page 110

6. F

7. B

8. I

9. A

10. H

11. C

Answers

Chapter 5 Answer Key

Test, Form 1B
Page 111

1. D
2. H
3. C
4. H
5. D

Test, Form 1B ***(continued)***
Page 112

6. I
7. C
8. H
9. A
10. H
11. D

Chapter 5 Answer Key

Test, Form 2A
Page 113

1. D

2. I

3. C

4. H

5. B

Test, Form 2A ***(continued)***
Page 114

6. 75°

7. 20

8. 3,420°

9. $x^2 + 80^2 = 95^2$; 51.2 yd

10. 108°

11. 25.2 ft

12. 17.8 units

Answers

Chapter 5 Answer Key

Test, Form 2B
Page 115

1. D
2. G
3. C
4. G
5. C

Test, Form 2B ***(continued)***
Page 116

6. 115°
7. 65
8. 2,880°
9. $x^2 + 200^2 = 220^2$; 91.7 ft
10. 540°
11. 46.6 ft
12. 12.6 units

Chapter 5 Answer Key

Test, Form 3A
Page 117

1. $x^2 + 30^2 = 37^2$; 21.7 ft

2. $8^2 + 20^2 = x^2$; 21.5 in

3. 5.3 units

4. 10.8 units

5. 72

6. 32

7. 30°, 60°, 90°

Test, Form 3A *(continued)*
Page 118

8. Sample answer: The sum is 90°

9. 48 in.

10. 40°

11. 5,040°

12. 360°

13. No; $10^2 + 24^2 \neq 28^2$

14. Corresponding angles have equal measures.

Answers

Chapter 5 Answer Key

Test, Form 3B
Page 119

1. $x^2 + 20^2 = 25^2$; 15 ft

2. $10^2 + 18^2 = x^2$; 20.6 in

3. 5.6 units

4. 6.7 units

5. 24

6. 48

7. 30°, 30°, 120°

Test, Form 3B ***(continued)***
Page 120

8. Sample answer: They are all 60°.

9. 12 in.

10. 36°

11. 8,640°

12. 210°

13. Yes; $10^2 + 24^2 = 26^2$

14. Corresponding angles have equal measures.

Chapter 6 Answer Key

Are You Ready?—Review
Page 121

1. 160
2. −86
3. 89
4. 7
5. −44
6. 14
7. −48
8. 50
9. 70

Are You Ready?—Practice
Page 122

1. −6
2. −1
3. −13
4. 5
5. −11
6. −1
7. −4
8. 5
9. −4
10. −8
11. Sample answer: *G*(0, 0)

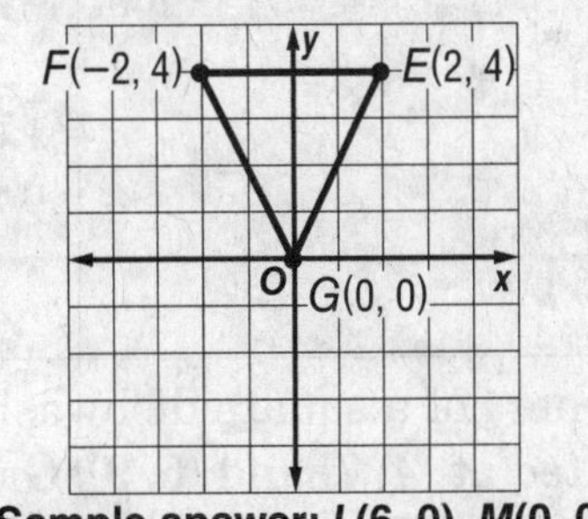

12. Sample answer: *L*(6, 0), *M*(0, 0)

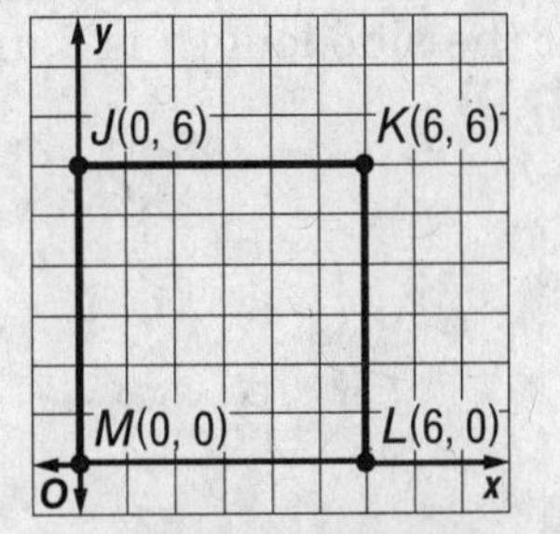

13. *C*(4, 4), *D*(4, 0)

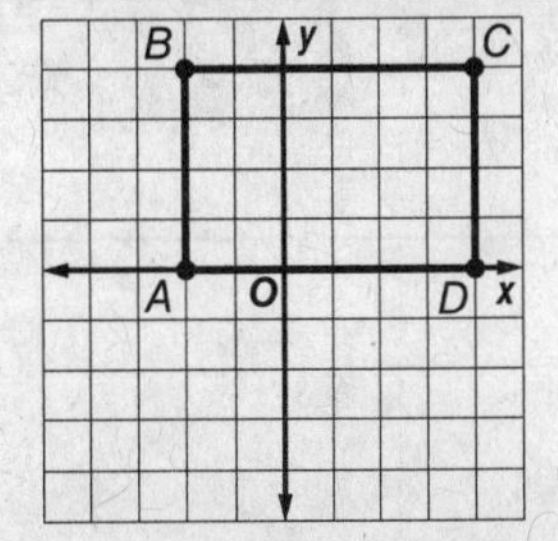

Answers

Chapter 6 Answer Key

Are You Ready?—Apply
Page 123

INTERIOR DESIGN Rudell is designing his kitchen. He wants to place the appliances in his kitchen on a coordinate grid to see how they will look.

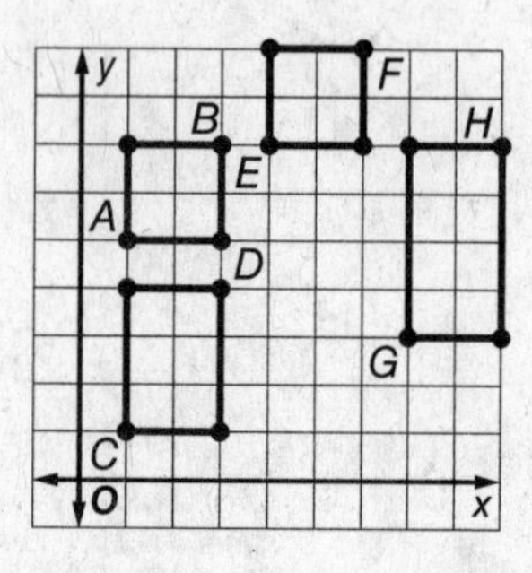

1. Two corners of a square stove are located at (1, 5) and (3, 7). Graph the figure and find the missing vertices if the side length is 2 units. **(3, 5), (1, 7)**	**2.** Two corners of a rectangular refrigerator are located at (1, 1) and (3, 4). Graph the figure and find the missing vertices if the side length is 3 units. **(3, 1), (1, 4)**
3. Two corners of a square dishwasher are located at (4, 7) and (6, 9). Graph the figure and find the missing vertices if the side length is 2 units. **(6, 7), (4, 9)**	**4.** Two corners of a rectangular cabinet are located at (7, 3) and (9, 7). Graph the figure and find the missing vertices if the side length is 4 units. **(9, 3), (7, 7)**

Chapter 6 Answer Key

Diagnostic Test
Page 124

1. −3
2. −7
3. −2
4. 4
5. −10
6. −6
7. −3
8. 1
9. 0
10. −2
11. Sample answer: $Z(0, 6)$

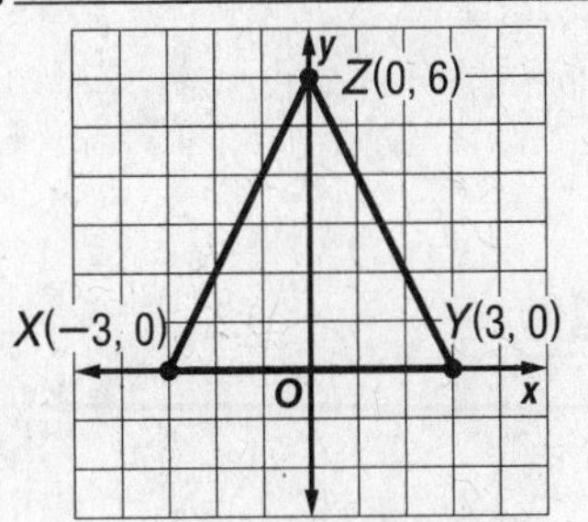

12. Sample answer: $C(3, 3)$, $D(0, 3)$

y
D(0, 3) C(3, 3)
A(0, 0) B(3, 0)
O
x

13. Sample answer: $G(2, 0)$, $H(0, 0)$

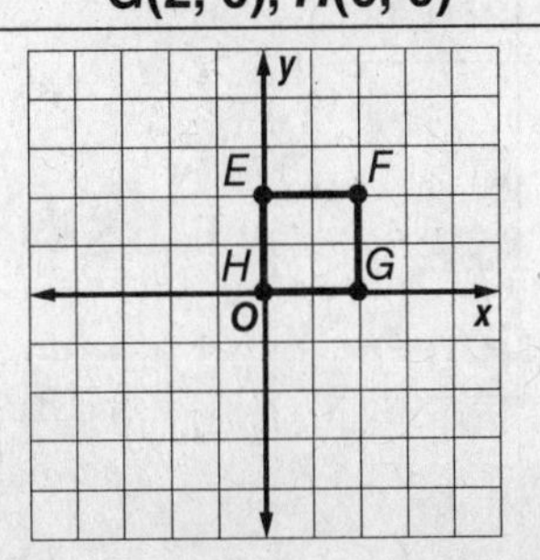

Pretest
Page 125

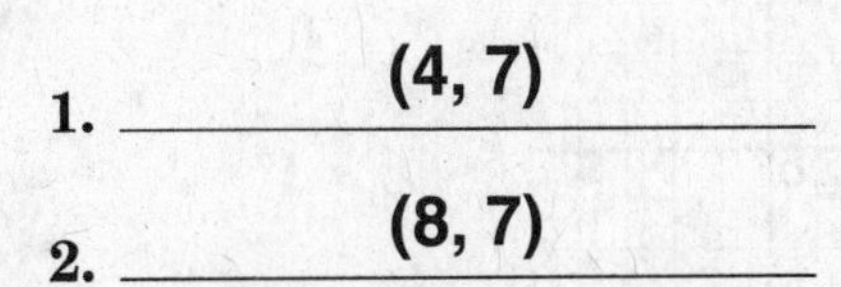

1. (4, 7)
2. (8, 7)
3.

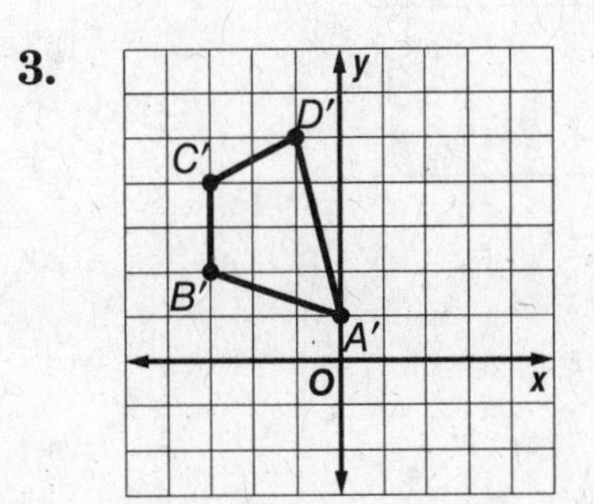

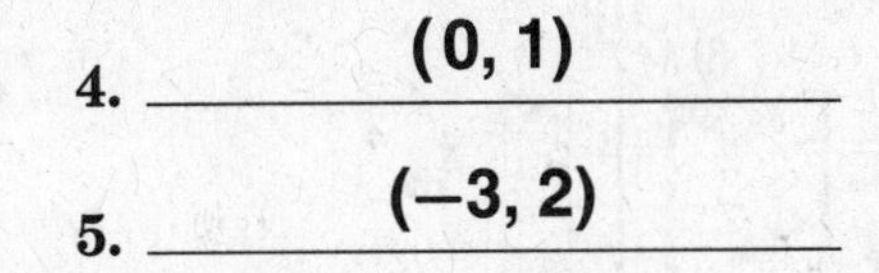

4. (0, 1)
5. (−3, 2)
6. $H'(-1, 0)$, $I'(-4, 0)$, $J'(-2, -1)$, $K'(1, -1)$
7.

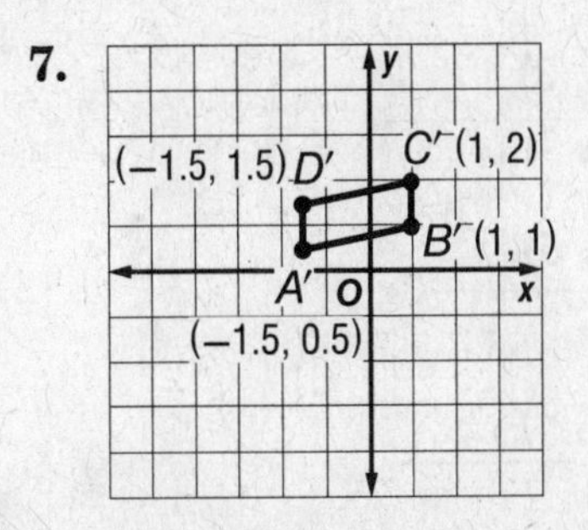

Answers

Chapter 6 Answer Key

Chapter Quiz
Page 126

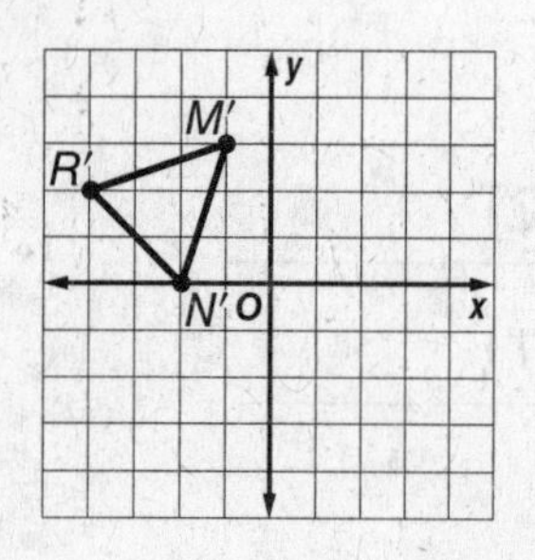

1. ______

2. $M'(-1, 3)$

3. $N'(-2, 0)$

4. $R'(-4, 2)$

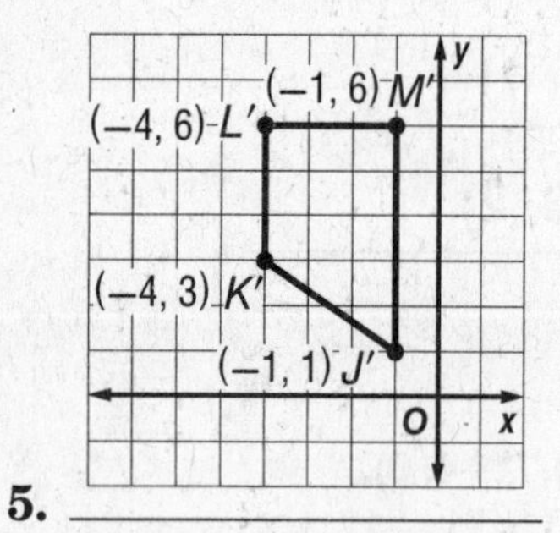

5. ______

6. $J'(-1, 1)$

7. $K'(-4, 3)$

8. $L'(-4, 6)$

9. $M'(-1, 6)$

Vocabulary Test
Page 127

1. transformation

2. preimage

3. rotation

4. image

5. reflection

6. dilation

7. congruent

8. translation

Chapter 6 Answer Key

Student Recording Sheet, Page 130

Use this recording sheet with the Standardized Test Practice pages.

Fill in the correct answer. For gridded-response questions, write your answers in the boxes on the answer grid and fill in the bubbles to match your answers.

1. Ⓐ ● Ⓒ Ⓓ

2. Ⓕ Ⓖ ● Ⓘ

3. Ⓐ Ⓑ ● Ⓓ

4. reflections or translation

5.

2	1	.	9	6	
	⊘	⊘	⊘	⊘	
⊙	⊙	●	⊙	⊙	⊙
⓪	⓪	⓪	⓪	⓪	⓪
①	●	①	①	①	①
●	②	②	②	②	②
③	③	③	③	③	③
④	④	④	④	④	④
⑤	⑤	⑤	⑤	⑤	⑤
⑥	⑥	⑥	⑥	●	⑥
⑦	⑦	⑦	⑦	⑦	⑦
⑧	⑧	⑧	⑧	⑧	⑧
⑨	⑨	⑨	●	⑨	⑨

6. Ⓕ ● Ⓗ Ⓘ

7. Ⓐ Ⓑ ● Ⓓ

8.

5	7	.	6	0	
	⊘	⊘	⊘	⊘	
⊙	⊙	●	⊙	⊙	⊙
⓪	⓪	⓪	⓪	●	⓪
①	①	①	①	①	①
②	②	②	②	②	②
③	③	③	③	③	③
④	④	④	④	④	④
●	⑤	⑤	⑤	⑤	⑤
⑥	⑥	⑥	●	⑥	⑥
⑦	●	⑦	⑦	⑦	⑦
⑧	⑧	⑧	⑧	⑧	⑧
⑨	⑨	⑨	⑨	⑨	⑨

9. Ⓕ ● Ⓗ Ⓘ

10. (4, 6)

11. Ⓐ Ⓑ ● Ⓓ

12. Ⓕ Ⓖ ● Ⓘ

13a. $X'(-1, -3)$, $Y'(4, -1)$, and $Z'(0, 2)$

13b. $X'(1, 0.5)$, $Y'(3.5, 1.5)$, $Z'(1.5, 3)$

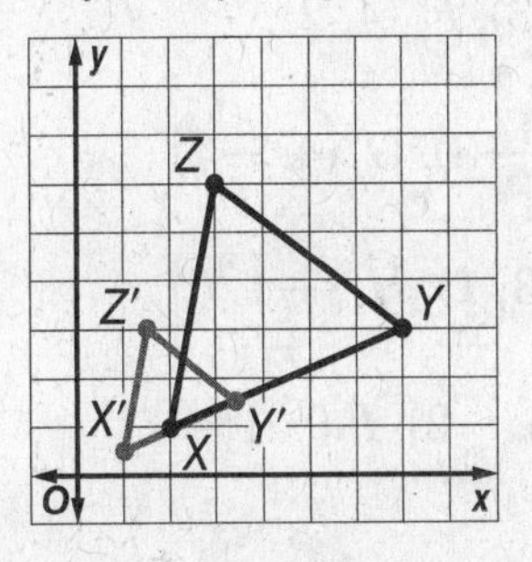

13c.

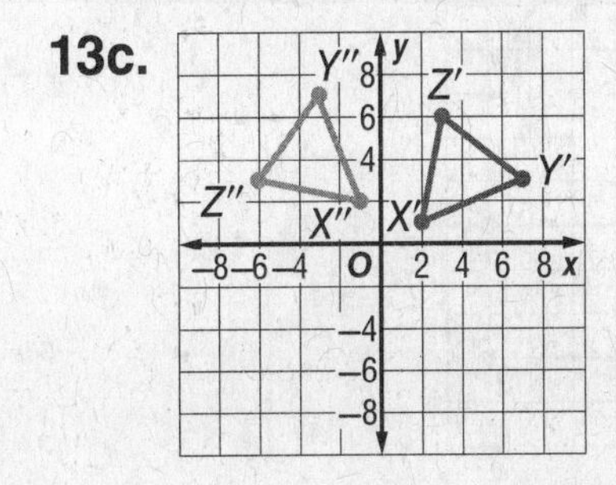

Extended Response

Record your answers for Exercise 13 on the back of this paper.

Answers

Chapter 6 Answer Key

Extended-Response Test, Page 131
Sample Answers

In addition to the scoring rubric, the following sample answers may be used as guidance in evaluating extended response assessment items.

1. The two transformations are the same. Conjecture: Reflecting a figure across the y-axis and then across the x-axis will give the same results as rotating the figure 180° about the origin.

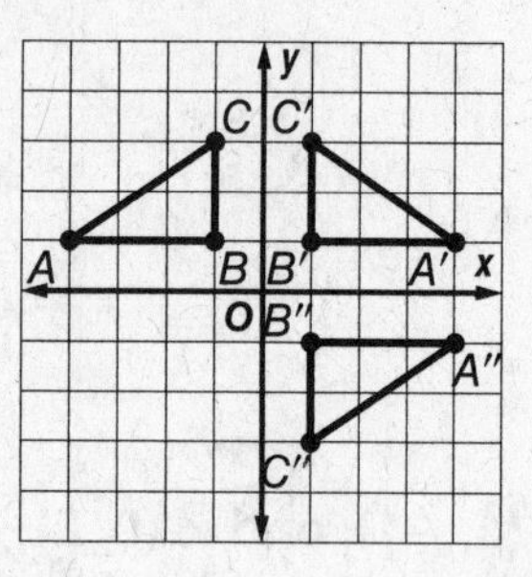

2. **a.** F is at (2, 3), F' (4, 2)

   b. G is at (2, −3), G' (4, −4)

   c. H is at (−3, 1), H' (−1, 0)

   d. I is at (−3, −2), I' (−1, −3)

3.

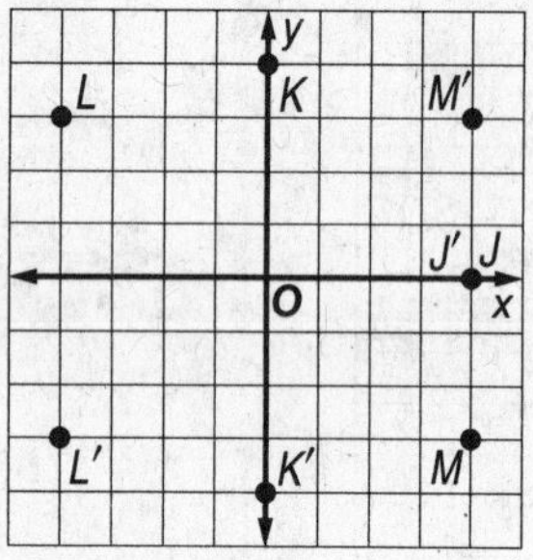

4. **a.** A translation slides a figure up, down, to the left, and/or to the right. A reflection flips the figure across a line. A rotation turns the figure. A dilation enlarges or reduces the figure.

   b. Slide the figure 2 units to right and 3 units down.

   c. Multiply each x-and y-coordinate by 2. Then plot the points.

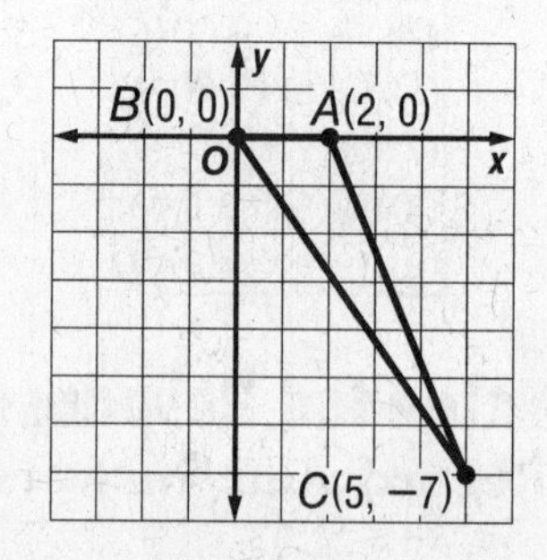

Chapter 6 Answer Key

Test, Form 1A
Page 133

1. C
2. I
3. B
4. H
5. A
6. H

Test, Form 1A ***(continued)***
Page 134

7. D
8. G
9. C
10. G
11. B

Chapter 6 Answer Key

Test, Form 1B
Page 135

1. B

2. G

3. D

4. I

5. D

6. H

Test, Form 1B ***(continued)***
Page 136

7. A

8. I

9. A

10. G

11. C

Chapter 6 Answer Key

Test, Form 2A
Page 137

1. B
2. G
3. D
4. G
5. D
6. F

Test, Form 2A ***(continued)***
Page 138

7. $\frac{1}{2}$
8. reduction
9. $H'(6, -1)$
10. $K'(-4, -1)$
11. $P'(1, 5)$
12. $A'(0, 8)$

Answers

Chapter 6 Answer Key

Test, Form 2B
Page 139

1. D
2. F
3. B
4. G
5. D
6. H

Test, Form 2B ***(continued)***
Page 140

7. $\frac{1}{3}$
8. reduction
9. (4, –3)
10. (3, –4)
11. (1, 8)
12. (1, –2)

Chapter 6 Answer Key

Test, Form 3A
Page 141

1. $N(-2, -1)$, $L(2, -3)$, $M(4, 0)$

2. 2, enlargement

3. $P'(1, -3)$, $Q'(4, -1)$, $R'(-5, -2)$

4. They are the same shape and size.

5. $(x - 4, y + 2)$

Test, Form 3A *(continued)*
Page 142

6. $(-2, -5)$

7. 35 in.

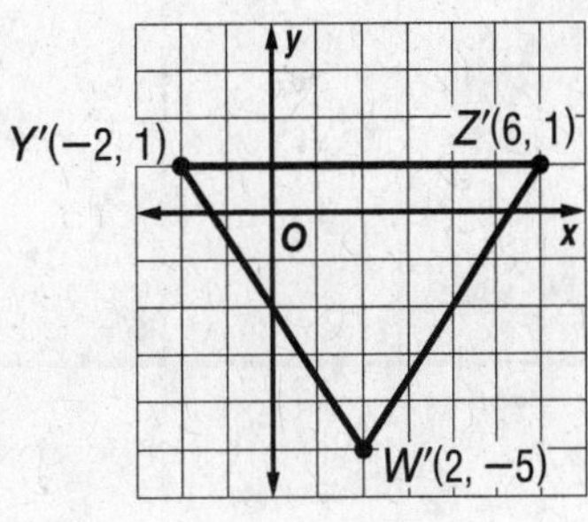

8. ______

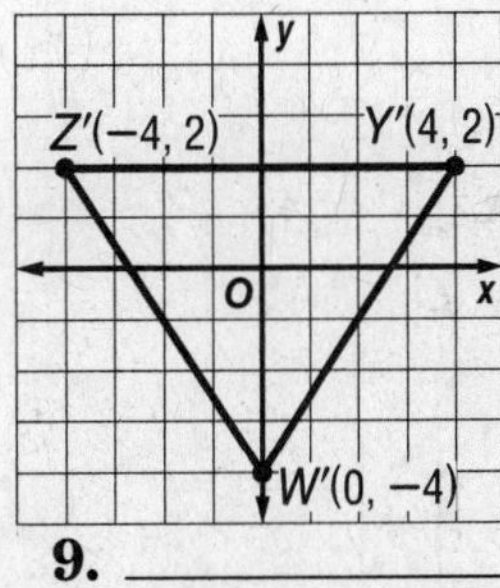

9. ______

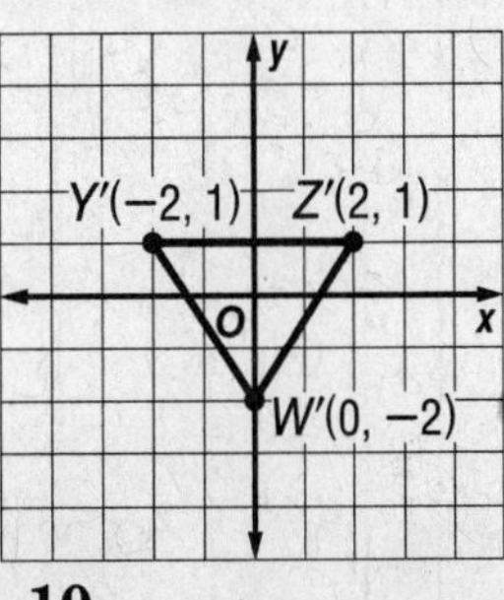

10. ______

Answers

Chapter 6 Answer Key

Test, Form 3B
Page 143

1. $R(2, -3)$, $K(4, -1)$, $G(1, 1)$

2. $\frac{1}{2}$; reduction

3. $L'(1, -3)$, $P'(-2, 1)$, $T'(0, 3)$

4. They are the same shape and size

5. $(x - 6, y + 2)$

Test, Form 3B ***(continued)***
Page 144

6. $Y'(2, -5)$

7. 10 mm

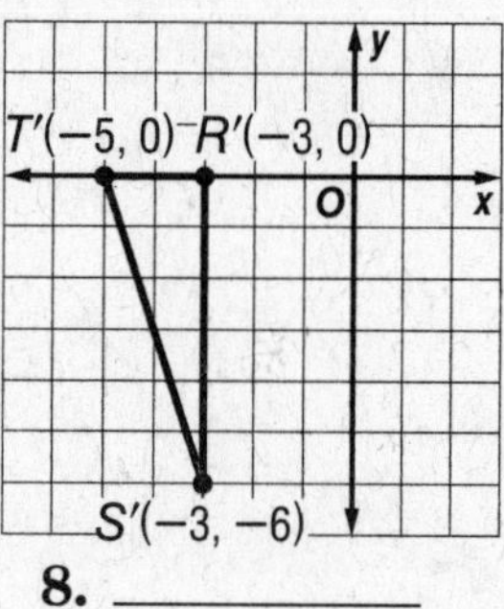

8.

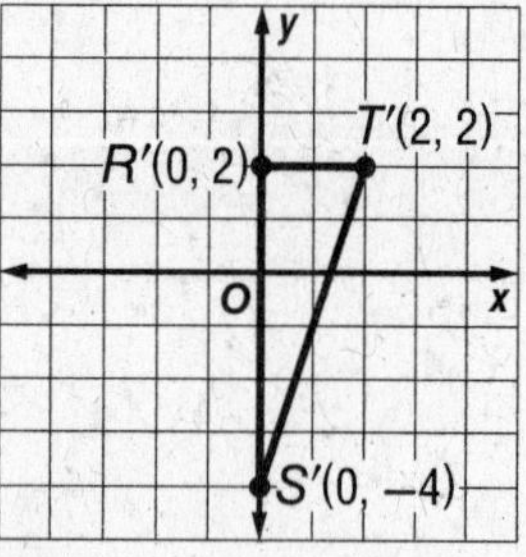

9.

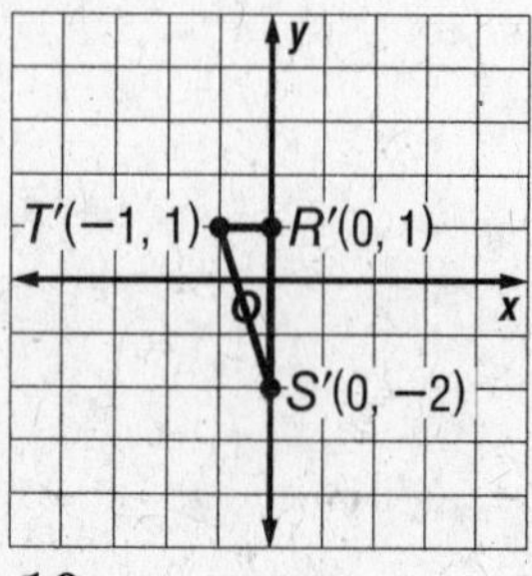

10.

Chapter 7 Answer Key

Are You Ready?—Review
Page 145

1. 90
2. 55
3. 39
4. 14
5. 9.5
6. 20
7. 20
8. 18
9. 6.8
10. 7.5

Are You Ready?—Practice
Page 146

1. 7
2. 6
3. 11
4. 42
5. 3
6. 48
7. 11.8
8. 36
9. 12 parts
10. 36 min
11. 2
12. $\frac{1}{2}$
13. $-\frac{1}{5}$

Answers

Chapter 7 Answer Key

Are You Ready?—Apply
Page 147

1. BAKERY In the window at a local bakery the ratio of muffins to cookies is 3 to 5. If there are 75 cookies in the window, how many muffins are in the window? **45 muffins**	**2.** FIELD TRIPS For the field trip to the museum, the ratio to adults to students is 2 to 9. If there are 108 students going on the field trip, how many adults will there be? **24 adults**
3. BLOOD TYPES For every 10 people that donated blood, 3 had blood type A positive. If 51 people who had type A positive donated blood, how many people donated blood? **170 people**	**4.** RECIPES A recipe for punch calls for 2.5 parts cranberry juice to 3 parts orange juice. If you use 12 parts orange juice, how many parts of cranberry juice do you need? **10 parts**
5. DRIVING A family drove 585 miles in 9 hours. At this rate, how many miles will the family drive in 7 hours? **455 mi**	**6.** TEXTING Jin can type 90 words in 2.5 minutes on his cell phone. At this rate, how many words can he type in 7.5 minutes? **270 words**

Chapter 7 Answer Key

Diagnostic Test
Page 148

1. 24
2. 18
3. 90
4. 10
5. 30
6. 18
7. 2.4
8. 1.8
9. 18 parts
10. 11 mi
11. 2
12. $\frac{1}{2}$
13. $\frac{3}{4}$

Pretest
Page 149

1. congruent; Sample answer: reflection followed by a translation
2. 3
3. 10 m
4. 16 in.
5. 1,296 in^2

Answers

Chapter 7 Answer Key

Chapter Quiz
Page 150

1. congruent; a rotation

2. congruent; a reflection followed by translation

3. not congruent

4. $\angle X \cong \angle F$, $\angle Y \cong \angle G$, $\angle Z \cong \angle J$ $\overline{XY} \cong \overline{FG}$; $\overline{YZ} \cong \overline{GJ}$; $\overline{ZX} \cong \overline{JF}$

Vocabulary Test
Page 151

1. similar polygons

2. similar

3. corresponding parts

4. scale factor

5. Sample answer: a sequence transformations

6. Sample answer: Finding a measurement using similar figures of an object that is too difficult to measure directly.

Chapter 7 Answer Key

Student Recording Sheet, Page 154

Use this recording sheet with the Standardized Test Practice pages.

Fill in the correct answer. For gridded-response questions, write your answers in the boxes on the answer grid and fill in the bubbles to match your answers.

1.

2. Ⓐ Ⓑ ● Ⓓ

3. 20.6

4. Ⓕ ● Ⓗ Ⓘ

5. 27.5

6. ● Ⓑ Ⓒ Ⓓ

7. Ⓕ Ⓖ ● Ⓘ

8. Ⓐ ● Ⓒ Ⓓ

9. $b + r = 15$, $b = r + 3$; $b = 9$, $r = 6$

10. Ⓕ Ⓖ ● Ⓘ

11. Ⓐ ● Ⓒ Ⓓ

12. Ⓕ Ⓖ Ⓗ ●

13. Ⓐ Ⓑ ● Ⓓ

14a. congruent

14b. The figure was reflected.

Extended Response

Record your answers for Exercise 14 on the back of this paper.

Answers

Chapter 7 Answer Key

Extended-Response Test, Page 155
Sample Answers

In addition to the scoring rubric, the following sample answers may be used as guidance in evaluating extended response assessment items.

1. a. Two polygons are similar if their corresponding angles are congruent and their corresponding sides are in proportion.

b. $\overline{AB}$ and $\overline{EF}$; $\overline{BC}$ and $\overline{FG}$; $\overline{CD}$ and $\overline{GH}$; and $\overline{DA}$ and $\overline{HE}$; You can find the ratio of the lengths of one pair of corresponding sides where both lengths are known. Because the ratios of the lengths of all the corresponding sides are equal, you can use the ratio you found to form proportions to find the lengths of the sides whose lengths are not given—$\overline{BC}$, $\overline{EF}$, and $\overline{GH}$.

c. Length of $\overline{BC}$:

$\frac{6}{4} = \frac{x}{9}$

$4x = 54$

$x = 13.5$

Length of $\overline{EF}$:

$\frac{6}{4} = \frac{12}{x}$

$6x = 48$

$x = 8$

Length of $\overline{GH}$:

$\frac{6}{4} = \frac{9}{x}$

$6x = 36$

$x = 6$

d. Sample answer:

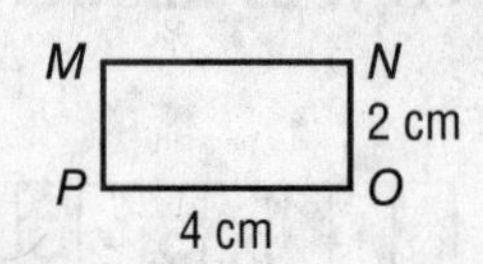

Q R

6 cm

T S

12 cm

Corresponding Parts:

$\overline{MN}$ and $\overline{QR}$

$\overline{NO}$ and $\overline{RS}$

$\overline{OP}$ and $\overline{ST}$

$\overline{PM}$ and $\overline{TQ}$

2. a. It is a combination of two or more of transformations.

b.

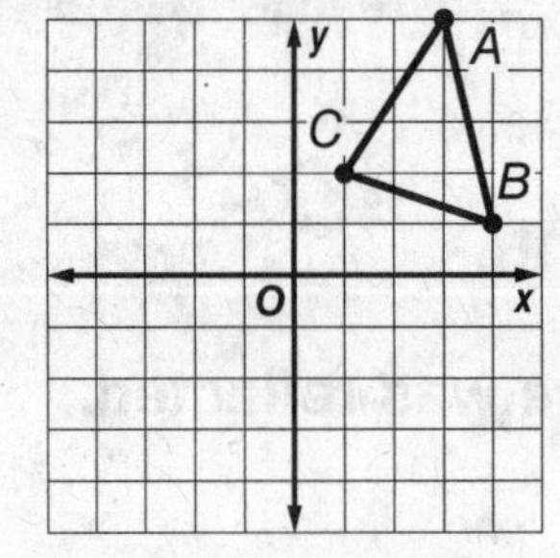

c.

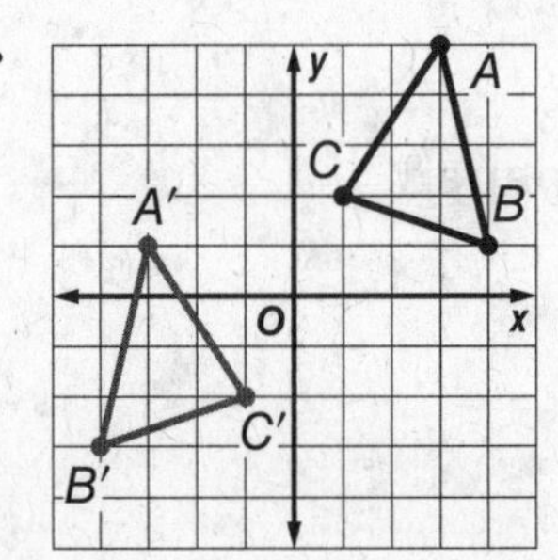

d. yes; Reflections and translations do not change the size or shape of the figure. The figure is congruent to the original figure.

Chapter 7 Answer Key

Test, Form 1A
Page 157

1. C

2. G

3. D

4. G

5. C

Test, Form 1A *(continued)*
Page 158

6. F

7. D

8. H

9. D

Answers

Chapter 7 Answer Key

Test, Form 1B
Page 159

1. B

2. G

3. B

4. F

5. C

Test, Form 1B ***(continued)***
Page 160

6. I

7. C

8. H

9. B

Chapter 7 Answer Key

Test, Form 2A
Page 161

1. C

2. G

3. D

4. G

5. D

Test, Form 2A ***(continued)***
Page 162

6. 6 cm

7. similar; $\triangle ABC \sim \triangle DBE$

8. congruent; figure reflected then translated

9. similar; figure dilated then translated

10. $\frac{AB}{BC} = \frac{CD}{DE}$
$\frac{-2}{2} = \frac{-4}{4} = -1$

Answers

Chapter 7 Answer Key

Test, Form 2B
Page 163

1. D

2. H

3. C

4. G

5. B

Test, Form 2B ***(continued)***
Page 164

6. 7 cm

7. similar; $\triangle DEF \sim \triangle HEJ$

8. congruent; figure rotated then translated

9. similar; figure dilated then translated

10. $\frac{BC}{CD} = \frac{DE}{EF}$
$\frac{2}{3} = \frac{4}{6}$

Chapter 7 Answer Key

Test, Form 3A
Page 165

1. 64 tatami

2. similar; The corresponding angles are congruent and the corresponding sides are proportional.

3. 10 ft

4. 78 cm

5. similar; $\triangle LMN \sim \triangle OPQ$

6a. $\angle A \cong \angle F$; $\angle B \cong \angle G$; $\angle C \cong \angle H$; $\overline{AB} \cong \overline{FG}$; $\overline{BC} \cong \overline{GH}$; $\overline{AC} \cong \overline{FH}$

6b. Sample answer: reflection then translation

Test, Form 3A ***(continued)***
Page 166

7a, b.

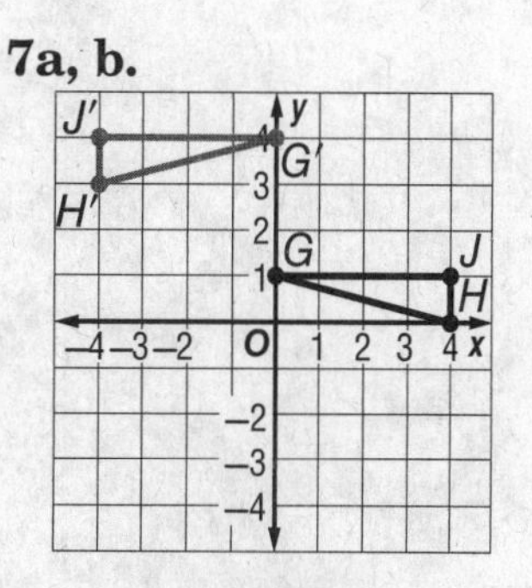

7c. yes; A reflection followed by a translation will map $\triangle GHJ$ onto $\triangle G'H'J'$.

8. not similar; The ratios comparing the corresponding side lengths are not equal so a dilation did not occur.

9. Sample answer: $\frac{AB}{BC} = \frac{-2}{2} = -1$ and $\frac{FG}{GH} = \frac{-4}{4} = -1$

10. 1,200 m^2

Chapter 7 Answer Key

Test, Form 3B
Page 167

1. 12 tatami

2. similar; The corresponding angles are congruent and the corresponding sides are proportional.

3. 31.25 ft

4. 318 in.

5. similar; $\triangle PQR \sim \triangle TUV$

6a. $\angle A \cong \angle F$; $\angle B \cong \angle G$; $\angle C \cong \angle H$; $\overline{AB} \cong \overline{FG}$; $\overline{BC} \cong \overline{GH}$; $\overline{AC} \cong \overline{FH}$

6b. rotation then translation

Test, Form 3B *(continued)*
Page 168

7a, b.

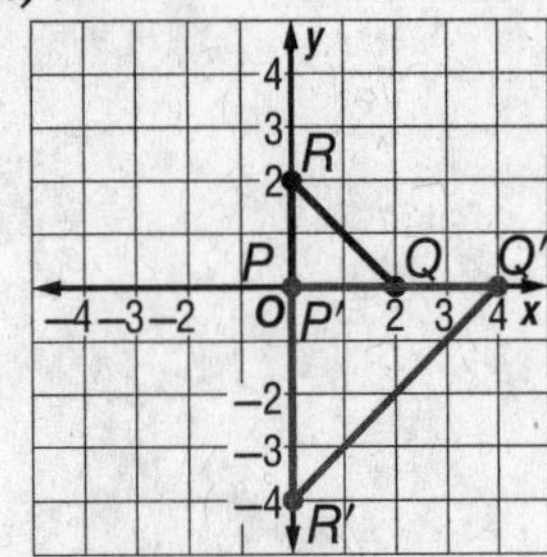

7c. no; The dilation enlarged the figure.

8. not similar; The ratios comparing the corresponding side lengths are not equal so a dilation did not occur.

9. Sample answer: $\frac{AB}{BC} = \frac{-1}{1} = -1$ and $\frac{FG}{GH} = \frac{-2}{2} = -1$

10. 320 m^2

Chapter 8 Answer Key

Are You Ready?—Review
Page 169

1. 310.5 cm^2
2. 360 ft^2
3. 270 mm^2
4. $1,200 \text{ ft}^2$
5. 234 in^2
6. 24 yd^2

Are You Ready?—Practice
Page 170

1. 380 cm^2
2. 464 in^2
3. 120 cm^2
4. 240 ft^2
5. 67.24 m^2
6. 512 yd^2
7. 7 ft

Chapter 8 Answer Key

Are You Ready?—Apply
Page 171

1. **FLAG** A flag of Florida measures 12 inches by 18 inches. What is the area of the flag? **216 in^2**

2. **TILES** Rondell covered a table top with 36 square tiles. Each tile measures 8 inches by 8 inches. What is the area of the table top? **2,304 in^2**

3. **ROAD SIGN** Cullen saw the sign below on his walk. The triangle is 12 inches on each side and the height is 10.4 inches. Find the area of the sign. **62.4 in^2**

4. **HEXAGONS** The patio shown below is made up of hexagons, which are two trapezoids joined at one base. If each trapezoid has a height of 1.7 feet and base lengths of 2 feet and 4 feet, what is the area of one hexagon? **10.2 ft^2**

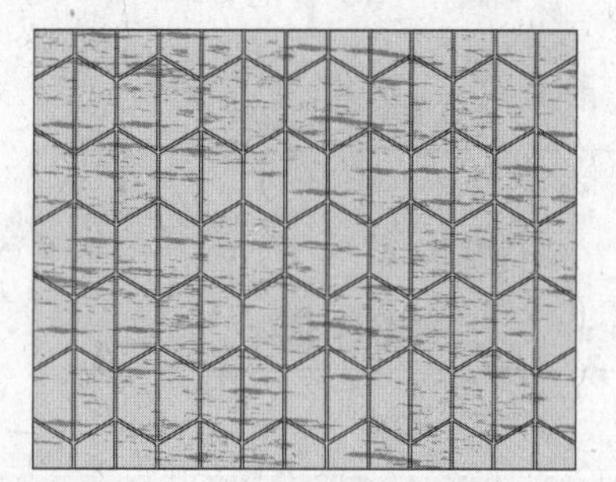

5. **LOGO** Josh used the figure below for a logo contest. If the height of the figure is 1.8 inches and the base is 3.8 inches, what is the area of the figure? **6.84 in^2**

6. **DESIGN** Russell made the design below on his computer. Each triangle has a height of 12 millimeters and a base of 42 millimeters. Find the area of the design. **1,764 mm^2**

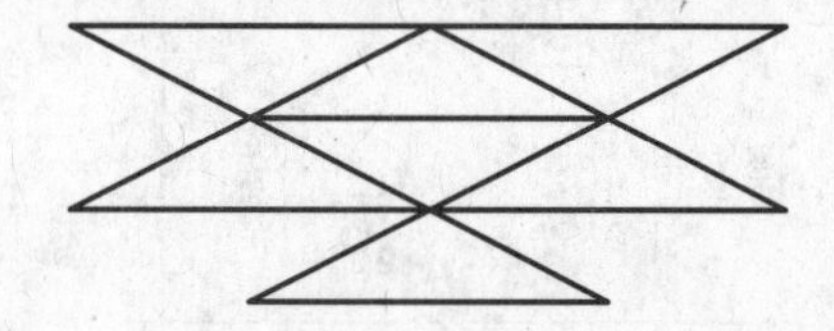

Chapter 8 Answer Key

Diagnostic Test
Page 172

1. 262.5 ft^2
2. 155 ft^2
3. 300 in^2
4. 608 cm^2
5. 57.76 yd^2
6. 414 mm^2
6. 15 ft

Pretest
Page 173

1. 33.5 in^3
2. 4,778.4 cm^3
3. 282.7 ft^3
4. 4,712.4 cm^3
5. 100.5 ft^2
6. 2,035.8 m^2
7. 102.5 in^2
8. 66.8 yd^2

Answers

Chapter 8 Answer Key

Chapter Quiz
Page 174

1. 3,455.8 cm^3
2. 340.3 ft^3
3. 706.9 cm^3
4. 3.7 m^3
5. 735.6 in^3
6. 812.9 cm^3

Vocabulary Test
Page 175

1. cylinder
2. polyhedron
3. composite solid
4. hemisphere
5. sphere
6. volume
7. cone
8. Sample answer: area of the non bases of a 3-D figure
9. Sample answer: solids that have the same shape and their corresponding measures are proportional

Chapter 8 Answer Key

Student Recording Sheet, Page 178

Use this recording sheet with the Standardized Test Practice pages.

Fill in the correct answer. For gridded-response questions, write your answers in the boxes on the answer grid and fill in the bubbles to match your answers.

1. ● Ⓑ Ⓒ Ⓓ

2. ● Ⓖ Ⓗ Ⓘ

3. 12 ft

4. 8.2

5. Ⓐ Ⓑ ● Ⓓ

6. Ⓕ ● Ⓗ Ⓘ

7. Ⓐ ● Ⓒ Ⓓ

8. 49

9. Ⓕ Ⓖ Ⓗ ●

10. Ⓐ Ⓑ ● Ⓓ

11. Ⓕ Ⓖ Ⓗ ●

12. Ⓐ Ⓑ ● Ⓓ

13a. Container A; The volume of Container A is about 452.4 in^3, while the volume of Container B is 332.75 in^3.

13b. Container B; The surface area of Container B is 272.25 in^2, while the surface area of Container A is about 276.5 in^2.

Extended Response

Record your answers for Exercise 13 on the back of this paper.

Answers

Chapter 8 Answer Key

Extended-Response Test, Page 179
Sample Answers

In addition to the scoring rubric, the following sample answers may be used as guidance in evaluating extended response assessment items.

1. a. Container A is a cylinder since it has two parallel bases that are circles. Container B is a prism since it has two parallel bases that are polygons (rectangles). Container C is a prism since it has two parallel bases that are polygons (trapezoids).

b. 6 faces; 12 edges; 8 vertices; Sample answer: the solid has three pairs of parallel faces, so it has 3×2 or 6 faces.

c. To find the area of a circle, square the radius of the circle and multiply the result by π.

$A = \pi r^2$

$A = \pi \cdot 2.5^2$

$A \approx 19.6$

The area of the circular base is about 19.6 in^2.

d. 41.5 in^2

e. For both cylinders and prisms, the volume is computed by multiplying the area of the base by the height.

A: $V = Bh$

$V = 19.6 \cdot 8$ or 156.8

157.1 in^3

B: $V = Bh$ or $b \cdot w \cdot h$

$V = 6 \cdot 2.5 \cdot 8$ or 120

120 in^3

C: $V = Bh$

$V = 24 \cdot 7$ or 168

168 in^3

Container C will hold the most snack mix since it has the greatest volume.

f. A: $S.A. = 2\pi r^2 + 2\pi rh$

$S.A. = 2\pi(2.5)^2 + 2\pi(2.5)(8)$

$S.A. \approx 164.9$

164.9 in^2

B: $S.A. = Ph + 2B$

$S.A. = [2(2.5) + 2(6)]8 + 2(2.5)(6)$

$S.A. = 166$

166 in^2

Container A has less surface area.

g. To determine the cost, multiply the surface area by \$0.0015.

$166 \cdot 0.0015 = 0.249$

Rounded to the nearest cent, the cost is \$0.25.

Chapter 8 Answer Key

Test, Form 1A
Page 181

1. A

2. G

3. B

4. H

5. C

6. F

Test, Form 1A *(continued)*
Page 182

7. B

8. H

9. B

10. F

11. D

12. F

Chapter 8 Answer Key

Test, Form 1B
Page 183

1. B
2. G
3. C
4. G
5. B
6. F

Test, Form 1B ***(continued)***
Page 184

7. C
8. G
9. C
10. F
11. D
12. G

Chapter 8 Answer Key

Test, Form 2A
Page 185

1. C
2. H
3. C
4. H
5. C
6. H

Test, Form 2A *(continued)*
Page 186

7. A
8. G
9. 488.8 in^2
10. 560 in^3
11. 15 m
12. 20.0 in.
13. $1,072.3 \text{ cm}^3$

Answers

Chapter 8 Answer Key

Test, Form 2B
Page 187

1. B

2. H

3. C

4. F

5. C

6. G

Test, Form 2B ***(continued)***
Page 188

7. C

8. F

9. 194.5 cm^2

10. 1,152 in^3

11. 27 m

12. 14 in.

13. 1,526.8 cm^3

Chapter 8 Answer Key

Test, Form 3A
Page 189

1. $50.3\ ft^3$

2. $14{,}137.2\ in^3$

3. $16{,}755.2\ m^3$

4. $161.2\ m^2$

5. $5{,}428.7\ in^2$

6. $163.4\ yd^2$

7. $271.4\ m^2$

8. $189.1\ cm^2$

9. $904.8\ in^3$

10. $377\ in^2$; $L.A. = \pi r l$ $= \pi \cdot 6 \cdot 20$ ≈ 377

11. 15.9 cm

Test, Form 3A *(continued)*
Page 190

12. 8 units

13. $251.3\ ft^2$

14. $754.0\ in^2$

15. 15 cakes

16. $1{,}069.3\ cm^3$

17. $145.33\ in^2$

18. 10

19. $851.97\ in^3$

Answers

Chapter 8 Answer Key

Test, Form 3B
Page 191

1. 57,905.8 cm^3

2. 804.2 yd^3

3. 56,548.7 in^3

4. 10.5 m^3

5. 2,243.1 in^2

6. 452.4 cm^2

7. 290.3 m^2

8. 846.7 ft^2

9. 1,436.8 in^3

10. 283 in^2; $L.A. = \pi r l$ $= \pi \cdot 5 \cdot 18$ ≈ 283

11. 12.2 yd

Test, Form 3B *(continued)*
Page 192

12. 7.5 units

13. 263.9 ft^2

14. 439.8 in^2

15. 48 arrangements

16. 2,010.0 cm^3

17. 132.8 cm^2

18. $\frac{1}{10}$

19. 718.75 in^3

Chapter 9 Answer Key

Are You Ready?—Review
Page 193

1. 18
2. 29
3. 33.8
4. 33.7
5. 9.1

Are You Ready?—Practice
Page 194

1. Sample answer: Most students live within 10 miles of the school. 14 students
2. Sample answer: Most games had 31–40 points scored. 6 games
3. 8
4. 18
5. 28.8
6. 34.5
7. 128

Answers

Chapter 9 Answer Key

Are You Ready?—Apply
Page 195

1. **GOLF** Sky is a member of her school's golf team. The table shows her scores from each of her matches. What was her average golf score? Round to the nearest tenth if necessary. **70.2**

Sky's Golf Scores		
72	69	72
71	71	69
68	72	68

2. **AUTUMN** Karly owns a lawn service. During the month of October she averaged 42 customers the first week, 38 the second week, 55 the third week, and 65 the fourth week. How many customers did she average per week? **50 customers**

3. **SALES** The number of bags of popcorn sold at a certain movie theater for several days was 27, 17, 24, 38, 47, and 21. Find the average number of bags of popcorn sold during these days. **29 bags**

4. **CUSTOMERS** Big Mart keeps track of the number of customers entering the store in the first hour of each day. In the first four days of April, Big Mart had 215, 125, 118, and 214 customers in the first hour. Find the average number of customers Big Mart had in the first hour during those four days. **168 customers**

5. **HEIGHT** The histogram below shows the heights of students in a classroom. Describe the histogram. Then find the number of students who are more than 59 inches tall.

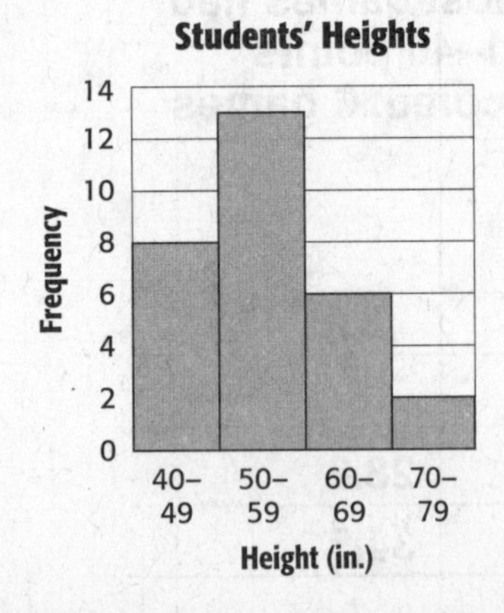

Sample answer: Most students have a height that is between 40 and 59 inches; 8 students.

6. **CHOIR** Timmy belongs to the choir in his hometown. The histogram below shows the ages of the choir members. Describe the histogram. Then find the number of members who are less than 31 years old.

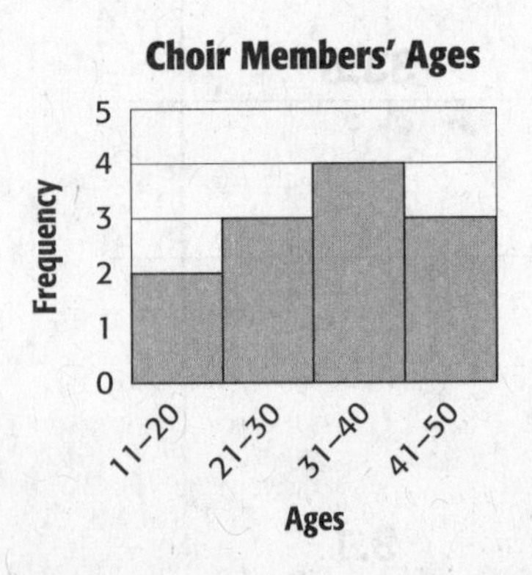

Sample answer: Most members are between the ages of 31 and 40; 5 members.

Chapter 9 Answer Key

Diagnostic Test
Page 196

1. Sample answer: Most zoos had an attendance of less than 2 million.; 10 zoos

2. Sample answer: The team usually scored at least 41 points.; 18 times

3. 30

4. 38

5. 19

6. 23.3

7. 32.6

8. 41.9

9. 62

Pretest
Page 197

1. mean: 34, median: 29, mode: 62

2. 14

3. minimum: 13
Q1: 15
median: 23.5
Q3: 42
maximum: 71

4. 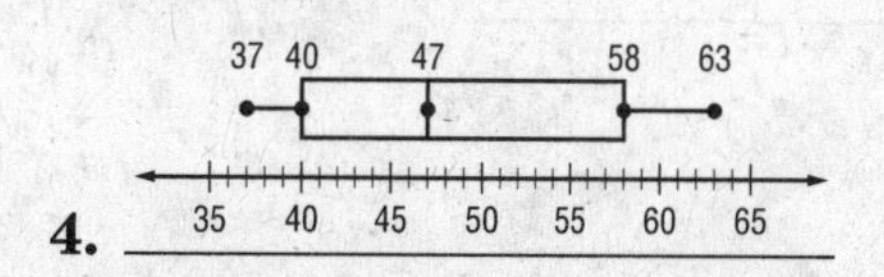

5. min.: 24; Q1: 27; median: 35; Q3: 44; max.: 46

Points Scored per Team

24 27 35 44 46

20 25 30 35 40 45 50

6. 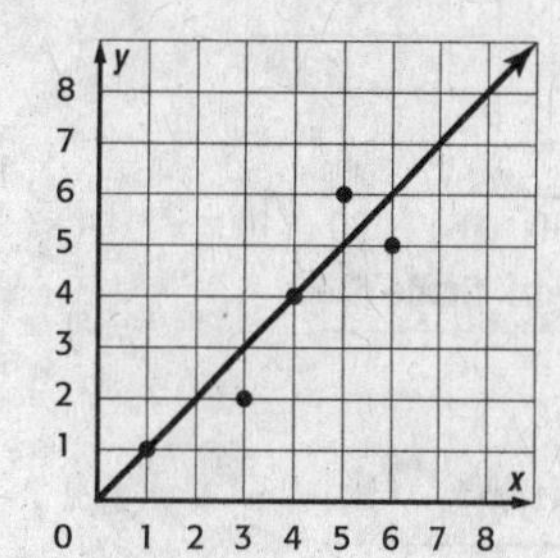

Answers

Chapter 9 Answer Key

Chapter Quiz
Page 198

1. 35 people

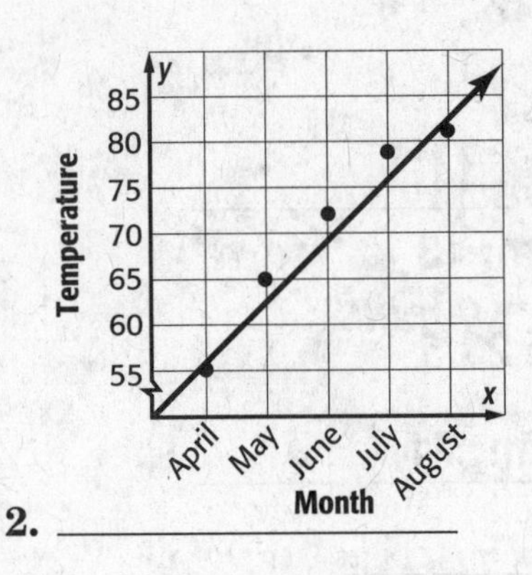

2.

3. Sample answer: $y = 5x + 50$

4. 80°

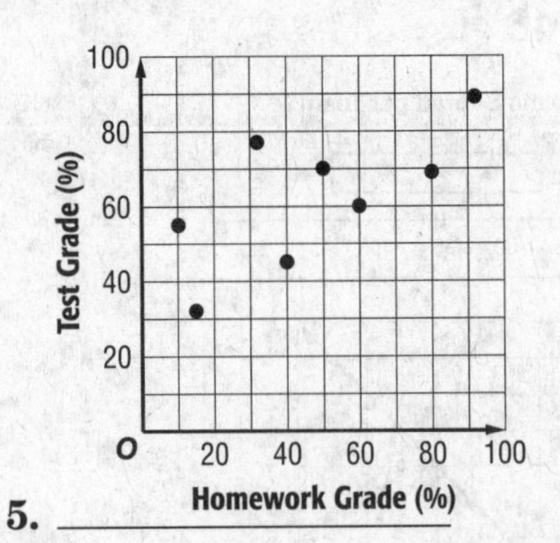

5.

6. positive

7. 0.83

Vocabulary Test
Page 199

1. mean absolute deviation

2. scatter plot

3. relative frequency

4. standard deviation

5. two-way table

6. distribution

7. line of best fit

8. Sample answer: If both sides of the distribution looks the same.

9. Sample answer: Data that can be measured.

Chapter 9 Answer Key

Student Recording Sheet, Page 202

Use this recording sheet with the Standardized Test Practice.

Fill in the correct answer. For gridded-response questions, write your answers in the boxes on the answer grid and fill in the bubbles to match your answers.

1. ● Ⓑ Ⓒ Ⓓ

2. Ⓕ Ⓖ ● Ⓘ

3. 

4. ● Ⓑ Ⓒ Ⓓ

5. $36.03

6. Ⓕ Ⓖ ● Ⓘ

7. ● Ⓑ Ⓒ Ⓓ

8. 24 square units

9. ● Ⓖ Ⓗ Ⓘ

10. 5.0 in.

11a. Sample answer: It shows individual pieces of data while emphasizing trends in the data.

11b. 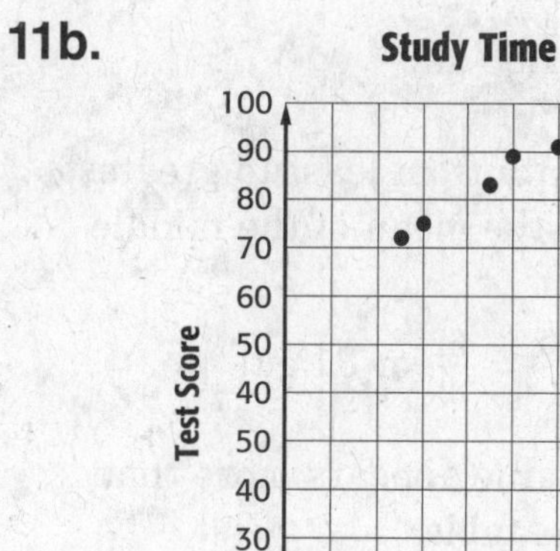

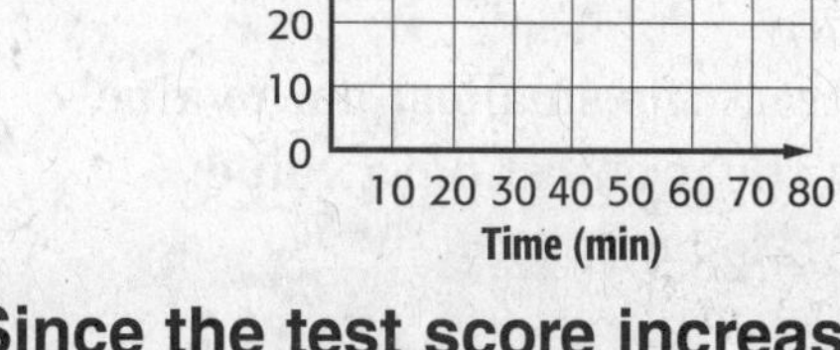

Since the test score increased as the study time increased, this is a positive relationship.

11c. 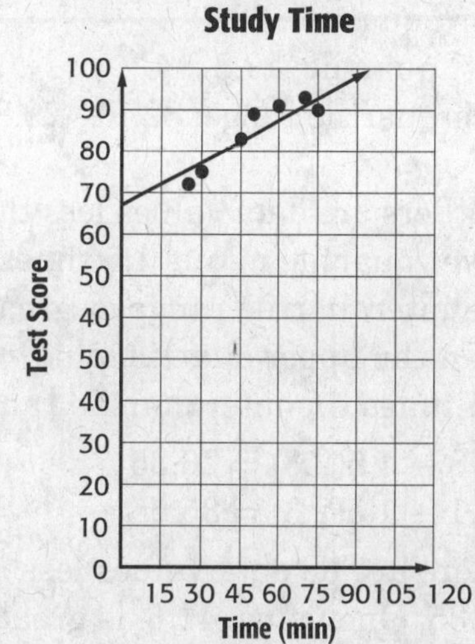

Sample answer: $y = \frac{1}{3}x + 67$

11d. 97

Extended Response

Record your answers for Exercise 11 on the back of this paper.

Chapter 9 Answer Key

Extended-Response Test, Page 203
Sample Answers

In addition to the scoring rubric, the following sample answers may be used as guidance in evaluating extended response assessment items.

1. a. Add the numbers of gallons of gasoline for all ten weeks, and then divide by 10.

$\frac{318.8}{10} = 31.88$ gal

b. List the data from least to greatest. Then find the mean of the middle two values.

$\frac{30.5 + 31.5}{2} = \frac{62}{2}$ or 31 gal

c. No data value appears more than once in the table.

d. Subtract the smallest data value from the greatest data value.

$41.6 - 28.9 = 12.7$

e. minimum: 28.9
maximum: 41.6
median: 31
third quartile: 32.1
first quartile: 29.8
interquartile range: 32.1 – 29.8 or 2.3

f. Outliers are data values less than the lower quartile minus 1.5 times the interquartile range or greater than the upper quartile plus 1.5 times the interquartile range.

$29.8 - 1.5(2.3) = 26.35$
$32.1 + 1.5(2.3) = 35.55$

There are no data values less than 26.35. One value, 41.6, is greater than 35.55. The only outlier is 41.6.

g.

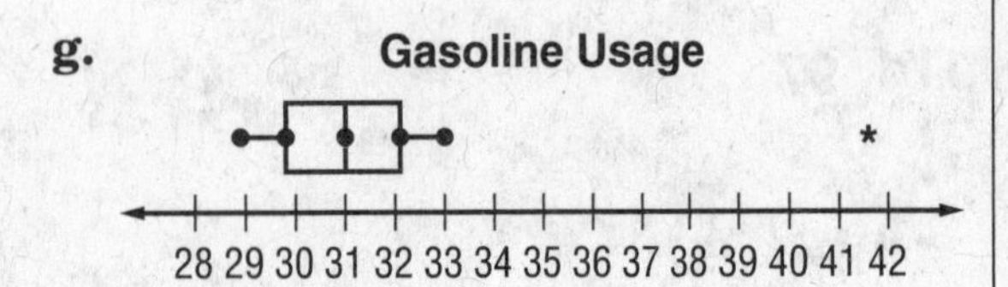

2. a.

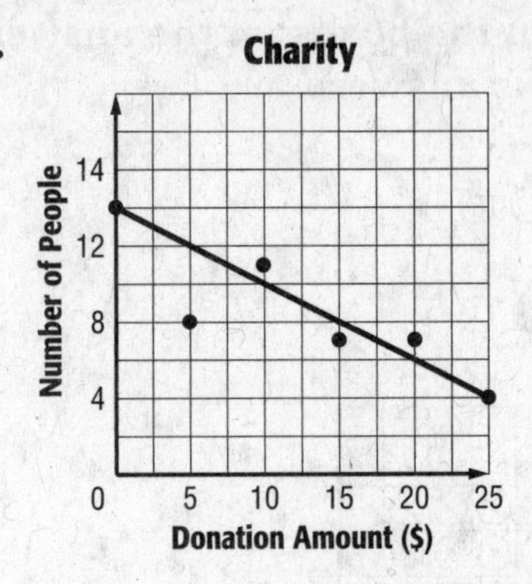

b. Negative; as the donation amount increases, the number of people decreases.

c. For graph, see part **a**. A best-fit line is a line that is very close to most of the data points.

d. Let $(x_1, y_1) = (0, 13)$ and $(x_2, y_2) = (25, 4)$.

$$m = \frac{y_2 - y_1}{x_2 - x_1}$$
$$= \frac{4 - 13}{25 - 0}$$
$$= \frac{-9}{25} \text{ or } -0.36$$

e. The y-intercept is 13; the line intersects the y-axis where $y = 13$.

Chapter 9 Answer Key

Test, Form 1A
Page 205

1. A

2. H

3. D

4. I

5. B

Test, Form 1A ***(continued)***
Page 206

6. I

7. B

8. I

9. B

10. H

11. C

Answers

Chapter 9 Answer Key

Test, Form 1B
Page 207

1. B

2. F

3. A

4. I

5. C

Test, Form 1B ***(continued)***
Page 208

6. H

7. C

8. G

9. D

10. H

11. A

Chapter 9 Answer Key

Test, Form 2A
Page 209

1. C
2. G
3. D
4. H
5. A
6. H
7. D

Test, Form 2A *(continued)*
Page 210

8a, b.

	Like Fishing	Dislike Fishing	Total
Male	35; 0.63	15; 0.38	50
Female	21; 0.38	24; 0.62	45
Total	56; 1.00	39; 1.00	95

8c. Most people that like fishing are males. Most that dislike fishing are females.

9.

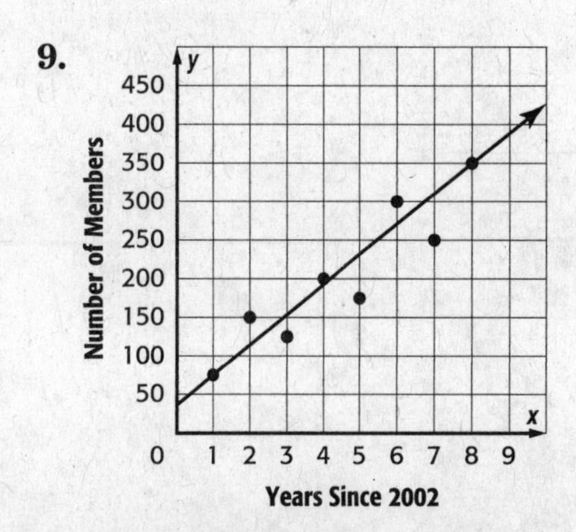

10. Sample answer: All of the points are close to the line.

11. Sample answer: $y = 40x + 36$

12. Sample answer: There are about 40 new members per year. In 2002, there were 36 members.

13. about 396 members

Chapter 9 Answer Key

Test, Form 2B
Page 211

1. A

2. G

3. B

4. H

5. B

6. I

7. C

Test, Form 2B *(continued)*
Page 212

8a, b.

	Like Skiing	Dislikes Skiing	Total
Male	25; 0.71	10; 0.29	35
Female	20; 0.40	30; 0.60	50
Total	45	40	85

8c. Most males like skiing. Most females dislike skiing.

9a.

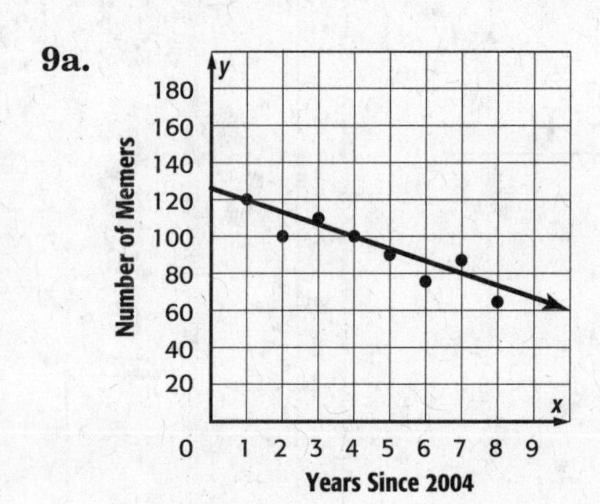

10. Sample answer: Most of the points are close to the line.

11. Sample answer: $y = -7x + 125$

12. Sample answer: The club loses 7 members per year. In 2004 there were 125 members.

13. 55 members

Chapter 9 Answer Key

Test, Form 3A
Page 213

1\.

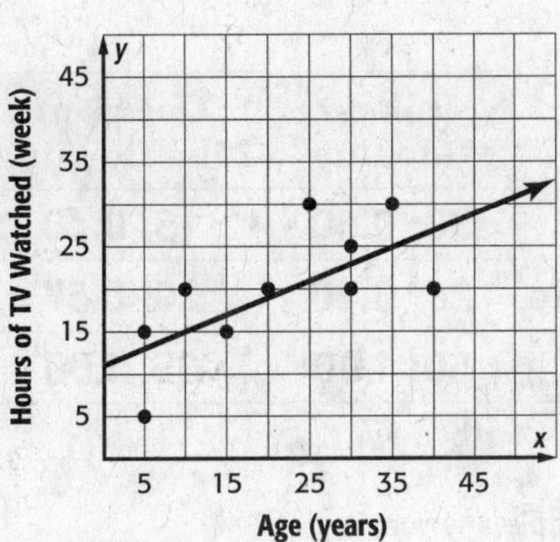

2\. Sample answer: The scatter plot shows a positive linear association with no clusters or outliners.

3\. Sample answer: Most of the points are close to the line.

4\. Sample answer: $y = 0.4x + 11$

5\. 33 hours

6\. minimum: 16; Q_1: 18; median: 23.5; Q_3: 30; maximum: 33

7\.

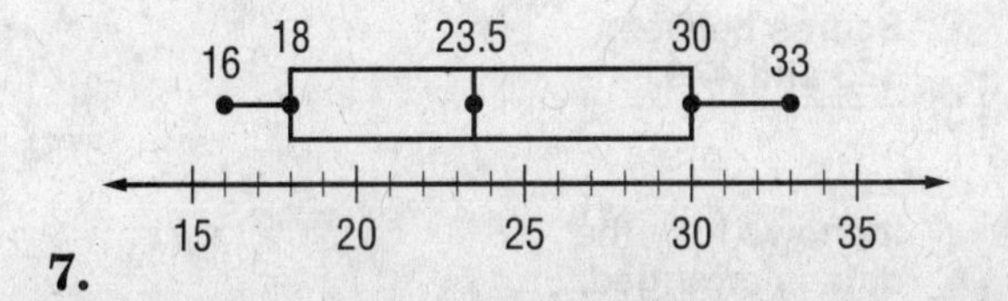

Test, Form 3A *(continued)*
Page 214

8a, b.

	Likes Soccer	Does Not Like Soccer	Total
Male	20; 0.80	5; 0.20	25; 1.00
Female	24; 0.80	6; 0.20	30; 1.00
Total	44	11	55

8c. Sample answer: Most males and females like soccer.

8d. No; Sample answer: 80% of girls like soccer so most girls like soccer.

9\. 22.75

10\. 4.9

11\. The average number of points each student's score is from the mean is 4.9.

12\. Scores between 17.15 and 28.35.

13\. When the distribution of a data set looks the same on both the right and left sides.

Answers

Chapter 9 Answer Key

Test, Form 3B
Page 215

1.

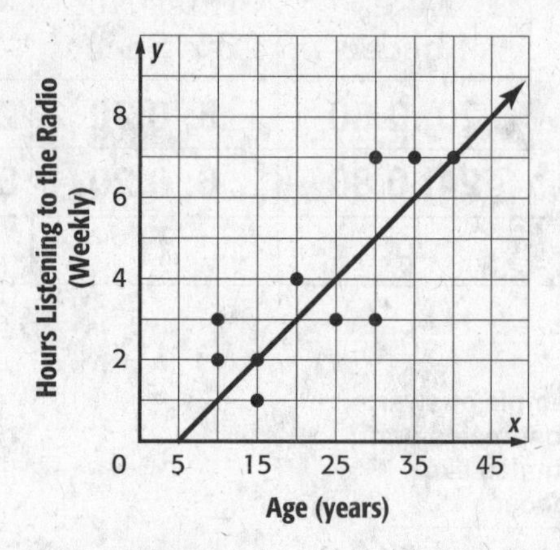

2. Sample answer: The scatter plot shows a positive linear correlation. Three appears to be a cluster between 10 years and 30 years.

3. Sample answer: Most of the points are close to the line.

4. Sample answer: $y = \frac{1}{5}x - 1$

5. 11 h

6. minimum: 59; Q_1: 65.5; median: 76; Q_3: 84; maximum: 90

7.

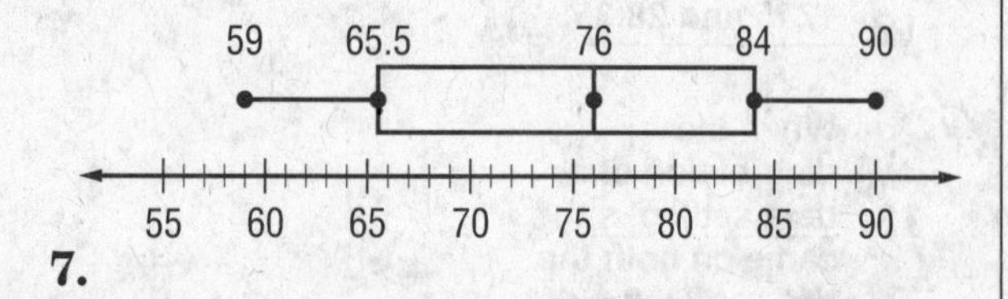

Test, Form 3B ***(continued)***
Page 216

8a, b.

	Likes Bowling	Does Not Like Bowling	Total
Male	30; 0.60	15; 0.40	45
Female	20; 0.40	20; 0.57	40
Total	50; 1.00	35; 1.00	85

8c. Sample answer: Males like bowling more often than females.

8d. No; Sample answer: Only 40% of students that like bowling were girls.

9. 42.6

10. 4.5

11. The average number of points each student's score is from the mean is 4.5.

12. Scores between 37.3 and 47.9.

13. It shows how the data is arranged.

NAME ________________ DATE ________ PERIOD ______

Course 3 Benchmark Test – First Quarter

1. The average distance from the Earth to the moon is about 384,000 kilometers. What is this number written in scientific notation?

   A. 384×10^5

   B. 384×10^3

   C. 3.84×10^6

   *D. 3.84×10^5

2. **SHORT ANSWER** Marc is finding the product of the monomials $3c^2d^4$ and $-4c^3d$. His work is shown below. What error did he make?

   Marc
   $3c^2d^4(-4c^3d)$
   $= 3(-4)(c^2c^3)(d^4d)$
   $= -12c^6d^4$

   He multiplied the exponents instead of adding them.

3. Which point on the number line shows $\sqrt{45}$?

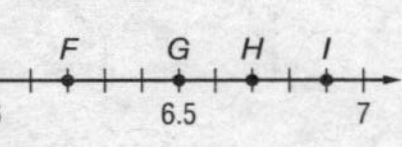

   F. point F

   G. point G

   *H. point H

   I. point I

4. A moving company charges $40 plus $0.25 per mile to rent a van. Another company charges $25 plus $0.35 per mile to rent the same van. For what number of miles will the rental cost be the same for both companies?

   *A. 150 miles

   B. 180 miles

   C. 260 miles

   D. 650 miles

5. A taxicab service charges $3.75 plus $0.40 per mile. Molly takes a taxicab from the hotel to the airport. If the total charge was $10.95, which equation could be used to determine the number of miles from the hotel to the airport?

   F. $3.75m + 0.4 = 10.95$

   *G. $3.75 + 0.4m = 10.95$

   H. $4.15m = 10.95$

   I. $3.35m = 10.95$

6. Which value is equivalent to 4^{-3}?

   A. -12

   B. -1

   C. $-\frac{1}{64}$

   *D. $\frac{1}{64}$

NAME ________________ DATE ________ PERIOD ______

Course 3 Benchmark Test – First Quarter *(continued)*

7. **SHORT ANSWER** The Venn diagram shows the real number system. Write the names of the missing sets of numbers.

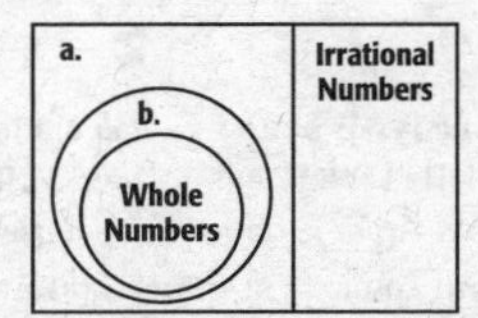

   a. Rational Numbers; b. Integers

8. Which of the following does *not* represent a rational number?

   F. -250

   G. $\frac{11}{39}$

   *H. $\sqrt{60}$

   I. $12.0\overline{982}$

9. The school marching band has 36 members. The band director wants to arrange the band members into a square formation. How many band members should be in each row?

   A. 8

   *B. 6

   C. 5

   D. 4

10. Which expression is equivalent to the expression below?

    $a \cdot a \cdot a \cdot b \cdot a \cdot b \cdot b \cdot a \cdot b \cdot a$

    *F. a^6b^4

    G. $a^{-6}b^{-4}$

    H. $(ab)^{10}$

    I. $(ab)^2$

11. What is the solution to the equation below?

    $-\frac{2}{3}p + \frac{1}{6} = \frac{7}{10}$

    A. $-\frac{13}{10}$

    *B. $-\frac{4}{5}$

    C. $-\frac{26}{45}$

    D. $-\frac{16}{45}$

12. Solve the equation below for t.

    $3t - 5 = -21 + t$

    F. -52

    G. -32

    H. -13

    *I. -8

Answers

NAME ________________ DATE ________ PERIOD ________

Course 3 Benchmark Test – First Quarter *(continued)*

13. The distance from the Sun to Earth is about 1.5×10^{11} meters. Suppose light travels at a speed of 3×10^{8} meters per second. About how long does it take light from the Sun to reach Earth?

A. 4.5×10^{19} seconds

B. 1.503×10^{11} seconds

C. 5×10^{3} seconds

***D.** 5×10^{2} seconds

14. What is the value of b in the equation below?

$$4(b - 1) = 2b + 10$$

F. 4

G. 5.5

***H.** 7

I. 11.5

15. The table shows the populations of several states. What is the population of Ohio written in scientific notation?

State	Population
Georgia	9,400,000
Illinois	12,900,000
Ohio	11,500,000
California	36,900,000

A. 1.15×10^{-8}

B. 1.15×10^{-7}

***C.** 1.15×10^{7}

D. 1.15×10^{8}

16. Which of the expressions below is *not* equivalent to the other three?

F. 0.015625

***G.** 15.625%

H. 4^{-3}

I. $\frac{1}{64}$

17. **SHORT ANSWER** What is the result when the monomial $-5x^3y^2z$ is raised to the third power?

$-125x^9y^6z^3$

18. The area of a square living room is 169 square feet. What is the perimeter of the room?

Area = 169 ft²

A. 13 ft

B. 17 ft

***C.** 52 ft

D. 68 ft

NAME ________________ DATE ________ PERIOD ________

Course 3 Benchmark Test – First Quarter *(continued)*

19. Between which two integers does $\sqrt{88}$ lie on the number line?

3 4 5 6 7 8 9 10 11

F. between 6 and 7

G. between 7 and 8

H. between 8 and 9

***I.** between 9 and 10

20. Which of the following symbols results in a true number sentence when placed in the blank?

$$\sqrt{12.96} \ \_\_\_ \ 3\frac{3}{5}$$

***A.** =

B. >

C. <

D. ×

21. **SHORT ANSWER** The area of an equilateral triangle is given by the expression $\frac{s^2\sqrt{3}}{4}$, where s is the side length of the triangle. What is the area of triangle below? Round to the nearest tenth.

5 cm, 5 cm, 5 cm

10.8 cm²

22. Which of the following numbers has the least absolute value?

F. 3.5×10^{-5}

***G.** 8.75×10^{-7}

H. 5.62×10^{3}

I. 1.002×10^{12}

23. Which equation shows the following relationship?

Seven less than four times a number is equal to 5.

A. $7 - 4n = 5$

***B.** $4n - 7 = 5$

C. $7n - 4 = 5$

D. $4 - 7n = 5$

24. Which equation is equivalent to the equation below?

$$5(n + 6) = 2(n - 3) + 4$$

F. $5n + 6 = 2n + 1$

G. $5n + 6 = 2n - 2$

H. $5n + 30 = 2n + 1$

***I.** $5n + 30 = 2n - 2$

NAME ________________ DATE ________ PERIOD ________

Course 3 Benchmark Test – First Quarter *(continued)*

25. SHORT ANSWER Juanita has saved $65 for vacation. She plans to save an additional $5 per week. How many weeks will it take for Juanita to save a total of $125? Write and solve an equation.

$65 + 5n = 125$; 12 weeks

NAME ________________ DATE ________ PERIOD ________

Course 3 Benchmark Test – Second Quarter

1. The table shows how much Addison earns for working various numbers of hours at a part-time job.

Hours, x	Earnings ($), y
10	72.50
15	108.75
20	145.00

Which of the following describes the constant rate of change?

A. 5 hours per dollar

B. $5.00 per hour

C. 7.25 hours per dollar

***D.** $7.25 per hour

2. Let n represent the figure number in the pattern below.

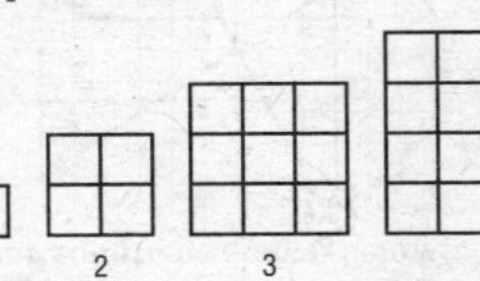

Which function represents the number of squares in each figure?

***F.** $f(n) = n^2$

G. $f(n) = 2n$

H. $f(n) = n^3$

I. $f(n) = 4n$

3. Which systems of linear equations has a solution of $(-2, 1)$?

***A.** $2x + 3y = -1$
$x - y = -3$

B. $2x + 3y = 1$
$x - y = 3$

C. $2x + 3y = -1$
$x - y = 3$

D. $2x + 3y = 1$
$x - y = -3$

4. What is the solution to the system of equations below?

$$3x - 2y = 7$$
$$-3x + 5y = 5$$

F. (3, 1)

G. (0, 1)

***H.** (5, 4)

I. no solution

5. SHORT ANSWER Missy walked around the school track to warm up. Then she ran several laps before walking to cool down. Sketch a graph to represent Missy's distance run over time.

Sample answer:

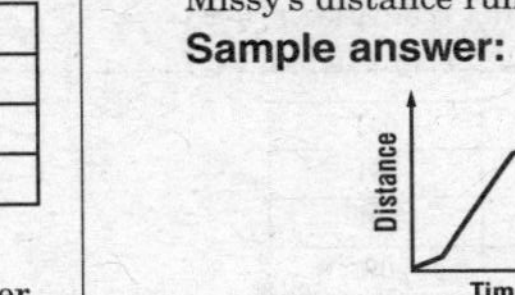

6. Which term describes the function shown below?

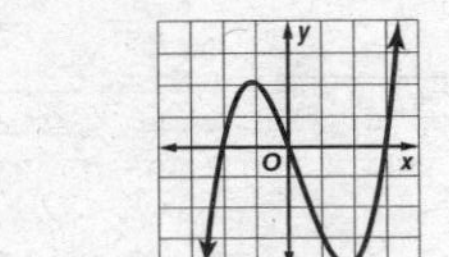

A. constant

B. linear

***C.** nonlinear

D. quadratic

NAME ________________ DATE ________ PERIOD ______

Course 3 Benchmark Test – Second Quarter *(continued)*

7. What is the equation of the quadratic function shown in the graph?

F $y = x^2 + 2$

***G** $y = x^2 - 2$

H $y = 2x^2$

I $y = \frac{1}{2}x^2$

8. **SHORT ANSWER** Find the x- and y-intercepts of the linear equation below.

$$4x - 5y = 20$$

(5, 0), (0, –4)

9. What is the slope of the line that passes through $M(-6, 1)$ and $N(2, 5)$?

A 2

***B** $\frac{1}{2}$

C $-\frac{1}{2}$

D –2

10. What is the domain of the function shown in the table?

x	–4	–2	0	2	4
y	–3	7	5	0	–1

F. all real numbers

G. all even integers

H. {–3, –1, 0, 5, 7}

***I.** {–4, –2, 0, 2, 4}

11. What are the slope and y-intercept of the linear equation below?

$$y = -5x + 2$$

A. slope: 2, y-intercept: (0, –5)

B. slope: 2, y-intercept: (–5, 0)

***C.** slope: –5, y-intercept: (0, 2)

D. slope: –5, y-intercept: (2, 0)

12. A tank contains 550 gallons of water. When the valve is opened, the tank drains at a rate of 12 gallons per minute. Which function shows the relationship between the time t the valve is opened and the amount of water in the tank?

***F.** $A(t) = -12t + 550$

G. $A(t) = 12t + 550$

H. $A(t) = 12 + 550t$

I. $A(t) = -12 + 550t$

NAME ________________ DATE ________ PERIOD ______

Course 3 Benchmark Test – Second Quarter *(continued)*

13. Which relation is *not* a function?

A.

x	–2	0	2	4	6
y	3	3	3	3	3

***B.**

x	–3	0	2	–3	1
y	–5	4	2	0	–1

C.

x	1	2	3	4	5
y	1	2	3	4	5

D

x	–4	1	2	–3	4
y	0	3	–1	–2	3

14. What is the solution to the system of linear equations shown below?

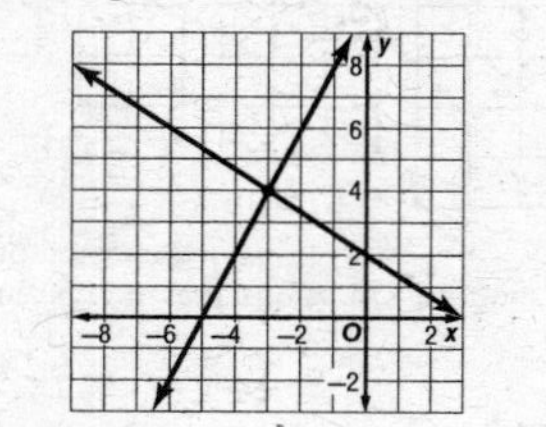

F. (4, –3)

G. (–4, 3)

***H.** (–3, 4)

I. (3, –4)

15. **SHORT ANSWER** What is the equation in slope-intercept form of the line that passes through (–2, 17) and (3, –13)?

$y = -6x + 5$

16. Which linear function has the steepest slope?

A. $y = \frac{1}{2}x - 5$

B. $y = -\frac{2}{5}x + 3$

C. $y = 4x - 2$

***D.** $y = -6x + 1$

17. The table shows the cost of renting a van from a moving company for different numbers of miles driven.

Miles, m	Cost, C
50	\$42.50
100	\$65.00
150	\$87.50
200	\$110.00

Construct a function that relates the cost of renting a van to the number of miles driven.

F. $C(m) = 0.85m$

G. $C(m) = 0.85m + 10$

H. $C(m) = 0.45m$

***I.** $C(m) = 0.45m + 20$

18. Which two points form a line that has a slope of –3?

A. (–5, 3) and (2, 4)

***B.** (1, –6) and (–4, 9)

C. (–4, –3) and (5, 0)

D. (2, 8) and (–1, –1)

NAME ______________ DATE ________ PERIOD ______

Course 3 Benchmark Test – Second Quarter *(continued)*

19. What are the x– and y–intercepts of the linear equation below?

$6x - 2y = 12$

***F.** (2, 0) and (0, −6)

G. (0, 2) and (−6, 0)

H. (−6, 0) and (2, 0)

I. (0, 2) and (0, −6)

20. The quadratic function $h(t) = -16t^2 + 120$ represents the height of an object in feet t seconds after when it falls from a height of 120 feet. What is the height of the object after 1.5 seconds?

A. 58 ft

***B.** 84 ft

C. 92 ft

D. 156 ft

21. SHORT ANSWER The table below shows the number of teams remaining in each round of a tournament. Is the number of teams a linear function of the number of rounds? Explain.

Round	Teams
1	32
2	16
3	8
4	4
5	2

No; Sample answer: there is not a constant rate of change.

22. What is the constant rate of change of the function represented in the table below?

x	y
−5	23
−1	7
3	−9
7	−25

F. 16

G. 4

***H.** −4

I. −16

23. The slope of a line is $-\frac{1}{5}$ and the y-intercept is (0, 6). What is the equation of the line in slope-intercept form?

A. $x + 5y = 30$

B. $x - 5y = 30$

C. $y = -\frac{1}{5}x - 6$

***D.** $y = -\frac{1}{5}x + 6$

24. Which of the following equations represents a horizontal line?

F. $y = x$

G. $y = -x + 1$

***H.** $y = -12$

I. $x = 5$

NAME ______________ DATE ________ PERIOD ______

Course 3 Benchmark Test – Second Quarter *(continued)*

25. SHORT ANSWER The graph below shows the length of Michael's hair as a function of time. Describe the change in the length of Michael's hair over time.

Length

Time

Michael's hair grows at a steady rate until he gets it cut. This cycle is continually repeated.

NAME ______________________ DATE ____________ PERIOD ________

Course 3 Benchmark Test – Third Quarter

1. **SHORT ANSWER** Alfonso leans a 20-foot long ladder against a wall with the base of the ladder 6 feet from the wall. How far up the wall does the ladder reach? Round to the nearest tenth if necessary.
about 19.1 ft

2. What is the sum of the measures of the interior angles of a pentagon?

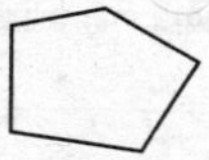

A. 900°

B. 720°

*C. 540°

D. 450°

3. What is the distance between points A and B shown on the coordinate plane?

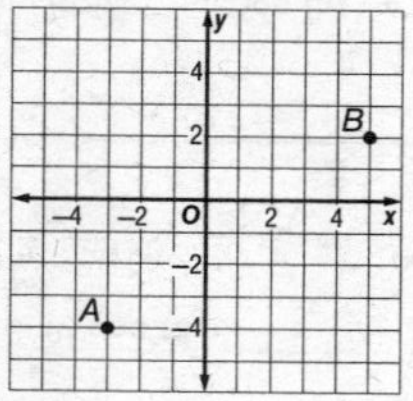

F. 8 units

*G. 10 units

H. 12 units

I. 14 units

4. Which of the following figures show a 90° clockwise rotation of the figure shown below?

A.

B.

*C.

D.

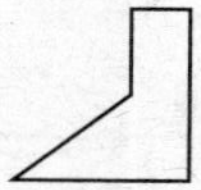

NAME ______________________ DATE ____________ PERIOD ________

Course 3 Benchmark Test – Third Quarter *(continued)*

5. If point $H(-6, 2)$ is translated 4 units up and 3 units right, what are the coordinates of the translated image?

F. $H'(-2, 5)$

*G. $H'(-3, 6)$

H. $H'(-9, -2)$

I. $H'(-9, 6)$

6. The dilation of $\overline{CD}$ is shown below. What is the scale factor of the dilation?

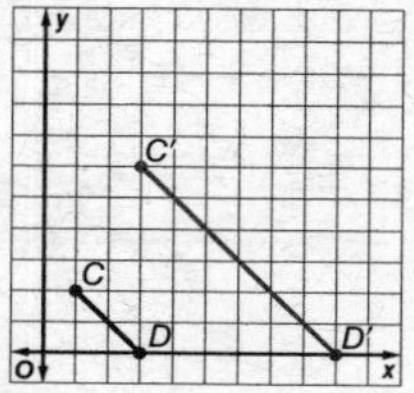

A. $\frac{1}{3}$

B. $\frac{1}{2}$

C. 2

*D. 3

7. Which of the following terms describes two lines that intersect to form right angles?

F. parallel

*G. perpendicular

H. skew

I. straight

8. What is the measure of angle 3?

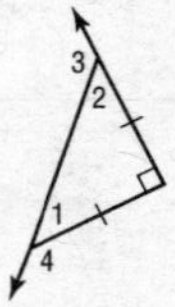

A. 45°

B. 90°

*C. 135°

D. 225°

9. **SHORT ANSWER** Determine whether the following figure is a right triangle. Justify your answer.

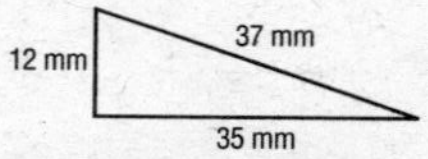

Yes, the figure is a right triangle because the sides satisfy the Pythagorean Theorem: $12^2 + 35^2 = 37^2$.

10. Point $N(6, -5)$ is reflected across the x-axis. What are the coordinates of the image?

F. $N'(-6, -5)$

G. $N'(-5, 6)$

H. $N'(5, -6)$

*I. $N'(6, 5)$

NAME ______________________ DATE ____________ PERIOD ________

Course 3 Benchmark Test – Third Quarter *(continued)*

11. Parallel lines l and m are intersected by transversal t as shown below. Which of the following angles are *not* congruent?

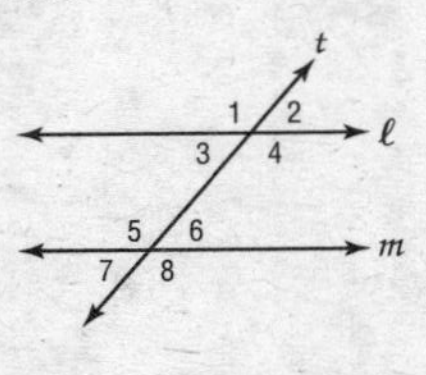

***A.** 1 and 2

B. 2 and 3

C. 3 and 6

D. 4 and 8

12. Suppose triangle RST shown on the coordinate grid is reflected across the y-axis. Which ordered pair does *not* represent a vertex of the reflected triangle?

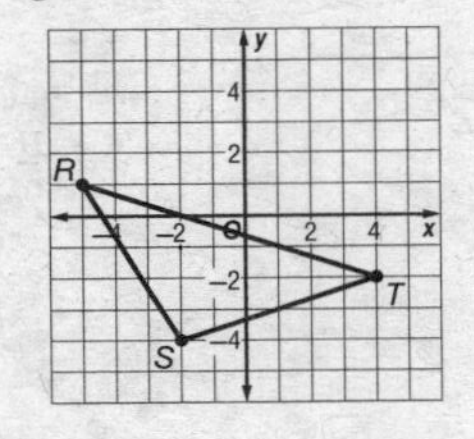

F. (5, 1)

G. (−4, −2)

H. (2, −4)

***I.** (−2, 4)

13. **SHORT ANSWER** Using the figure below, write a paragraph proof to show that $m\angle a = m\angle b = 45°$.

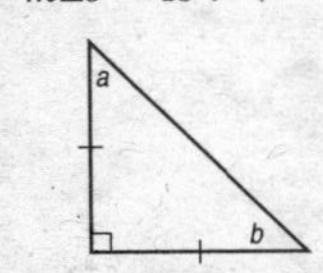

Sample answer: The triangle is isosceles, so $m\angle a = m\angle b$. The sum of the angles of a triangle is 180°. So, $a + b + 90 = 180$, or $a + b = 90$. Since $a = b$, $a + a = 90$, or $a = 45$. So, $m\angle a = m\angle b = 45°$.

14. What is the approximate distance between points $W(-4, 1)$ and $Z(3, 7)$? Round to the nearest tenth.

A. 10.8 units

***B.** 9.2 units

C. 8.3 units

D. 6.1 units

15. What is the value of n in the triangle below?

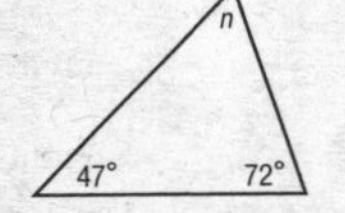

F. 47°

G. 51°

***H.** 61°

I. 72°

NAME ______________________ DATE ____________ PERIOD ________

Course 3 Benchmark Test – Third Quarter *(continued)*

16. What is the measure of an interior angle of a regular hexagon?

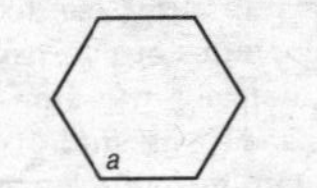

***A.** 120°

B. 135°

C. 720°

D. 810°

17. Which rotation best describes the transformation shown below?

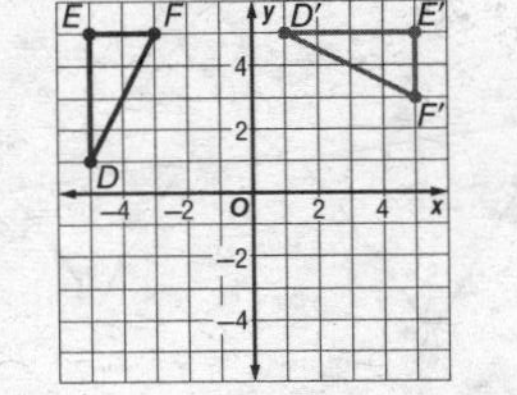

F. 90° counterclockwise rotation

G. 270° clockwise rotation

H. 180° rotation

***I.** 90° clockwise rotation

18. Which set of numbers could be the sides of a right triangle?

A. 6, 8, 12

***B.** 8, 15, 17

C. 4, 12, 16

D. 9, 11, 21

19. What is the approximate length of $\overline{NP}$ with endpoints $N(7, 3)$ and $P(-6, -2)$? Round to the nearest tenth.

F. 5.7 units

G. 6.5 units

H. 10.2 units

***I.** 13.9 units

20. **SHORT ANSWER** What is the length of the diagonal of a square with 8-foot sides? Round to the nearest tenth.

d
8
8

11.3 ft

21. Which transformation does *not* result in an image congruent to the original figure?

A. translation

B. rotation

C. reflection

***D.** dilation

Answers

NAME ______ DATE ______ PERIOD ______

Course 3 Benchmark Test – Third Quarter *(continued)*

22. What is the value of x in the figure below?

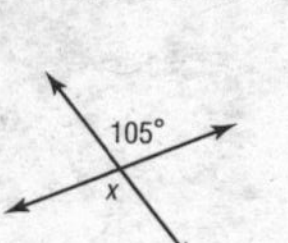

F. 75°

G. 85°

***H.** 105°

I. 115°

23. Mary enlarged a 4- by 6-inch photo to a 10- by 15-inch photo. What is the scale factor of the dilation?

A. 2

***B.** 2.5

C. 6

D. 9

24. The legs of a right triangle measure 7 units and 24 units. What is the measure of the hypotenuse? Round to the nearest tenth if necessary.

F. 17 units

G. 20.4 units

H. 23.0 units

***I.** 25 units

25. **SHORT ANSWER** Prove that triangle ABC is an isosceles triangle.

Sample answer: Using the distance formula, $AB = \sqrt{58}$ and $AC = \sqrt{58}$. Since $AB = AC$, the triangle is isosceles.

NAME ______ DATE ______ PERIOD ______

Course 3 Benchmark Test – End of Year

1. The area of a figure is 64 square centimeters. Suppose the sides of the figure are doubled. What will be the new area of the similar figure?

A. 16 square centimeters

B. 32 square centimeters

C. 128 square centimeters

***D.** 256 square centimeters

2. Triangle MNO is similar to triangle WXY. Which of the following statements is not necessarily true?

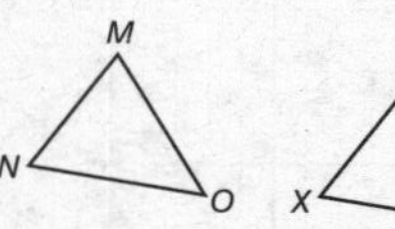

F. $\angle Y = \angle O$

***G.** $\frac{MO}{MN} = \frac{WX}{WY}$

H. $\angle N = \angle X$

I. $\frac{MN}{NO} = \frac{WX}{XY}$

3. **SHORT ANSWER** A moving company charges \$30 plus \$0.15 per mile to rent a moving van. Another company charges \$15 plus \$0.20 per mile to rent the same van. For how many miles will the cost be the same for the two companies? Write and solve an equation.

$30 + 0.15m = 15 + 0.2m$; 300 miles

4. A marching band has 64 members. The band director wants to arrange the band members into a square formation. How many band members will be in each row?

***A.** 8

B. 7

C. 6

D. 5

5. Between which two integers does $\sqrt{42}$ lie on the number line?

3 4 5 6 7 8 9 10 11

F. between 5 and 6

***G.** between 6 and 7

H. between 7 and 8

I. between 8 and 9

6. What are the slope and y-intercept of the linear equation below?

$$y = \frac{2}{3}x - 1$$

***A.** slope: $\frac{2}{3}$, y-intercept: $(0, -1)$

B. slope: $\frac{2}{3}$, y-intercept: $(-1, 0)$

C. slope: -1, y-intercept: $\left(0, \frac{2}{3}\right)$

D. slope: -1, y-intercept: $\left(\frac{2}{3}, 0\right)$

NAME ________________ DATE ________ PERIOD ________

Course 3 Benchmark Test – End of Year *(continued)*

7. What is the equation of the quadratic function shown in the graph?

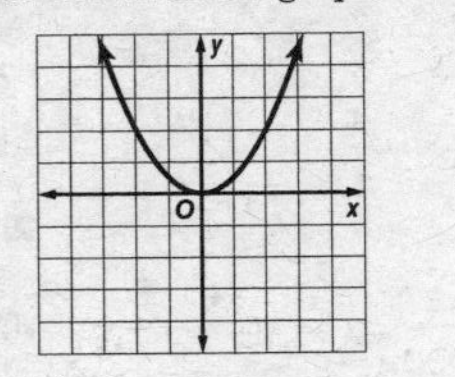

F. $y = x^2$

G. $y = -x^2$

H. $y = 2x^2$

***I.** $y = \frac{1}{2}x^2$

8. What is the volume of a sphere with a radius of 9 inches?

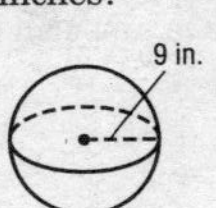

A. 1016π in^3

***B.** 972π in^3

C. 486π in^3

D. 324π in^3

9. What are the *x*- and *y*-intercepts of the linear equation below?

$$-5x + 3y = -15$$

***F.** (3, 0) and (0, –5)

G. (0, 3) and (–5, 0)

H. (–5, 0) and (3, 0)

I. (0, 3) and (0, –5)

10. **SHORT ANSWER** The two-way table shows the number of boys and girls in the school band and choir. Is there a greater percentage of girls in the school band or in the choir? Explain.

	Band	Choir
Boys	14	5
Girls	12	9

choir; The band is about 46% girls, but the choir is about 64% girls.

11. What is the sum of the measures of the interior angles of a hexagon?

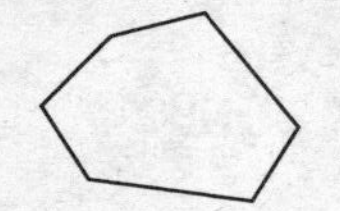

A. 540°

***B.** 720°

C. 900°

D. 1,080°

12. **SHORT ANSWER** Determine whether the following figure is a right triangle. Justify your answer.

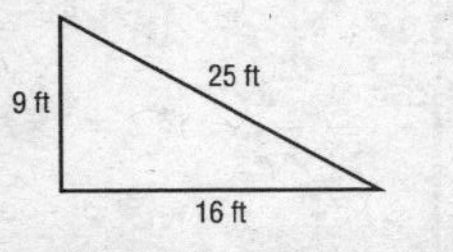

No, the figure is not a right triangle because the sides do not satisfy the Pythagorean Theorem; $9^2 + 16^2 \neq 25^2$.

NAME ________________ DATE ________ PERIOD ________

Course 3 Benchmark Test – End of Year *(continued)*

13. A soup can has a diameter of 8 centimeters and a height of 15 centimeters. About how much soup does the can hold? Use 3.14 for π. Round to the nearest tenth.

F. 376.8 cm^3

***G.** 753.6 cm^3

H. 1028.7 cm^3

I. 3014.4 cm^3

14. **SHORT ANSWER** The table shows the number of goals scored by the Cougars so far this soccer season.

Game	1	2	3	4	5
Goals Scored	3	2	6	5	4

What is the mean absolute deviation?

1.2

15. Parallel lines *l* and *m* are intersected by transversal *t* as shown below. Which of the following angles are alternate interior angles?

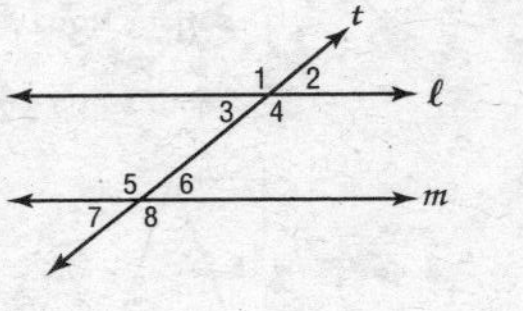

A. 1 and 5

B. 4 and 6

C. 2 and 7

***D.** 3 and 6

16. What is the distance between points $L(-5, 7)$ and $M(3, -8)$?

F. 9 units

G. 13 units

H. 15 units

***I.** 17 units

17. The slope of a line is –3 and the *y*-intercept is (0, 4). What is the equation of the line in slope-intercept form?

A. $y = -\frac{1}{3}x + 4$

B. $y = \frac{1}{3}x - 4$

C. $y = 3x + 4$

***D.** $y = -3x + 4$

18. What is the value of *n* in the triangle below?

30° 44° *n*

F. 68°

G. 74°

H. 96°

***I.** 106°

Answers

NAME ______________________ DATE ____________ PERIOD ________

Course 3 Benchmark Test – End of Year *(continued)*

19. Suppose the dimensions of a rectangular prism are enlarged by a factor of 3. By what scale factor will the volume of the prism be scaled?

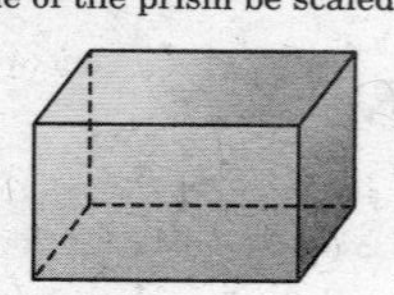
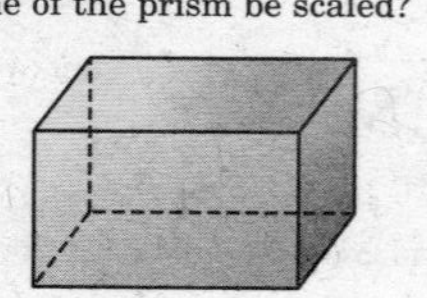

A. $\frac{1}{3}$

B. 3

C. 9

***D.** 27

20. What is the measure of an interior angle of a regular octagon?

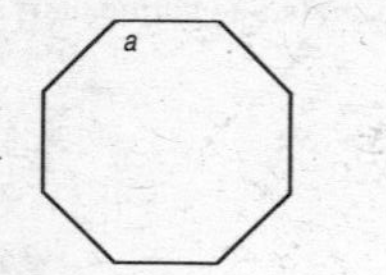

F. 1,080°

G. 720°

H. 540°

***I.** 135°

21. **SHORT ANSWER** What is the expression $(3x^2y^3)^3$ simplified?

$27x^6y^9$

22. Which equation is equivalent to $3x + 2y = -2$?

A. $y = -\frac{2}{3}x - 5$

B. $y = \frac{3}{2}x + 7$

***C.** $y = -\frac{3}{2}x - 1$

D. $y = \frac{2}{3}x + 4$

23. Which of the following symbols when placed in the blank results in a true number sentence?

$1.7\overline{3}$ ____ $\sqrt{3}$

F. =

***G.** >

H. <

I. ×

24. What type of relationship is shown in the scatter plot below?

A. positive

***B.** negative

C. skewed

D. no relationship

NAME ______________________ DATE ____________ PERIOD ________

Course 3 Benchmark Test – End of Year *(continued)*

25. About how much water can the paper drinking cup shown below hold? Use 3.14 for π. Round to the nearest tenth.

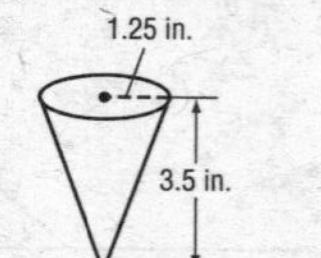

F. 17.2 cubic inches

G. 9.2 cubic inches

***H.** 5.7 cubic inches

I. 4.8 cubic inches

26. **SHORT ANSWER** Determine if the two figures below are congruent by using transformations. Explain your reasoning.

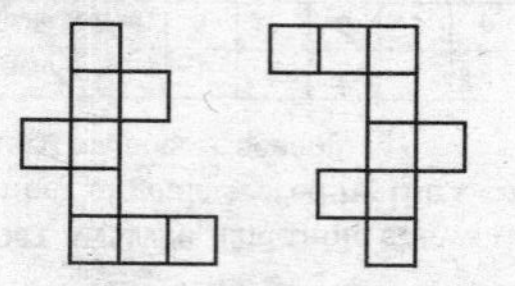

congruent; A rotation of 180° maps one figure exactly onto the other figure.

27. Which two points form a line that has a slope of $\frac{5}{2}$?

***A.** (3, 6) and (−1, −4)

B. (−4, 2) and (7, −1)

C. (−4, 7) and (−9, 5)

D. (3, −7) and (8, 4)

28. What is the constant rate of change of the function represented in the table below?

x	y
−6	−7
−3	−1
0	5
3	11

***F.** 2

G. 3

H. 5

I. 6

29. **SHORT ANSWER** What is the equation of the line that passes through (−6, −6) and (12, 9)?

$y = \frac{5}{6}x - 1$

30. Which transformations could have been used to map Figure A onto Figure B?

Figure A

Figure B

***A.** dilation, translation

B. dilation, reflection

C. reflection, rotation

D. translation, rotation

NAME ______________________ DATE ____________ PERIOD ________

Course 3 Benchmark Test – End of Year *(continued)*

31. Katie is 5 feet tall. She casts a 3-foot long shadow at the same time that a flagpole casts an 18-foot long shadow.

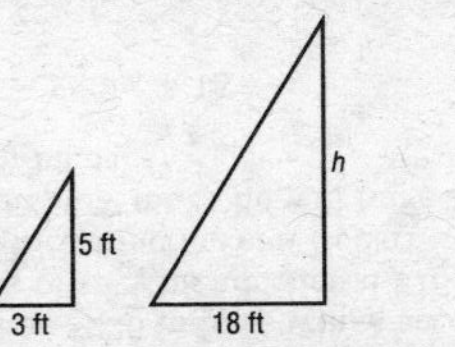

What is the height of the flagpole?

F. 10.8 ft

G. 22.4 ft

H. 28 ft

***I.** 30 ft

32. What is the approximate surface area of a cylinder with a height of 12 meters and a base radius of 2 meters? Use 3.14 for π. Round to the nearest tenth if necessary.

A. 242.1 m^2

***B.** 175.8 m^2

C. 150.7 m^2

D. 124.5 m^2

33. The distance from the Sun to Venus is about 1.08×10^{11} meters. If light travels at a speed of 3×10^8 meters per second, about how long does it take light from the sun to reach Venus?

***F.** 3.6×10^2 seconds

G. 4.2×10^2 seconds

H. 1.083×10^{11} seconds

I. 3.24×10^{19} seconds

34. Which of the following is equivalent to 2^{-4}?

A. −16

B. −8

C. $\frac{1}{32}$

***D.** $\frac{1}{16}$

35. What is the range of the function shown in the table?

x	−7	−5	−3	−1	1
y	4	6	1	−2	−3

F. all integers

G. all odd integers

***H.** {−3, −2, 1, 4, 6}

I. {−7, −5, −3, −1, 1}

36. **SHORT ANSWER** The area of a square patio is 225 square feet. What is the perimeter of the patio?

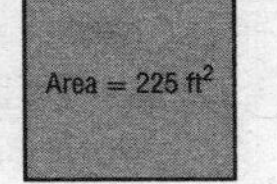

60 ft

37. A cone has a height of 24 inches, a slant height of 25 inches, and a diameter of 14 inches. What is the surface area of the cone?

A. $1{,}176\pi$ in^2

B. 392π in^2

***C.** 224π in^2

D. 178π in^2

NAME ______________________ DATE ____________ PERIOD ________

Course 3 Benchmark Test – End of Year *(continued)*

38. A hotel shuttle service charges $7.50 plus $0.85 per mile. A customer hires a shuttle, and the total charge is $12.60. Which equation can be used to determine the number of miles from the hotel to the airport?

***F.** $0.85m + 7.5 = 12.6$

G. $7.5m + 0.85 = 12.6$

H. $8.35m = 12.6$

I. $6.65m = 12.6$

39. **SHORT ANSWER** What is the relationship between the slope of the line and the side lengths of the triangles?

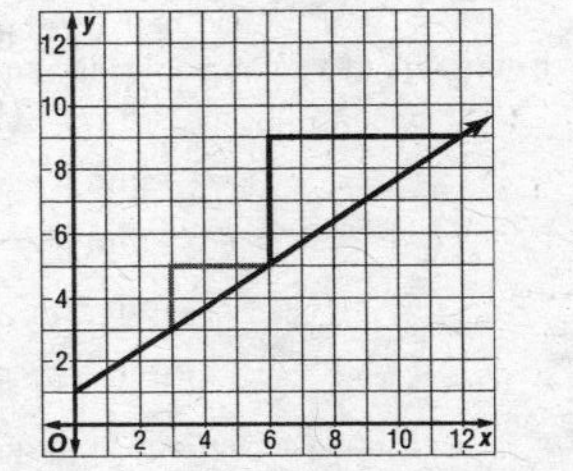

The slope of the line is $\frac{2}{3}$. This is the ratio of the side lengths of the legs of each triangle.

40. The population of the United States is about 3.1×10^8 people. What is this number written in standard form?

A. 3,100,000

B. 31,000,000

***C.** 310,000,000

D. 3,100,000,000

41. Which expression is equivalent to the expression below?

$$c \cdot c \cdot c \cdot c \cdot d \cdot d \cdot c \cdot d \cdot c \cdot c \cdot d$$

F. $(cd)^3$

G. $c^{-7}d^{-4}$

H. $(cd)^{11}$

***I.** c^7d^4

42. What is the solution to the system of linear equations shown below?

A. (0, 3)

B. (5, 6)

C. (−5, −4)

***D.** no solution

43. Jasmine determines figure $ABCD \cong$ figure $FGHI$. If $AB = 14$ meters, $BC = 11$ meters, $CD = 9$ meters, and $AD = 17$ meters, what is the length of $\overline{GH}$?

F. 9 m

***G.** 11 m

H. 14 m

I. 17 m

NAME ______________ DATE ________ PERIOD ______

Course 3 Benchmark Test – End of Year *(continued)*

44. SHORT ANSWER Twenty years ago, Mr. Williams purchased a classic car for $65,000. The table below shows the value of the car over time. Write an equation that represents the data.

Years from Purchase	Value (thousands)
0	$65
5	$67.5
10	$70
15	$72.5
20	$75

What will be the value of the car when it has been 30 years since he purchased it?

Sample answer: $y = 0.5x + 65$; about $80,000

45. What is the slope of the line that passes through points $R(0, 2)$ and $T(-3, -4)$?

*A. 2

B. $\frac{1}{2}$

C. $-\frac{1}{2}$

D. −2

46. Robert has $220 in his savings account. He plans to save an additional $15 each week. Which function can Robert use to determine how much he will have saved s after m months?

F. $s(m) = 220m + 15$

G. $s(m) = 235m$

*H. $s(m) = 15m + 220$

I. $s(m) = 15m$

47. What type of transformation is represented by the figures below?

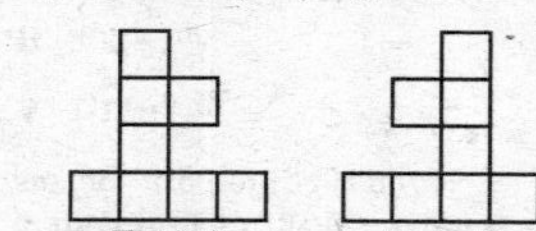

A. dilation

*B. reflection

C. rotation

D. translation

48. Which of the following equations represents a vertical line?

F. $y = x$

G. $y = x + 10$

H. $y = 4$

*I. $x = 5$

49. Which series of transformations can be used to prove that triangle RST is similar to triangle LMN?

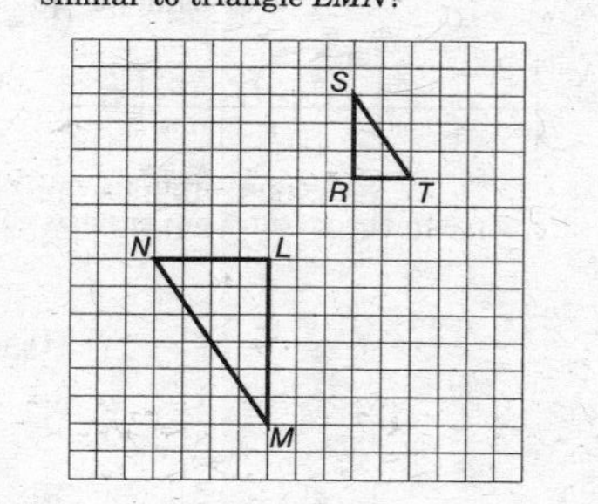

A. reflection, dilation

B. 90° rotation, dilation

C. translation, dilation

*D. 180° rotation, dilation

NAME ______________ DATE ________ PERIOD ______

Course 3 Benchmark Test – End of Year *(continued)*

50. Which of the following statements about a line of best fit is *not* true?

F. Most of the data points are close to the line.

G. About half of the points are above the line.

*H. All of the data points have to be on the line.

I. The line can be used to make conjectures.

51. The endpoints of $\overline{AR}$ are $A(8, -2)$ and $R(-4, 1)$. What is the length of $\overline{AR}$? Round to the nearest tenth.

*A. 12.4 units

B. 11.2 units

C. 7.5 units

D. 4.0 units

52. What is the value of x in the figure below?

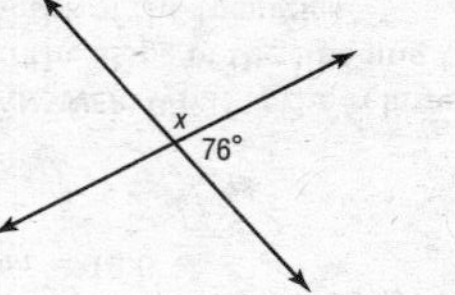

F. 114°

*G. 104°

H. 86°

I. 76°

53. SHORT ANSWER Is a triangle with side lengths of 33 inches, 56 inches, and 65 inches a right triangle? Explain your reasoning.

Yes, the triangle is a right triangle because the side lengths satisfy the Pythagorean Theorem: $33^2 + 56^2 = 65^2$.

54. Which set lists the values below from least to greatest?

$$3^{-2}, \sqrt{3}, 1.3 \times 10^{-1}, \frac{1}{3}$$

A $\left\{\sqrt{3}, \frac{1}{3}, 1.3 \times 10^{-1}, 3^{-2}\right\}$

B $\left\{\sqrt{3}, 1.3 \times 10^{-1}, \frac{1}{3}, 3^{-2}\right\}$

*C $\left\{3^{-2}, 1.3 \times 10^{-1}, \frac{1}{3}, \sqrt{3}\right\}$

D $\left\{3^{-2}, \frac{1}{3}, 1.3 \times 10^{-1}, \sqrt{3}\right\}$

55. SHORT ANSWER The table below shows the prices of digital cameras at an electronics store. Summarize the data.

Prices of Digital Cameras ($)					
75	115	95	105	115	95
100	100	70	80	105	75
120	95	115	175	105	110

Sample answer: Most of the prices are $120 or less and centered around $105. There is a small cluster of prices at $80 or less.

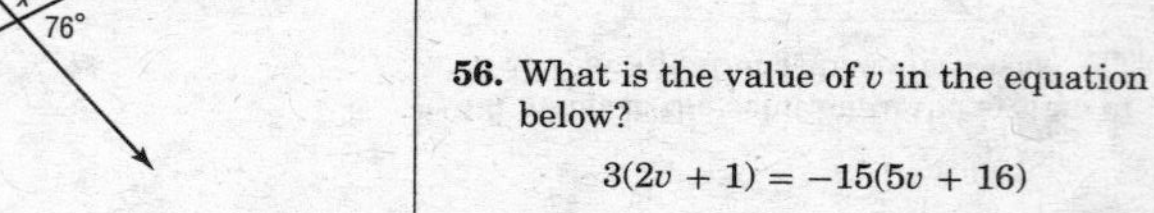

56. What is the value of v in the equation below?

$$3(2v + 1) = -15(5v + 16)$$

F. $\frac{13}{81}$

G. $\frac{5}{27}$

H. −2

*I. −3

57. What is the solution to the equation below?

$$0.4p + 0.1 = 1.15$$

A. 3.125

*B. 2.625

C. 0.5

D. 0.42

NAME ______________________ DATE ____________ PERIOD ________

Course 3 Benchmark Test – End of Year *(continued)*

58. Solve the system of equations below.

$$7x + 6y = -10$$
$$-2x + y = 11$$

***F.** $(-4, 3)$

G. $(-5, 1)$

H. $(7, 9)$

I. no solution

59. The quadratic function $h(t) = -16t^2 + 90$ represents the height, in feet, of an object t seconds after it begins falling from a height of 90 feet. What is the height of the object after 2 seconds?

A. 22 ft **C.** 58 ft

***B.** 26 ft **D.** 154 ft

60. Let n represent the figure number in the pattern below.

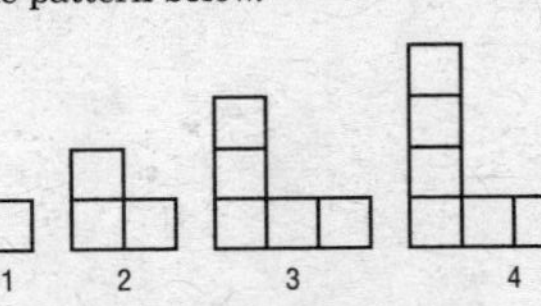

Which function represents the number of squares used to create each figure?

F. $f(n) = n^2$

G. $f(n) = n^2 - 1$

***H.** $f(n) = 2n - 1$

I. $f(n) = 2n + 1$

61. By what factor would you need to multiply the dimensions of a polygon in order for the resulting image to have a perimeter that is equal to $\frac{1}{4}$ the original perimeter?

***A.** $\frac{1}{4}$ **C.** 2

B. $\frac{1}{2}$ **D.** 4

62. A rectangular-shaped school courtyard has a length of 280 feet and a width of 150 feet wide. What is the approximate length of a diagonal of the courtyard to the nearest tenth?

F. 430.0 ft ***H.** 317.6 ft

G. 395.4 ft **I.** 295.1 ft

63. **SHORT ANSWER** Does the data in the table represent a linear or nonlinear function? Explain your reasoning.

x	y
−7	−37
−2	−7
1	11
5	35
7	47

Sample answer: linear function; There is a constant rate of change of $\frac{6}{1}$ or 6.

64. What is the scale factor of the dilated figure shown below?

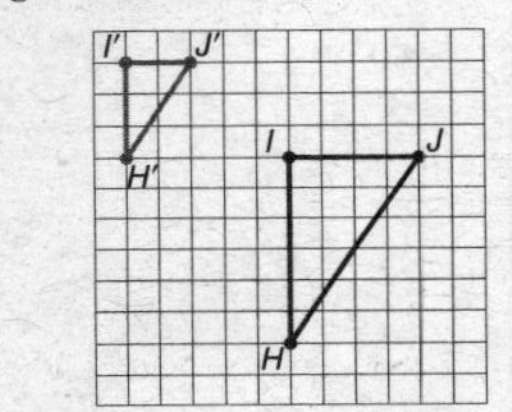

A. 0.25 **C.** 2

***B.** 0.5 **D.** 4

65. Point $A(-7, -3)$ is reflected across the y-axis. What are the coordinates of the image?

F. $A'(3, -7)$ **H.** $A'(-3, -7)$

G. $A'(-7, 3)$ ***I.** $A'(7, -3)$